Metametaphysics and the Sciences

This collection addresses metaphysical issues at the intersection between philosophy and science. A unique feature is the way in which it is guided by history of philosophy, by interaction between philosophy and science, and by methodological awareness. In asking how metaphysics is possible in an age of science, the contributors draw on philosophical tools provided by three great thinkers who were fully conversant with and actively engaged with the sciences of their day: Kant, Husserl, and Frege.

Part I sets out frameworks for scientifically informed metaphysics in accordance with the metametaphysics outlined by these three self-reflective philosophers. Part II explores the domain for co-existent metaphysics and science. Constraints on ambitious critical metaphysics are laid down in close consideration of logic, meta-theory, and specific conditions for science. Part III exemplifies the role of language and science in contemporary metaphysics. Quine's pursuit of truth is analyzed; Cantor's absolute infinitude is reconstrued in modal terms; and sense is made of Weyl's take on the relationship between mathematics and empirical aspects of physics.

With chapters by leading scholars, *Metametaphysics and the Sciences* is an in-depth resource for researchers and advanced students working within metaphysics, philosophy of science, and the history of philosophy.

Frode Kjosavik is Professor of Philosophy at the Norwegian University of Life Sciences. He was a group leader in philosophy at the Centre for Advanced Study at the Norwegian Academy of Science and Letters, 2015/16. He is the co-editor, with Christian Beyer and Christel Fricke, of *Husserl's Phenomenology of Intersubjectivity: Historical Interpretations and Contemporary Applications* (Routledge, 2019).

Camilla Serck-Hanssen is Professor of Philosophy at the University of Oslo, Norway. She was a group leader in philosophy at the Centre for Advanced Study at the Norwegian Academy of Science and Letters, 2015/16. She is now the Scientific Director of the Centre. She is also co-leader of ConceptLab at the University of Oslo.

Routledge Studies in Metaphysics

Neo-Aristotelian Perspectives in Metaphysics
Edited by Daniel D. Novotný and Lukáš Novák

Nominalism about Properties
New Essays
Edited by Ghislain Guigon and Gonzalo Rodriguez-Pereyra

Substance and the Fundamentality of the Familiar
A Neo-Aristotelian Mereology
Ross D. Inman

Philosophy of Time and Perceptual Experience
Sean Enda Power

Fundamental Causation
Physics, Metaphysics, and the Deep Structure of the World
Christopher Gregory Weaver

Eliminativism, Objects, and Persons
The Virtues of Non-Existence
Jiri Benovsky

Thisness Presentism
An Essay on Time, Truth, and Ontology
David Ingram

Metametaphysics and the Sciences
Historical and Philosophical Perspectives
Edited by Frode Kjosavik and Camilla Serck-Hanssen

For more information about this series, please visit: www.routledge.com/Routledge-Studies-in-Metaphysics/book-series/RSM

Metametaphysics and the Sciences

Historical and Philosophical Perspectives

Edited by Frode Kjosavik and Camilla Serck-Hanssen

Routledge
Taylor & Francis Group
NEW YORK AND LONDON

First published 2020
by Routledge
52 Vanderbilt Avenue, New York, NY 10017

and by Routledge
2 Park Square, Milton Park, Abingdon, Oxon OX14 4RN

Routledge is an imprint of the Taylor & Francis Group, an informa business

Library of Congress Cataloging-in-Publication Data
A catalog record for this book has been requested

ISBN: 978-0-367-26369-0 (hbk)
ISBN: 978-0-429-29295-8 (ebk)

Typeset in Sabon
by Apex CoVantage, LLC

Contents

Preface

This volume is based on the research project 'Disclosing the Fabric of Reality. The Possibility of Metaphysics in the Age of Science.' The project was hosted by the Centre for Advanced Study (CAS) at the Norwegian Academy of Science and Letters during the academic year 2015–16. The research team was led by Professor Frode Kjosavik and Professor Camilla Serck-Hanssen. Among the participants were experts on Kant, Husserl, and Frege, as well as on logic, philosophy of mathematics, and philosophy of science. Some additional scholars kindly accepted our invitation to contribute to the volume. We are grateful for the financial and administrative support from CAS. Special thanks go to Andrew Weckenmann at Routledge for his encouragement and constructive suggestions.

Introduction

Frode Kjosavik and Camilla Serck-Hanssen

Metaphysical questions are deep questions about the nature of reality, such as: To what extent is the world constructed rather than mind-independent? What is the relation between the physical "outside" and the mental "inside" of a being that has both? Can humans choose freely and act autonomously? What is the infinite, and how does it appear from the perspective of human finitude? Are the objects of mathematics invented, or are they discovered, and therefore already there, with a richness of properties, prior to our conceptions? What are we to make of what is unobservable yet postulated by our physical theories?

As these examples bear out, *questions* of a metaphysical character have not been rendered obsolete by modern science. Rather, there remain pressing metaphysical issues both in dealing with foundational scientific problems and, more generally, in dealing with problems that the human condition is entangled with. The entry point for all the chapters in this volume is the firm belief that *metaphysical questions are unavoidable even in an age of science* and that metaphysical aspects of scientific and other discourse ought to be made explicit and scrutinized.

How, then, are metaphysical questions to be *answered?* Our volume presents frameworks for and contributions to a *critical* metaphysics that is *shaped by philosophical reflection on science*. In our discussion of what ought to be the methodology for metaphysics, we rely on both the history of philosophy as well as on contemporary science. In doing so, we bring metaphysics and science together without substituting the one for the other. We thereby place ourselves well within the current dialectic between metaphysics and epistemology, between pure philosophy and empirical science, and between methodologies that rely on rational and those that rely on empirical evidence.

The following three main hypotheses were central in the research project that this volume grew out of.

1. *Metaphysical questions require a philosophical treatment.*

Scientific disciplines cannot on their own determine what is the boundary of their domain and what are their "proper" objects of investigation, methodological principles and standards of evidence. These are normative meta-questions that are continuous with fundamental questions within a discipline but which cannot be handled solely within these boundaries, and which therefore require an extended, methodological and metaphysical discussion. Furthermore, in life more generally, there are metaphysical questions that impose themselves upon us. These questions are similarly unavoidable and arguably philosophical in nature.

2. *Metaphysical questions can be given substantive answers.*

After a period of decline in the first half of the twentieth century, metaphysics has again become a respected area within philosophy itself. There is currently, e.g., a strong interest in the topic of metaphysical grounding—in questions concerning what depends on what, and in whether reality has a fundamental layer. Also, many metaphysical issues concerning the status of information (e.g., biological), structures (e.g., mathematical and physical), essences (fixed kinds to be found in nature), and constructions (social and contingent) are now seen as highly relevant to the foundations of particular sciences.

Under the label "metaphysics" one finds, however, a variety of approaches, which can be more or less science oriented. In particular, one has witnessed certain tendencies to lower metaphysical aspirations to a very modest level. For instance, metaphysical problems have been recast as questions about our conceptual framework rather than about the world. Mere discussion of the ontological commitments of theories or of conceptual schemes that go together with theories would not constitute metaphysics in the substantive sense that we are after.[1] Some have also explicitly rejected the more ambitious metaphysics,[2] arguing that it rests on mere armchair intuitions, the real task of metaphysics being that of unifying scientific theories.[3]

Still, there are recently more ambitious approaches, which set out to solve the questions without deflating them into empirical or conceptual issues, or to the systematization of these.[4] We believe that this understanding of metaphysics is defensible and that metaphysical questions concern objects and their mode of existence, not conceptual schemes or the relations between theories. We also think that there is no principled reason why answers to such questions should not lie within the reach of human minds.[5] We therefore think one should opt for an ambitious rather than a modest metaphysics.

3. *An ambitious metaphysics requires a meta-level investigation.*

To legitimize an ambitious metaphysics it does not suffice to point to the fact that metaphysical questions tend to be unavoidable or to refute anti-metaphysical arguments. Rather, we believe one must provide some positive account of how substantive metaphysics is to be possible. Without such methodological awareness, ambitious metaphysics runs the risk of being blind to its own tacit assumptions.[6]

To summarize, despite the recent wave of interest in philosophical meta-issues[7] much of the ambitious metaphysics now flourishing fails to go deeply enough into these meta-issues and thereby falls short of showing *how* metaphysics is possible.[8] What tends to be neglected is how a critical metaphysics ought to be anchored in a well-developed *metametaphysics*, i.e., a full-fledged study of what metaphysics is and how metaphysical theories are to be established as justified. We believe that a systematically worked out metametaphysics will be a valuable and distinct addition to contemporary ambitious metaphysics and its meta-issues. Thus, our volume is metametaphysical in its overall aim of providing frameworks for a critical metaphysics as well as metaphysical in its more concrete investigations of the issues at hand.

In our pursuit of a methodology for metaphysics, we are looking to sources of metaphysical insight outside semantics, rather than, say, for a revival of Carnap's linguistic frameworks and an orientation towards semantics.[9] Beyond attentiveness to methodological meta-issues, we believe in utilizing rich resources from different philosophical traditions that assign a very important role to science. Accordingly, many chapters in this volume deal with the possibility of metaphysics at the very intersection between philosophy and science. Moreover, they do so by drawing on the rich traditions from two thinkers that were both fully conversant with and actively engaged with the sciences of their day, namely, Immanuel Kant and Edmund Husserl. Historically, Husserl relates explicitly to Kant, and he offers his own version of transcendental philosophy, in the form of transcendental phenomenology. Common to both thinkers is first that they opt for an explicitly critical approach to naïve metaphysics, which offers no answers to the "How is metaphysics possible?" question; second, they are themselves both scientists and philosophers, and display an unusually high degree of methodological awareness and self-criticism even within their philosophical work; and finally, they attempt to bring together philosophy and science, while respecting the scope and limits of both.

Another philosopher who fits very well with this picture is Gottlob Frege. In Part I of the present volume, on the very possibility of metaphysics, the final chapter is on Frege and metaphysics. The chapter contrasts nicely with the broader takes on metametaphysics to be found in the preceding chapters on Kant and Husserl. Frege is also important when it comes to Husserl interpretations offered in this volume. Furthermore, in Part II, on the scope and limits of metaphysics, the need for a proper meta-theory and logical framework for critical metaphysics is highlighted. Frege—the founder of modern logic—is thus highly relevant here. One chapter in this part deals with the logic of the eighteenth century and what Kant considered to be its proper use. Another chapter deals with transcendental arguments. Two subsequent chapters bring in the logical tradition of the nineteenth century. In Part III, one chapter is devoted to Willard van Orman Quine—a philosopher and logician of the twentieth century.

There are important historical connections between the philosophies of Kant, Husserl, and Frege. Husserl and Frege related to each other, and both related to Kant. Common to all three is also that they took the science of their own time very seriously—indeed, we consider them to be exemplary in that regard. Furthermore, they all—explicitly or effectively—considered science and metaphysics to be interdependent, or so we claim. Thus, we believe that we can learn about the possibility and method of metaphysics in an age of science from the accounts of the possibility and method of metaphysics provided by these three self-reflective philosophers. We think that such an approach is a promising alternative to present trends in philosophy that are less ambitious or less science oriented.

As far as we know, Kant, Husserl, and Frege have never been brought into dialogue with each other in a systematic attempt to develop forms of ambitious metaphysics. Since their positions have been used in the past both to inflate and deflate metaphysical claims, one has to tread cautiously here, pursuing the middle path of being neither over-ambitious and naive nor too modest. Any acceptable form of metaphysics must thus be sufficiently guided by sound methodological principles.

In the research project underlying the present volume, our three most important working hypotheses concerning the methodology of an ambitious metaphysics were very general:

1. Metaphysics has its own methodological principles, just as mathematics and the empirical sciences, and these have to pass a test of coherence, adequacy, and fruitfulness with regard to metaphysical questions.
2. An important methodological principle in metaphysics consists in reflection upon the conditions of the possibility of mathematical and empirical science, and thereby in explication of a metaphysical framework that is implicitly there in the sciences themselves.
3. The reflection upon the conditions of possibility of mathematical and empirical science must be guided by an investigation into some logical framework or other as the most fundamental condition on the judgments and inferences made in any of the sciences.

In developing forms of critical metaphysics, bringing together history of philosophy and contemporary science as two major sources of critique suggests a general approach to metametaphysics. Indeed, such an approach should be considered as exemplified rather than exhausted by our more specific choices of thinkers, perspectives, and topics. While many of the contributors are situated within "transcendental philosophy" broadly construed, the present volume is not to be seen as a mere collection of interpretive pieces on key figures within the history of philosophy. That being said, we do intend to provide more careful interpretations of the philosophers that are central in this volume—interpretations that are sufficiently attentive both to their metaphysical insights as well as to their own sensitivity to science.

What is characteristic of critical metaphysics as pursued in most of the chapters in this volume is the following. It goes straight to a subject matter, asks fundamental questions of the type "How is x possible?" and seeks answers to questions of that kind by identifying enabling conditions for x. Kant asked how metaphysics, mathematics, and natural science are possible. Husserl even asked how logic is possible. In this sense, the basis for most of the contributions to the present volume lies within a "broadly transcendental" approach. However, this approach itself falls within a more general approach through which the possibility, scope, and limits of metaphysics are explored by way of combining input from the history of philosophy and from contemporary science. This more general ambition of our research project underlies the framing and organization of the entire collection, which aspires to bring out the relevance of "transcendental philosophy," broadly construed, for the general debate on metaphysics, beyond the circles of Kant and Husserl scholarship. In fact, the three chapters in Part III illustrate this, in that they do not belong to Kant or Husserl scholarship but display the more general orientations of philosophy of language (Quine), philosophy of mathematics (Cantor), and philosophy of physics (Weyl), respectively.

Part I: Metametaphysics—The Very Possibility of Metaphysics

This first part of the volume discusses the philosophies of Kant, Husserl, and Frege. Not only are frameworks thereby laid for forms of critical metaphysics, but, in doing so, the interpretations presented of these three thinkers go well beyond what is offered by the standard pictures. After all, Kant's restrictions on metaphysics may seem too strict. He is often taken to be concerned only with "domesticated" metaphysics, as it were, which is thought to be too close to epistemology. Our aim is to show that this understanding of Kant is misguided and that he also has adequate resources to inquire into the kind of metaphysical questions that still impose themselves upon us. Indeed, Kant placed the method of metaphysics on the philosophical agenda in an unprecedented manner. Through his method, Kant thinks that he can (1) secure mathematical and empirical science against speculative and skeptical objections; (2) provide a systematic overview of the interconnections between the sciences; and (3) demarcate the scope of validity of science and protect human reasoning against scientism, i.e., the view that all knowledge can be subsumed under the sciences.

As in the case of Kant's transcendental philosophy, Husserl's phenomenology is sometimes taken to be anti-metaphysical. However, Husserl is influenced by Kant's critical philosophy. We believe it is more correct to view his project as a methodologically clarified attempt to engage in metaphysical investigations that he calls 'First Philosophy.' In our view, Husserl's phenomenology is particularly apt for handling the kind of metaphysical questions that are of interest and relevance to science. This is so because (1) Husserl introduces a method for studying the ontological premises of special sciences, i.e., 'eidetic variation,' and through his notion of 'eidetic seeing' he expands the concept of "intuition" in Kant so that it becomes a more comprehensive source of evidence; (2) Husserl brings subjectivity and objectivity together in a very precise way, in that they presuppose each other, since objects are invariants of variation in subjective perspectives and "constitute themselves" in intentional consciousness; and (3) Husserl develops a rich conception of the *life-world*, i.e., the world that we take for granted in our daily lives, and argues that ultimately all sciences have the life-world as their source of meaning and justification.

Frege has been very influential with regard both to the idea that metaphysical questions are to be handled through some form of analysis of language, and also to the idea that reflections within philosophy of science are of great metaphysical relevance. His famous distinction between sense and reference brings out an important insight in Husserl's phenomenology, namely, that objects come with multiple aspects and are only grasped aspect-wise. Like Kant, Frege pursued both scientific and metaphysical questions. He founded modern logic, and thought that this provides maximally general laws for all the sciences. Frege raises a metaphysical question about what kinds of objects there are, and like Husserl claims that in addition to physical and mental objects, there are also abstract objects, i.e., objects that are not in space-time and which are causally inert. Indeed, logic itself comes with a domain of logical objects. The importance of Frege in the present volume lies in (1) the two previously mentioned ideas and their methodological role within analytic metaphysics; (2) his acute awareness of methodology in dealing with scientific foundational issues; (3) his revolutionizing of logic, which bears on the task of clarifying the relation between logic and metaphysics.

In Chapter 1, Frode Kjosavik sets out to account for the methodology of metaphysics, in 'Kant on Method and Evidence in Metaphysics.' Philosophizing is in some way continuous with critical thinking in everyday life, and it is subject to maxims. Transcendental philosophy, conceived of as transformed general metaphysics, even has a distinctive critical method of philosophizing and a distinctive source of evidence in the form of rational insight. The author suggests that there is an abductive route to Kant's system of transcendental philosophy—in the form of an inference to the best explanation of the possibility of mathematics and the natural sciences. However, there is also a route that is intrinsic to reason itself by way of an architectonic "schema" for metaphysics. After all, metaphysical concepts are "given," i.e., they are fixed and can be discovered by us as such. Mathematical concepts, by contrast, are introduced freely through real genetic definitions, say, the definition of a circle, which is an original construction in pure intuition. Mathematical properties then "flow" from arbitrary syntheses of intuitive manifolds. Thus, once the concept of a circle is introduced many properties can be proved of the circle, including properties concerning intersecting chords, etc. Hence, mathematics relies on a combination of invention and discovery. Physics, on the other hand, has both a special metaphysical part, a mathematical part, and an empirical part. On the basis of this, the discussion aims at clarifying how metaphysics is related to mathematics and physics in Kant—also by tracing the evolution from the early writings to his mature works—and what we might learn from this concerning (1) what should be the methodology of metaphysics; (2) to what extent should metaphysics provide a framework for mathematics and empirical science; (3) what are the scope and limits of further developments within the framework of metaphysics itself.

In Chapter 2, Houston Smit delves into the issue of 'Essence, Nature, and the Possibility of Metaphysics' in Kant, with a special view to natural science. He aims to explain how, according to Kant's critical metametaphysics, "all true metaphysics is taken from the essence of our capacity of thinking itself" and "contains the pure acts of thinking" that make experience possible (Ak. 4: 472). This characterization of metaphysics is explained in a way that clarifies Kant's Copernican revolution. The essence of our capacity of thinking itself can serve as that from which a true metaphysics is to be taken only if it is an essence that we can cognize synthetically a priori. However, we cannot cognize any real essence, or nature. Geometric figures also do not have any real essence, or nature. What is proposed, then, is that the essence in question stands to the capacity of which it is the essence as the essence of a geometric figure stands to bodies that have shapes that conform to it. This interpretive proposal is pursued by sketching, if only in general outline, a reading of the Transcendental Deduction of the Categories on which Kant identifies the principle of the original synthetic unity of apperception as the non-real essence of our capacity of thinking itself.

In the *Metaphysical Foundations of Natural Science*, Kant aims at deriving the metaphysics that provides foundations for natural science. Transcendental philosophy thus comes to ground another science and serve as 'First Philosophy.' The metaphysics of nature is concerned with cognition *of things* under purely a priori principles. For this to be possible, empirical concepts of matter are incorporated into the insight into the formal possibility of our experience in general that Kant achieved in the Transcendental Analytic of the *Critique of Pure Reason*. Objects of these empirical concepts are subsumed under the principles of pure understanding (including, crucially, the Analogies of Experience). There is *metaphysical* cognition of objects ["Gegenstände"] whose real possibility we can prove merely by appeal to the actuality of these objects, which is attested by the experiences from which we derive these concepts. This contrasts with the manner in which traditional dogmatic metaphysics strived to achieve metaphysical cognition of things. Accordingly, the incorporation of empirical concepts is gained through insight into how, e.g., matter being the subject of attractive and repulsive powers that balance each other is not only necessary, but would, together with other conditions, be sufficient for matter to occupy a determined region of space, ultimately under the principles of pure understanding.

Christian Beyer turns to Husserl in his piece 'Toward a Husserlian (Meta-)Metaphysics,' in Chapter 3. The account begins by explaining Husserl's conception of the constitution of reality by the *pure Ego*. This requires one—to some extent—to go into the motives that led Husserl to "transcendental idealism." This position is compared to some aspects of Hilary Putnam's "internal realism." The author also provides a brief explanation of Husserl's analysis of the constitution of nature as conceived of in physical (as well as cognitive) science. Husserl developed this analysis in book II of *Ideas*.

As for the author's approach, the following serves as a contrasting background within contemporary philosophy. Many recent contributors to so-called analytic metametaphysics (e.g., Sider, Fine, Schaffer) assume that ontological claims made in a philosophical tone of voice are substantial, rather than trivial or "internal" (Carnap), and that if some of these claims are true, then there are *basic* objects or properties—"referential magnets" as Lewis called them—whose existence helps make these claims true. These fundamental truth-makers, or the corresponding ontological claims, are supposed to "carve out nature at its joints" (Lewis). As a consequence, philosophical statements regarding ontology are held to be made true by entities to be postulated by physical science, and metaphysics is supposed to investigate which of these entities are fundamental—which of them are "substances," as Neo-Aristotelians put it. Metaphysics goes hand in hand with science, and it is nature as investigated in science that metaphysical statements (that do not concern abstract objects) carve out at its joints, thereby teaching us *how* existing things exist.

From the viewpoint of Husserl's transcendental phenomenology, this metametaphysical picture is naïve. Here the author agrees with Putnam, who criticizes the natural properties or "elite classes" postulated by Lewis as "something 'spooky'" (Putnam, 'A Defence of Internal Realism,' 1990, 38). To be sure, Husserl also distinguishes between substances, which he refers to as "substrates," and other entities, but he has a complex notion of substrate embedded in his epistemological theory of the pre-predicative, or pre-judgmental, constitution of empirical reality. Husserl would agree that there are referential magnets, in the sense of fundamental truth-makers or their underlying basic referents, but deny that in the final analysis this is due to natural properties as conceived of in (present or future) physical science, as mainstream analytic (meta)metaphysics has it. Rather, referential magnetism is a "constitutive" achievement; it is due to the essential and epistemic structure of perception and, more generally, pre-predicative experience, to be studied in transcendental phenomenology, and to the structure of empirical reality thus "constituted."

The final chapter in this part, Chapter 4, is on Gottlob Frege. Leila Haaparanta writes on "Frege on 'Es gibt,' Being in a Realm and (Meta-) Ontology." Frege did not call any of his philosophical views by the name "ontology," let alone "metaontology." But the author discusses two philosophical positions that Frege held and that have metaontological relevance. They are the doctrine of the ambiguity of the word "is" and the doctrine of three realms. It is well known that Frege regarded existence as a second-order concept. Likewise, Frege's distinction between three realms has been studied in detail. However, the relation between the two doctrines has not been sufficiently worked out; instead, they are treated as separate themes that occur in Frege's works. It is true that Frege does not combine the two topics. The author seeks to demonstrate that the two doctrines are connected in an interesting way: The concept of existence that is expressed as a second-order concept in the formula language plays an important logical role, while the doctrine of three realms is a means to express a division between three modes of being or three ways to exist. It is argued that while the modes of being are predicated of objects, they do have a special, constitutive role in relation to those entities of which they are predicated. The word "exist" is thus an analogous rather than a completely ambiguous word. The author suggests that Frege's views on "being" and being contain elements both of Aristotelian and of Kantian approaches to metaphysics. The author also briefly addresses a claim to the effect that Frege was not really a metaphysician because he was not concerned with making sense of things but rather with making sense of sense (A. W. Moore).

Part II: Critical Metaphysics—The Scope and Limits of Metaphysics

This part of the volume is intimately linked to the first part. It elaborates on what ought to be the reach of an ambitious yet critical metaphysics—how far does it extend and which restrictions is it subject to? Underlying all the chapters is the idea that metaphysics is not only enabled but also constrained by the conditions of the possibility of mathematical and empirical science. The first two chapters are concerned with what belongs to metaphysics and how inclusive it might be. The remaining four chapters contain not only reflections upon the conditions of mathematics and empirical science; rather, they bring out how these reflections must themselves be guided by a careful examination of logical frameworks. After all, such frameworks are fundamental to the judgments and inferences made in any of the sciences.

The topic for Chapter 5 is the very domain of metaphysics—what belongs to it and what is the ground of its unity. Under the intriguing title 'Thinking-the-World,' Joseph Almog and Olli Koistinen maintain that metaphysics, science, and religion are all concerned with the same "world." In early modernity, this was a trio that existed in deep unity. If the one is divorced from the others, as in late modernity, the world itself is "lost." Here lies a source of critique of any science/metaphysics dualism—where metaphysics includes even caring for the world in moral or aesthetic terms. Rather than viewing Kant as one who segregates domains, he is considered by the authors as one who integrates them.

Kant is related to two rationalist thinkers that preceded him—René Descartes and Nicolas Malebranche. Descartes pointed out perceptively in the last stretches of his Third Meditation from 1641 that to think of one's own *limited* being (*Cogito, Sum*) is *inevitably* to think of one's being part of *unlimited* Being. Furthermore, Malebranche considered the infinitude of space as linked to the infinitude of God. It is argued that there is a distinct influence from the latter thinker on Kant's view of space. Indeed, the closest we get to a "world" that is given to us in Kant is in the form of space itself, which is said to be represented as an infinite given magnitude in the Transcendental Aesthetic, in the *Critique of Pure Reason*. In Malebranche, space is neither a substance that is separate from God nor is it an accident of God. Rather, space is ideal, since it is an idea God forms that expresses God's infinity. It is pointed out how Kant likewise ascribed ideality to space, and, how he, albeit in more speculative passages, suggested a link between the infinity of space and the infinity of God—not only in pre-critical writings but also in his lectures on rational theology of the 1780s. However, Kant was not able to justify that there is an intimate relationship between the infinity of space and the infinity of God within his own framework for theoretical philosophy.

Chapter 6 also presents a critical approach to what ought to be the domain of metaphysics—and the ground of its unity—but now with a view to the sharp distinction Kant himself draws within his system between theoretical and practical philosophy. In 'Kant's Metaphysics of Nature and Freedom,' Michael Friedman explores an approach to the relationship between nature and freedom in Kant that has not been sufficiently explored before. The focus is on Kant's conception of the best science of nature—articulated in his *Metaphysical Foundations of Natural Science*—and the relationship of this kind of science to his conception of the demands of morality is examined, that is, the demands of pure *practical* reason. This kind of science represents our ideal of *theoretical* knowledge, for Kant, and it is science in this sense that must be brought into a satisfactory relation with morally practical knowledge. It is argued that Kant's metaphysics of nature and freedom involves no mysterious division of reality into two different "worlds"—phenomenal and noumenal. It instead involves two different standpoints—theoretical and practical—on the same natural (phenomenal) world. These two standpoints are inextricably connected by a teleological evolutionary conception wherein the approximation of the Kingdom of Ends by humanity necessarily appears as a really possible continuation of the history of life on earth.

These two chapters, which bring together theoretical and practical philosophy in their discussion of what is the proper scope for metaphysics, are followed by four chapters where meta-theory and a logical framework play a central part in restricting any legitimate metaphysics.

In Chapter 7, Camilla Serck-Hanssen focuses on the limitations of logic, in 'From Nothing to Something—Why Metaphysics Cannot Be Reduced to Logic.' The topic is Kant's dismissal of the special metaphysics of the tradition, in particular his critique of rational psychology and rational cosmology. For the majority of interpreters, the Transcendental Dialectic's rejection of such metaphysics need not be studied in much detail because it simply follows from the outcome of the Transcendental Analytic. Since the Analytic has already established that the only legitimate use of the categories to yield cognitions is restricted to objects in space and time, it is this result that does the job and shows that any purported insights into transcendent objects such as the soul, the world, and God, must be dismissed, however appealing the arguments for these insights might be. The author deems this a misreading of Kant. Kant is to be taken seriously when he claims, e.g., that such proofs could indeed be a "stumbling block to his whole theory." What Kant does in the Dialectic is not to offer mere "therapy" in the face of metaphysical illusions but instead to deal in a serious and substantive way with the challenge that reason, which is not at all touched in the Analytic, can be the source of genuine metaphysical insights into objects that transcend what can be given in sensibility.

The hypothesis that is put forth by the author goes as follows. To understand the stance Kant takes towards the special metaphysics of the tradition, we should zoom in on the negative method ("apagogic," or *reductio ad absurdum* proofs) by means of which those who subscribe to special metaphysics (allegedly) proceed, as well as on the distinction between 'negative' and 'infinite' judgments. A line of critique to be found in Kant, according to the interpretation on offer, is that the proponents of traditional metaphysics commit a logical mistake: In predicating properties of the soul or the world, they treat S is P and S is non-P as contradictories and attempt to reason, say, from S is not non-P, to S is P, without worrying about the possible emptiness of S. If the subject term is empty, S is P cannot be established as a truth on the basis of the obvious falsehood of S is non-P. It is maintained that there is a line of critique here that runs through the entire Dialectic. While there are several other sources of critique of traditional special metaphysics in Kant, this one stands out, since it does not depend on Transcendental Idealism and its restrictions as such. Indeed, a critique that takes logical fallacies as its point of departure cannot be set aside as question-begging from a viewpoint that is external to Kant's own form of idealism.

In Chapter 8, Toni Kannisto shifts the perspective to one that is immanent to Kant's system of transcendental philosophy, in "Transcendentally Idealistic Metaphysics and Counterfactual Transcendental Arguments." Transcendental arguments have been widely discussed in analytic philosophy. Still, no formalization has been offered that can set them apart from other arguments either employing or seeking to ground necessity—e.g., analytic and mathematical. The author aspires to do precisely that by using counterfactuals to formalize a feature of such arguments, namely, the inverted order of inferring to a necessary condition rather than to a consequence of a given premise. The author's account lends support to Barry Stroud's claim that *metaphysical* or *objective* use of transcendental arguments must resort to idealism. However, *pace* Stroud, this is not to be seen as a sign of weakness. Rather, it is maintained that Kant considered transcendental arguments as the proper method for transcendentally idealistic metaphysics *that has no validity in non-idealistic metaphysics*.

The next two chapters are devoted to Husserl and metaphysics, and logic plays an important part here as well.

David Woodruff Smith addresses 'Phenomenology as Constitutive Realism' in Chapter 9. The heart of Husserl's conception of phenomenology is his theory of intentionality, analyzing the structure of our consciousness of things in the world around us. Husserl's phenomenology led to a form of "transcendental idealism," like in Kant, but the author thinks a better name for Husserl's position is "constitutive realism." Things in the world are "constituted" in our experience, and so experienced as "real," by virtue of the form of intentional content Husserl calls *noema*. Following his "transcendental" turn, Husserl made use of the method of epoché, "bracketing" the question of the existence of our surrounding world, as a technique for turning to the structure of our consciousness, including the noema.

In *Ideas* I, Husserl outlines a system of "formal" and "material" ontology within which he develops the fine details of his phenomenological theory of intentionality. His mature "transcendental" phenomenology thus emerges in interdependence with his complex ontology, each conditioning the other—according to Husserl's broadly logical *meta-theory*. Given his background in mathematics, Husserl's theory of noema and constitution reflects a broadly Platonistic view of meaning (in the spirit of Bolzano and later Frege) as well as an emerging view of metalogic or metamathematics (from Cantor and Hilbert onward toward Gödel and Tarski). Thus, noemata are both "*entertained*" in *everyday experiences* intending familiar objects (e.g., trees) and "*constituted*" in acts of *reflection* intending such experiences bearing noemata—that is, in phenomenological reflection via the practice of epoché. Husserl's "theory of theories," in the Prolegomena of the *Logical Investigations* thus prefigures a *meta-theory* in *Ideas* I as we move from everyday experience to higher-order reflection on everyday experience.

Furthermore, Husserl occasionally speaks of a "zigzag" movement forward and backward in our developing theory or explication of a given domain. In transcendental phenomenology, we may see such a movement between everyday lived experience and phenomenological reflection on everyday experience (to which we may return). Accordingly, the "constitution" of "real" objects in the world around us, by virtue of a manifold of noematic meaning, is revealed in phenomenological reflection on the given form of experience. In our analysis of intentionality, we normally zigzag back and forth between our surrounding *world* and our *experience* of that world. In this pattern of activity, we weave together our *metaphysics* of the world and our *phenomenology* of our consciousness of the world, following the very structure of intentionality. The foundations of our metaphysics lie in our lived *intentional relations* with the world. Accordingly, in Husserlian phenomenology we find a metaontology of "constitutive realism," defined by the very *structure of intentionality*.

Finally, in Chapter 10, Mirja Hartimo takes up another methodological issue in "Husserl on 'Besinnung' and Formal Ontology." Again, it is underlined that logic in Husserl is a *theory of science*, in particular of the norms of science. But the author now emphasizes that the philosopher's task is that of clarifying and renewing the "final sense" of logic towards which the scientists have been aiming. This aim should be pursued by means of the method of *Besinnung*. The employment of this method is thus a matter of "science first, philosophy second," rather than "philosophy first." To find out what the scientists are aiming at one should stand in, or enter into, a *community of empathy* with the scientists. Yet, Husserl's approach is not merely descriptive either. Husserl points out that "[r]adical Besinnung, as such, is at the same time criticism for the sake of original clarification" (*Formal and Transcendental Logic*, p. 10). The author discusses Husserl's notion of *formal ontology* as an outcome of radical *Besinnung*. It is argued that the goal of mathematics is to develop structuralist theories. However, objects as thus construed are too abstract to account for genuine formal ontology, which has to relate to what is possibly true. The goal of logic is *to seek truth*, and ultimately intuition of concrete objects, which leads to a more committal formal ontology than the thin one that is offered by formal mathematics. Husserl thus presents a judgment theory in which ontology is related to the actual world—an ontology that is universal but still "worldly."

Part III: Contemporary Metaphysics—The Role of Language and Science

This part of the volume contains three additional chapters in which the relation between contemporary metaphysics and modern science is central.

Chapter 11 brings up analytic metaphysics, which tends to be driven by philosophical intuitions. In Charles Parsons's 'Quine on Truth and Metaphysics,' this contemporary approach is discussed with a view to W. V. Quine's philosophy of language and his belief in the *continuity* between science and philosophy. The first part of the chapter undertakes to explicate Quine's view of truth. Quine holds that truth is a scientifically acceptable concept, while key notions of the theory of meaning are not. He maintains that the Tarski truth schema makes attributions of truth to sentences as clear as the sentences themselves. In *Word and Object*, Quine holds to the strict Tarskian schema by substituting non-context dependent expressions for indexicals and demonstratives. This coheres with the objective of giving a canonical language for science, contrasting with the program of Davidson and others for a semantics of natural language.

In this and other writings, Quine is concerned with immanent truth, applying to sentences in one's own language or theory. Otherwise, problems of translation arise. The grammar of "true" makes truth absolute, but the specter of relativism arises when two theories might be empirically equivalent but the usual canons of scientific method do not give rise to a decision between them. If they are intertranslatable, we can regard them as one theory. Otherwise, Quine offers different solutions. The author goes on to inquire whether Quine was a metaphysician and what he might have thought about the analytic metaphysics of today. On the first point, science does the work of metaphysics, with reservations due to ontological relativity. On the second point, an instance of contemporary analytic metaphysics is examined, namely, Theodore Sider's Four-Dimensionalism. The author speculates about what Quine would have thought about it.

The final two chapters exemplify a kind of metaphysics that is *distinctive* in its own right but which is very much scientifically informed. Cantor and Weyl are the two key figures from mathematics and physics, respectively, and 'infinity' and 'local symmetry' are the two main concepts from mathematics and physics, respectively. The chapters thus highlight more concretely how metaphysical investigations can be conducted in the restricted manner that is recommended in our volume, i.e., in a methodologically clarified way—by building on self-reflective traditions within the history of thought, as well as in a way that is informed by present-day science.

In Chapter 12, Øystein Linnebo presents "The Paradox of the Largest Number. From Aristotle to Cantor." The infinite is a central topic in metaphysics. The chapter is (a) historically informed, referring specifically to the views of Aristotle/Kant and Cantor; (b) scientifically informed, in relating to contemporary logic and mathematic; and (c) critical, in being shaped by concerns similar to those found in Aristotle's and Kant's infinity critique. The chapter is thus firmly placed within critical metaphysics.

The author starts out with a simple version of the paradox of the largest number. Is there a largest number? Suppose a young child asks the question. Most parents would answer "no." They might add the following explanation. Suppose a gazillion was the largest number. Then a gazillion and one would be an even larger number. So, there cannot be a largest number. The child might respond by asking whether infinity is not a largest number. She might reason as follows. There are infinitely many natural numbers. So, if we added one more thing, say a fraction, we would still have infinitely many things. So, there cannot be a number greater than infinity. And this means that infinity is the largest number. Although admittedly somewhat naive, this reasoning should not be dismissed as just childish confusion. On the contrary, it is argued, there are important lessons to be learnt from pondering the paradox.

To defend this claim, it is first shown how a less naive version of the paradox arises in the context of the ancient conception of infinity as the property of unboundedness. Next, it is explained how an elegant solution is provided by the Aristotelian concept of potential infinity. This remained the dominant conception of infinity in mathematics and philosophy for more than two millennia, until the Cantorian revolution in the second half of the nineteenth century, which ushered in the now-standard view. Although the Cantorian revolution solves many of the traditional paradoxes of infinity, it is shown that a version of the paradox of the largest number remains. To improve on the situation, a "successor concept" to the ancient concept of potential infinity is finally articulated. It is argued that this concept enables us to resolve the paradox. If this is indeed the case, the concept of potential infinity is not only of great historical importance but also of profound contemporary significance. This illustrates how the history of philosophy can serve as an inspiration and source of ideas for the contemporary philosophical debate.

Whereas Linnebo's chapter is primarily about modal metaphysics and pure mathematics, it is followed by a piece that turns to metaphysics of nature, in its current form, and, thus, to the intersection between mathematics and physics. In Chapter 13, Thomas Ryckman scrutinizes "Symbolic Construction from the 'Purely Infinitesimal': Gauge Invariance, Lie algebras and Metaphysics chez Hermann Weyl." The so-called "gauge principle" holds that the form of fundamental interactions (described by the Standard Model) is in some sense dictated by requirements of local (gauge) symmetry. The principle presents a mystery to contemporary philosophy of physics: How can apparently *a priori* mathematical degrees of freedom give rise to such astonishing empirical success? This author aims at shedding illumination on the issue by revisiting the transcendental phenomenological origin of the principle of local symmetry in the context of General Relativity by Hermann Weyl in 1918, and its subsequent development (by Weyl and others) in the quantum mechanical setting of matter waves. Reflection on this history illustrates the transcendental constitutive significance of Weyl's widely quoted statement, "As far as I know, all *a priori* statements in physics have their origin in symmetry."

This closing chapter of our volume thus also brings out a distinct influence from Kant's transcendental philosophy and from Husserl's transcendental phenomenology upon a scientist who made important contributions to present-day mathematics and physics.

Notes

1. For support of this view, cf. also Jessica Wilson, "Much Ado about 'Something'," *Analysis* 71/1, 2011.
2. Fine, Hawthorne, Lewis, and Lowe are among those who dominate the field of ambitious metaphysics.
3. This has been argued in Ladyman, James and Don Ross: *Every Thing Must Go: Metaphysics Naturalized* (Oxford, 2007).
4. As such, the project underlying this volume, although it draws on additional philosophical sources, is on a par with the approaches of Fine, Hawthorne, Lewis, and Lowe. However, it has more in common with some recent contributions to ambitious metaphysics, such as those of one of our contributors, Øystein Linnebo, and also some of Jessica Wilson's work.
5. For a different view, cf., e.g., Colin McGinn: *Problems in Philosophy: The Limits of Inquiry* (Blackwell, 1993).
6. For an account of the history of metaphysics that traces it to the present time, and also points to a certain lack of sensitivity to meta-issues in philosophical traditions, see A. W. Moore: *The Evolution of Modern Metaphysics: Making Sense of Things* (Cambridge, 2012).
7. One prominent example is Timothy Williamson: *The Philosophy of Philosophy* (Blackwell, 2007).
8. An example is E. J. Lowe: *The Possibility of Metaphysics. Substance, Identity, and Time* (Oxford, 1998). Despite its suggestive title, this work does not really address the problem: How is metaphysics possible?
9. For a strong reliance on semantics, cf. many of the papers in *Metametaphysics: New Essays on the Foundations of Ontology*, ed. by David J. Chalmers et al. (Oxford, 2009).

Part I

Metametaphysics

The Very Possibility of Metaphysics

1 Kant on Method and Evidence in Metaphysics

Frode Kjosavik

I. Introduction: Kant's Critical Ambition

In the *Critique of Pure Reason*, Kant comes up with his own "critical method" for philosophizing. In this chapter, I shall go into aspects of his methodology and what it entails for the scientific legitimacy and role of metaphysics. However, before turning to the origins and specifics of his proposed framework, it is perhaps good to be reminded of the following. There is a very general sense in which there are methodological maxims for critical thinking in Kant. In *What is Enlightenment?* (1784), Kant puts forth what he considers to be *the* maxim of enlightenment: Have the courage to make use of your own understanding! (Ak. 8: 35). In *What does it mean to orient oneself in thinking?* (1786), he states this as follows:

> *Thinking for oneself* [Selbstdenken] means seeking the supreme touchstone [Probierstein] of truth in oneself (i.e., in one's own reason); and the maxim of always thinking for oneself is *enlightenment*.
>
> (Ak. 8: 146n)

Kant presents three related maxims in his *Anthropology* (1798, Ak. 7: 228): 1. Think for oneself; 2. Think into the place of the other (person) (in communication with human beings); 3. Always think consistently ["einstimmig"] with oneself. The rules are also given as "maxims of common human understanding" in the *Critique of Judgment* (1790, Ak. 5: 294). Any discursive thinker ought to adopt these as maxims for a *sound* way of thinking, which is without prejudice, in an expanded way, and self-consistent ["konsequente Denkungsart"], respectively.

Now, just as one ought to think for oneself in everyday life, one ought to do so in philosophy as well. Indeed, there can be no rational philosophizing which does not spring from the reason of the thinker. The continuity between critical thought in everyday life and critical thought in philosophy is particularly evident in the Architectonic, in the Doctrine of Method, in CPR (A836–A838/B864–B866). Whereas a proposition may *objectively* be a sufficient ground for knowledge, it can have a different status *subjectively*, with regard to how it is in fact known by a subject. Kant had drawn a distinction already in *Inquiry* (1764) between an objective and a subjective aspect of the certainty with which a proposition is known. The objective aspect is "the sufficiency in the characteristic marks of the necessity of a truth" (Ak. 2: 290–291), whereas the subjective aspect is the "intuitiveness" with which it is known. A proposition can be known with more intuition ["mehr Anschauung"], or with less. Both objectively and subjectively, there is a difference in kinds of certainty as well as in degree—at least when it comes to the metaphysics that has been developed so far.

The difference in kinds of certainty is transformed into a *methodological* distinction between rational cognition through concepts and rational cognition through *construction* of concepts in CPR. The difference in degrees of certainty, on the other hand, which now merely pertains to the subjective aspect, is itself developed into two *ways of knowing*—a *historical* and a *rational* way. There is only the rational way of knowing within mathematics, Kant thinks, because a mathematical proposition cannot be grasped without exhibition of its concepts in pure intuition. Within philosophy, on the other hand, one and the same proposition may be known subjectively either way. I know it historically if I know it from the outside, from others through instruction, as one may learn the system of Wolffian philosophy from one's teachers. I know it rationally if I know it from the inside, i.e., from my own reason. The latter does not rely on intuition, though, which is now taken to be conditioned by space and time as forms of human sensibility, but on a distinctive form of rational insight. In particular, there must then be consciousness of the necessity of a proposition—the necessity must be exhibited through the judgment itself. Only then can there be the *apodictic certainty* that is required by a proper rational science, like mathematics or metaphysics.

Kant goes on to claim that one cannot learn philosophy in the form of a system developed by others in the rational sense but only in the historical sense. One can learn rationally only *to philosophize*. The requirement of "intuitiveness" of which he spoke in *Inquiry* is now fulfilled to the highest degree by thinking as an activity that illuminates itself—as in a "critique of pure reason." What is offered thereby is nothing less than a "treatise on the method" of metaphysics (Bxxii). In the *Jäsche Logic*, the primacy of method in critical philosophy is described like this:

> For the sake of practice in thinking for ourselves, or philosophizing, we will have to look more to the *method* for the use of our understanding than to the propositions themselves at which we have arrived through this method.
>
> (Ak. 9: 26)

The motivation for being critical—and even make use of a "critical method"—is ultimately a moral one. The methodological ought reason prescribes for a metaphysics of nature is subservient to a "metaphysics of morals"—or to the practical postulates of immortality, freedom, and the existence of God. As Kant puts it in an often-quoted passage in the B-edition preface: "I had to deny *knowledge* in order to make room for *faith*" (Bxxx). However, apart from these supreme ends, there is also the virtue that consists in the soundness of reason, be it in philosophy or in everyday life—a virtue that pertains to critical thinking as such.

Against this background of moral interests and virtuous thought, Kant brings up the *conceptus cosmicus*, the concept of philosophy which concerns what is of interest to everyone, as an alternative to the scholastic concept. To be sure, transcendental philosophy as such is a "philosophy of pure, merely speculative reason" (A15/B29) and does not include moral philosophy. However, philosophy is still "the science of the relation of all knowledge to the essential ends of human reason (*teleologia rationis humanae*)," and the philosopher is himself "the lawgiver of human reason." (A839/B867). The "idea of his legislation [i.e., of the ideal philosopher] is to be found in that reason with which every human being is endowed" (A839/B867). "Now, the legislation of human reason (philosophy) has two objects, nature and freedom" (A840/B868). One deals with *what is*, the other with *what ought to be*. The final end ["Endzweck"] is the "whole vocation of man, and the philosophy which deals with it is entitled moral philosophy" (A840/B868). Practical reason thus has primacy over theoretical reason. Kant even thinks that ultimately theoretical and practical philosophy belong together in a single system. But it is theoretical reason and the systems—or subsystems—thereof that shall be our focus here.

II. Toward Critical Thinking in Metaphysics—The Contrast With Mathematics

As has been pointed out by the analytic metaphysician Lowe (1998, 24–25), there are some examples of traditional metaphysical arguments in Kant, not only in the pre-critical but also in the critical works. One example is his argument from "incongruent counterparts"—non-superimposable mirror images, like a left and a right hand, in some idealized version—as this was used against a relationist theory of space in *Concerning the ultimate ground of the differentiation of directions in space*, of 1768. Another example, from CPR, is when Kant argues that space and time cannot be substances, for such a non-entity ["Unding"] cannot exist in reality (CPR, A39–40, B56–57). Kant's main approach to metaphysics, however, is critical, even in his pre-critical writings. While there is not yet the critical method that comes with the Copernican turn, Kant realizes from the outset that metaphysics cannot be pursued without proper regimentation. He sees it as his task to bring out its distinctive perspective and methodology. Mathematics serves as a contrast in this regard. It is a highly regimented pure science. The relationship between mathematics and metaphysics remains central throughout his philosophical writings, from *Living Forces*, of 1746/7, published in 1749, to his very last work *Opus postumum*, containing material composed in the period 1796–1803. (Cf. Menzel 1911 on the pre-critical writings; Büchel 1987 on the critical and "post-critical" writings.)

In *Living forces*, Kant highlights a difference in perspectives between mathematics and metaphysics by distinguishing between mathematical and physical bodies (§ 115). There is also a correlated difference in methods. Whereas mathematics defines its concept of body by means of presupposed axioms (§ 114), metaphysics has to analyze given concepts and seek the principles from which given consequences follow (cf. § 89, Ak. 1: 96). The distinction between kinds of bodies bears some resemblance to the later distinction in CPR between *extensive* and *intensive* magnitudes, and a similar correlated difference in methodology. In CPR, Kant maintains that mathematics is concerned with the "synthesis of the homogeneous," and it is precisely extensive magnitudes that are homogeneous. Indeed, in *Inquiry* Kant had stated that mathematics deals with a sphere of homogeneity, and metaphysics with a sphere of heterogeneity. However, in CPR it is said that even intensive magnitudes, like colors, heat, and gravity (A169/B211) can be mathematized and quantified according to their degree. The difference is that they cannot be mathematized directly. Rather, each sense quality or empirical "reality" has to be mathematized according to its own nature. In *Prolegomena* (1783), § 24, Kant calls this "the second application of mathematics (*mathesis intensorum*) to the science of nature." There is still a contrast, then, between outer properties and relations, which can be dealt with directly mathematically, and inner natures, which require metaphysics for their mathematization.

In *Physical Monadology*, of 1756, the relationship between mathematics and metaphysics is presented as more problematic: "But how, in this business, can metaphysics be married to geometry, when it seems easier to mate griffins with horses than to unite transcendental philosophy with geometry?" (Ak. 1: 475). Kant here employs the term "transcendental philosophy" in the sense of general metaphysics, and basically in the way the term is used in Christian Wolff. Of course, Kant's conception of transcendental philosophy is to undergo significant changes later. In the *Critique of Pure Reason*, *transcendental* cognition is said to be "occupied not so much with objects but rather with our mode of cognition of objects insofar as this is to be possible a priori. A *system* of such concepts is to be called *transcendental philosophy*" (CPR, A11/B25). Kant thus develops his own *critical* conception of 'transcendental philosophy,' which can be seen as a turn towards the faculties of the subject and how these are constitutive of general traits in cognition, and, thus, away from the objects of cognition in their specificity. This critical turn is arguably motivated, in part, by the very attempt to account for how metaphysics and mathematics can be combined with each other.

Of particular importance in that regard is an "antinomy" presented in *Physical Monadology*, namely, the problem of squaring the infinite divisibility of space, which is a mathematical property thereof, with the metaphysical grounding of matter in ultimate constituents, like Leibnizian monads, or some descendant thereof. Kant finds a peculiar solution in that work, i.e., the activity sphere of the force of repulsion centered in the monad is infinitely divisible but not the monad itself. The problem reappears in the form of the Second Antinomy in CPR, i.e., within the Transcendental Dialectic rather than Analytic. It is there seen as a problem that emerges only when reason does not curb its pretensions and confines itself to metaphysics that is immanent to the human forms of possible experience. Furthermore, Kant no longer thinks that the infinite divisibility of matter is a metaphysical implication of the infinite divisibility of space. *Mathematics does not have such a special metaphysical implication in its own right.* As Kant makes clear in the *Metaphysical Foundations of Natural Science* (1786), the proof of the infinite physical divisibility of matter requires further steps that take us from mathematics to physics, i.e., any portion of matter fills a space through the force of repulsion, and, as such, it is independently movable (Cf. Friedman 2013, 143–154 on this.)

In *Inquiry*, Kant is very clear on the difference in existential commitments and methodology between pure mathematics and general metaphysics. Mathematics begins with definitions, since its concepts are generated through definitions, except for some fundamental ones, like that of space itself. Metaphysics, on the other hand, ends with definitions. The reason for this is that in mathematics, the definitions are *synthetic*, i.e., novel concepts are freely composed from marks, whereas in metaphysics they are *analytic*, in that given concepts are resolved into marks. Kant here emphasizes a contrast in how one relates to concepts within the two fields, and also notes that mathematics can be cognized in stages—from the elementary to the more advanced, in a way that metaphysics cannot.

This contrast in how one relates to concepts is taken further in CPR, where mathematical concepts are said to "contain" a "synthesis" beyond that of discursive marks. A concept is not a proper mathematical one if its synthesis of marks does not also represent the *synthesis of an intuitive manifold*. Thus, the concept of a figure enclosed within three straight lines is a proper mathematical concept, whereas that of a figure enclosed within two (Euclidean) straight lines is not. Both contain a synthesis of discursive marks, but only the former represents a synthesis of an intuitive manifold. Metaphysics, by contrast, is not concerned with arbitrary synthesis but with the synthetic unity of transcendental apperception as the most general ground for synthesis and with each category as a special ground, or as a "ground of a special unity" (Paton 1936, 277n), of a priori synthesis. While mathematical concepts introduce specific grounds of unity in pure syntheses as well, the categories are *given* grounds in the sense that they are fixed by the essence of the faculty of thought. It is the transcendental philosopher's task to exhibit the categories and principles of the pure understanding in a complete system. Thus, in the preface of *Metaphysical Foundations*, Kant says that "All true metaphysics is drawn from the essence of the faculty of thinking itself" (Ak. 4: 472). He goes on to state that the concepts and principles of the pure understanding are manifestations of its nature. (For an in-depth interpretation of this passage and a concomitant account of 'rational insight' in Kant, see Chapter 2 in the present volume.) Categories are not freely generated, unlike concepts in mathematics, nor are any special objects, with their own essences, given through them as such.

According to *Inquiry*, mathematics also offers a kind of evidence that is absent in metaphysics, namely, that of shapes that are directly given. In mathematics, geometrical figures or signs are set before one's eyes. Intuition of shapes, e.g., in the form of finitary numerals, later became the central source of evidence in Hilbert's contentual mathematics. However, in CPR Kant subtly shifts the locus of intuitive evidence from figures or signs, and their properties and relations, to the *construction of concepts*. To construct a concept is "to exhibit a priori the intuition corresponding to it" (A713/B741). With construction comes an aspect of practical certainty as well. The intuition of spatial or temporal forms is essentially pure, without sensory content, and non-essentially empirical, as when one seeks support from the senses through drawn or inscribed shapes.

Moreover, mathematical concepts are introduced through real genetic definitions which are themselves original constructions in intuition. To define in mathematics is "just to exhibit originally the exhaustive [ausführlich] concept of a thing within its boundaries" (A727/B755). Or, as Kant puts it in the letter to K. L. Reinhold, of May 19, 1789 (Ak. 11: 43): "If a circle is defined as a curve all of whose points are equidistant from a mid-point, is not this concept given in intuition?" Mathematical postulates, on the other hand, with their practical certainty, are mere practical corollaries of the definitions: "And this even though the practical proposition ["Satz"] that follows, viz., *to describe a circle* (as a straight line is rotated uniformly about a point), is not at all considered" (Ak. 11: 43; cf. CPR, A234/B287 for a less accurate description of postulates in mathematics).

The novel concepts that are generated in mathematics can thereby be considered as inventions. Through these, novel synthetic unities are established through which objective forms or formal objects are originally given. There are mathematical objects in Kant, then, but no mathematical "things in themselves," as Platonism will have it. After all, an *object* is "that in the concept of which the manifold of a given intuition is *united*" (B137). The concept of a circle merely represents the *synthetic unity* of points on a curve that are equidistant from a fixed point. After a mathematical concept has been generated and a class of mathematical objects falling under it has been given, one can discover various properties that are not constitutive marks in the very definition but derivative properties, e.g., of circles. Mathematics thus relies on a combination of invention and discovery. (Cf. Sandmel 2001, 414–416, on "schematic construction" and its "metaphysical shadow" in mathematics, as well as his critique of intuitionism in that regard, p. 541).

However, Kant does not have much to say about the discovery aspect in mathematics in CPR, as opposed to the invention aspect. In *The only possible argument [Beweisgrund]* (1763), on the other hand, he characterizes mathematical objects as having infinitely many properties. We get to know these properties by proving theorems about them, like proposition 35 in Euclid, Book III, concerning pairs of intersecting chords of a circle. However, the objects have an inexhaustible richness to them, so there is no upper bound on the number of theorems that can be proved about them, e.g., of a circle: "the figure of the circle . . . which has infinitely many properties of which only a small number is known" (Ak. 2: 95). The properties can be taken to flow from the essence of a circle. In *Prolegomena*, using the same example of Proposition 35 in Book III in Euclid, Kant is likewise concerned with "the unity of the *manifold properties* of geometrical figures under common laws" (Ak. 4: 321; my italics).

Metaphysics, on the other hand, is a science of *mere* discovery, and not of invention also. It aims at clarification and justification of what is already constituted. For that purpose, metaphysics has its own methods, including that of *isolation*:

> It is of utmost importance to *isolate* cognitions that differ from one another in their species and origin, and carefully to avoid mixing them together with others with which they are usually connected in their use. What chemists do in analyzing materials, what mathematicians do in their pure theory of magnitude, the philosopher is even more obliged to do, so that he can securely determine the proper value and influence of the advantage that a special kind of cognition has over the aimless use of the understanding. Hence human reason has never been able to dispense with a metaphysics, as long as it has thought, or rather reflected (. . .).
>
> (A842/B870)

What is a priori is isolated from what is a posteriori (A20–22/B35–6; B870), contributions from sensibility are isolated from contributions from the understanding (A22/B36; A62/B87), and a further question is raised as to whether pure reason can be isolated from the understanding (A305/B362).

Complementing such isolation is a kind of *reflection* that the transcendental philosopher has a duty to perform explicitly so as not to fall prey to deceptions. It leads to radical transformations of traditional distinctions in metaphysics. Kant brings it up in the Amphiboly in CPR (A263/B318–19). In *logical* reflection, concepts are related to each other with a view to objective judging, i.e., making universal or particular, affirmative or negative, etc. judgments. The concepts for such reflection, and, thus, for establishing truths, come in four pairs, those of identity vs. difference, of agreement vs. opposition, etc. However, *transcendental* reflection in the same aspects is required for comparing or connecting representations that compose cognitions of things. Leibniz "intellectualizes" such cognition when he does not acknowledge the constitutive contribution from sensibility, just as Locke sensualizes it by not acknowledging the constitutive contribution from the understanding. A philosopher's explicit reflection on cognitions should not be performed on mere intellectual or mere sensible surrogates for physical phenomena—by an "amphiboly," or equivocation, as it were.

Transcendental reflection on representational content, when performed with sufficient care, brings out the ontological implications of such content. According to Kant, neither are the objects of our experience mere objects of pure understanding nor are they mere objects of pure sensibility. Rather, of necessity they have dual origins, and, hence, exemplify or instantiate both intellectual and sensible a priori forms. Physical phenomena therefore have constitutive features that are represented through two distinctive species of representations—concepts and intuitions. Thus, exactly similar portions of sensible matter can still be distinct—two drops of water can be perfectly alike—by occupying different locations in space, which is a form of intuition. Empirical realities can occur on a par—e.g., by filling spatial regions—albeit with opposite "charge" relative to each other, like two moving forces of the same intensity that cancel each other, as mere "negative magnitudes" with opposite orientation in space. (This is a metaphysical topic that goes back to Kant's pre-critical work *Negative Magnitudes*, of 1763.)

These are the transcendental philosopher's explicit deliberations, and, according to Kant, they undermine rationalist doctrine, like the identity of indiscernibles or the principle that what is negative has no reality of its own but is a mere privation of what is positive, respectively. Two further central distinctions in traditional metaphysics run the same risk of a transcendental amphiboly—those between intrinsic vs. extrinsic properties and between matter vs. form. The latter distinction is of special significance, since the Kantian priority of a priori form over a posteriori matter plays a part in all transcendental reflection in which deceptions are removed, i.e., where sensible and intellectual forms of empirical phenomena also have been isolated from each other.

Correlated with the difference in methodology between mathematics and metaphysics is a difference in their sources of evidence. Mathematics relies on construction in intuition, and its evidence is therefore *intuitive*. While such construction involves intuition of shapes, be it of figures or of signs, this is not all intuition boils down to in mathematics, according to Kant. There are also *intuitive principles* in the form of axioms, e.g., "Between two points only one straight line is possible," (A163/B204), which are "immediately certain" (A732/B760) synthetic a priori principles. The axioms are "self-evident" (A733/B761), and could of course be taken to be immediately certain merely in that sense—they justify themselves (cf. A261/B317: "no further mark of truth can be given for them"). However, there is also a more specific sense of "immediate certainty" at stake here, namely that the concepts in such propositions are connected in the medium of pure intuition, and not by means of further conceptual marks. For that reason, propositions that come with intuitive certainty—which Kant here calls *mathemata*—include theorems as well. In one of his reflections on metaphysics, mathemata are said to comprise all "cognitions insight into which and certainty about which are *intuitive* but thereby still apodictic" (R5645, Ak. 18: 292; my italics). In *Prolegomena*, § 7, Kant draws a general distinction between synthetic a priori judgments that are *intuitive* and those that are *discursive*.

Kant thus ascribes not merely intuition of shapes to mathematical practice but intuitive insight and intuitive certainty as well. All three are interlinked, though. This is made clear in the Discipline, in the Doctrine of Method, when he speaks of a "chain of inferences that is always guided by intuition" (A717/B745) and of intuitive proofs which are accompanied by "intuitive certainty" or "self-evidence" (A734/B762). In algebra, one relies on the display of steps and patterns in a symbolic construction that "secures all inferences against mistakes by placing each of them before one's eyes" (A734/B762). Unlike philosophy, which lacks structural transparency of the intuitive kind, "mathematics can assess the universal *in concreto* (in the individual intuition [in der einzelnen Anschauung]) and yet through *a priori* intuition, where every false step becomes visible" (A734–35/B762–63). Furthermore, in the Architectonic, in the Doctrine of Method, it is said that "reason is here used *in concreto* though nevertheless *a priori*, founded, that is, in pure and therefore *error-free intuition*, and excludes all deception and error" (A837/B865; my italics).

In metaphysics, evidence does not lie in intuition as such, be it of shapes or scenes, or in the form of support for executions of operations. Within theoretical philosophy, apodictic propositions can also not be rationally grasped directly from concepts, the way the rationalists maintained. The synthetic principles of the understanding are not immediately certain and cannot be posited as mere *dogmata* either: "Philosophy thus has no axioms *and can never simply offer its a priori principles as such*, but must content itself with justifying their authority through a thorough deduction" (A733/B761n; my italics). Its evidence lies in rational insight—albeit not in *mere* rational insight. In the General Comment on the System of

Principles in CPR, Kant makes it clear that even rational insight into the possibility of a thing or into the objective reality of the categories (B288) requires reliance on the conditions of intuition, and it does not extend beyond these. Indeed, the "boundaries of all human insight" is set by the "field of possible experience" (A753/B781). We do not have insight into a thing in itself, or its essence, but we do have insight into the essence of a cognitive capacity. Insight into the objective reality of mathematical concepts is different. It requires only insight into an arbitrary synthesis of an intuitive manifold. There is no empirical reality to be grasped. The insight is therefore intuitive.

III. Abduction, the Actuality of Sciences and the Possibility of Metaphysics

The task of clarification and justification in metaphysics brings us to what I consider as *abduction*, or inference to the best explanation, in Kant's critical project. Timothy Williamson sees this method as the central one in philosophy: "I propose that philosophy *should* use a broadly abductive methodology. Indeed, to some extent it already does so. I propose that it should do so in a bolder, more systematic, more self-aware way" (2016, 268). Explanation is precisely Kant's concern in CPR, when taken in a wide sense that includes constitution analysis, which is performed in a "systematic" and "self-aware way," as Williamson proposes. Kant's theoretical philosophy is *transcendental* in the novel sense of being concerned with "How is x possible?" questions. This suggests abductive moves that account for the possibility or even actuality of x. I shall argue that there are at least two abductive routes to Kant's critical metaphysics. Neither of these constitute the final justification of Kant's critical metaphysics, though, as we shall see later, in Section IV. Furthermore, just as Williamson takes the evidence base for abduction in philosophy to be *total*, evidence from mathematics and the natural sciences is certainly included in Kant's evidence base—metaphysics is not a mere "armchair science"—even if pure philosophy also comes with its own distinctive kind of evidence.

In CPR, Kant does not ask "How is experience possible?" He asks, "How is metaphysics possible as science?" (B22), along with "How is pure mathematics possible?" (B20) and "How is pure natural science possible?" (B20). The primary objective of the *Critique of Pure Reason* is to clarify the theoretical foundations for rational sciences. It is not to offer a descriptive metaphysics of experience—the invariants of a "conceptual scheme" for it, or its necessary "minimal conceptual structure," as P.F. Strawson maintains (1992, 26). Furthermore, whereas mathematics and physics are already established as successful forms of rational inquiry—mathematics in antiquity and physics through the scientific revolution—this is not yet the case with metaphysics. This raises the pressing question whether metaphysics can be set on "the secure course of a science" (Bxiv) as well.

The critique as such is a rational meta-project that aims at answering this question through reason's self-examination. To be sure, any theoretical science which is not merely formal has its own domain of discourse and contains synthetic a priori principles. In laying down a framework for such principles, Kant's transcendental philosophy has the potential to offer a radical critique of its current foundations—a critique that could potentially alter its very framework. However, Kant does not propose radical reforms in mathematics or physics as such—his attitude to these sciences remains largely anti-revisionist. After all, they are very successful in their contemporary forms. With regard to pure general logic, he does not even ask how it is possible. It is an *analytic* science which does not theorize about any specific domains of objects, and, hence, it is not to be subjected to a transcendental investigation into its foundations. Still, Kant proposes nothing less than a revolution in metaphysics. He pins down "elements"—concepts and principles—that must account for the very possibility of metaphysics as a science.

An important aim of this reflexive project is *to draw a demarcation line between good and bad metaphysics*, or between metaphysics that respects its proper foundations and metaphysics which does not. Good metaphysics turns out to be metaphysics that is restricted by the conditions of possible experience, and, thus, by the faculties of the cognizing subject. The aim of such metaphysics is not to establish the possibility of experience as such, but to unveil how that very possibility warrants the objective reality of our concepts. One has to make use of the "critical method" (A856/B884) to develop a metaphysics that is neither too ambitious, i.e., not "dogmatic," in ascribing objective reality to concepts which are empty in a theoretical context; nor too modest, i.e., not "skeptical" by questioning whether anything like experience, in the thick sense of empirical cognition, can even be made intelligible.

Since metaphysics is a foundational science, to ask how metaphysics is possible as a science is also to ask how pure mathematics and pure natural science are possible. The questions are intimately related. Kant is of course not raising doubts about the possibility of the rational sciences which are already actual, i.e., mathematics and natural science. Section V of Dissertatio (1770) on the "method in metaphysics," § 23, is very illuminating in this regard. There Kant distinguishes between sciences in which "the principles are given intuitively" (Ak. 2: 410–11), be it in pure or empirical intuition, and sciences in which they are not. To the former belong mathematics and natural science. In these, "*use gives the method.*" The thought is that much has been developed from "intuitively given," in some sense, concepts and principles (axioms) even before the methodological clarification that is required for there to be fully developed exact sciences. In metaphysics, on the other hand, "*method precedes all science*." One does not even reach the fundamental concepts or principles, its starting points, without having a proper methodology.

Similarly, in CPR Kant speaks of "the reality [Wirklichkeit] of the scientific cognition a priori that we possess, that namely of pure mathematics and general natural science" (B128), and says that Locke's attempted "empirical derivation" of the categories is "refuted by the fact [das Factum]." There is a sense, then, in which these rational sciences have already rationally justified themselves. Against this background, we can see even more clearly how it is *metaphysics* that is put to a rational test in CPR in relation to sciences that are ruled by, or at least contain, intuitive principles.

Kant states explicitly that metaphysics deals with reason according to "its elements and highest maxims," and these must underlie (1) the possibility of some sciences and (2) the use of all sciences (cf. A851/B879). Physics is a science whose very possibility has to be accounted for in metaphysics. Mathematics is a science whose applicability certainly has to be accounted for in metaphysics. But even its possibility—albeit a thin formal one—calls for an account. After all, the principle of the axioms of intuition in CPR is not only that all appearances are extensive magnitudes but effectively that all synthesis in pure intuition is subjected to the quantitative categories. (Cf. Sutherland 2005 on this topic.) Kant's revolution in metaphysics certainly has implications for the grounding or validation of both mathematics and physics, as is clear from the analytic of principles in CPR. It even has implications for the application of formal logic—as can be seen in the use of "apagogic"—or *reductio ad absurdum* proofs—in the antinomies in the Dialectic of CPR and the ultimate suspension of this proof method in metaphysics (cf. Chapter 7 in the present volume).

As we have already seen, there is a sense in which mathematics makes up a semi-autonomous domain of rational thought that is not subject to the same pitfalls as metaphysics. The intuitive demonstrations and semi-formal proofs in mathematics are secure in their own right. Physics, on the other hand, has a special metaphysical part, in addition to its applied part. Of course, the very idea that there can be a part of physics that is a priori might seem very puzzling to us today, in particular if this is taken to entail immunity to falsification. In a paper on Kant's *Metaphysical Foundations*, Charles Parsons speaks of the "hardest nut in Kant's philosophy of science, his conception of an a priori or "pure" part of science" (Parsons 2012, 69). He discusses some of the intricacies involved in this. I shall go more into the relation between metaphysics and physics in Section IV.

At this point, we may note that it is precisely because physics is entangled with metaphysics that it leaves so much room for speculation that is not (yet) backed by evidence, unlike mathematics—as seen in the debates between the Leibnizians and the Newtonians, or in present-day debates on theoretical physics. But while Kant attempts to remove all dogmatism through his critique, he does not thereby seek to eradicate metaphysics from physics. Rather, he seeks to clarify and justify the metaphysical foundations of natural science and thereby properly restrain all speculation within it.

Kant does seem to think that there is a route through abduction from mathematics and physics as established sciences, then, to the establishment of transcendental philosophy as itself a science. Among the rival rationalist and empiricist accounts of the sciences, Kant's own transcendental philosophy is supposed to offer the best justification of their possibility and actualization. The presentation in *Prolegomena* is very much in accordance with this route of abduction, as it makes use of what Kant calls the *regressive* or *analytic* method, rather than the *progressive* or *synthetic* method, i.e., it moves from the conditioned—mathematics and natural science in their actuality—to the conditions, i.e., their grounds of possibility in the human sensibility and understanding, as well as their actualization in scientific practice, as guided by ideas of reason. (The latter are supplemented by reflective judgment in the *Critique of Judgment*.)

However, I shall suggest that there is another route of abduction in CPR as well. In the B-introduction, Kant speaks of "what we assume as the altered method of our way of thinking" and identifies this with his Copernican turn, i.e., "that we can cognize of things *a priori* only what we ourselves have put into them" (Bxviii). He goes on in an appended footnote to characterize the method as imitating experiments within the natural sciences, in that it seeks "the elements of pure reason in that *which admits of being confirmed or refuted through an experiment*." He notes that one cannot experiment with objects here in the way one can within natural science, but rather with "*concepts* and *principles* that we assume a priori." In physics, which is not wholly pure, or even wholly a priori, there has to be empirical input from empirical experiments. In metaphysics, on the other hand, we are clearly facing a mere thought experiment, which Kant also calls the "experiment of pure reason" (Bxxin).

This experiment consists in considering things as if the Copernican turn is true, and, thus, from a two-fold standpoint, or from two different sides ["Seiten"]—the standpoint of appearances and that of things in themselves, or as Kant puts it in the footnote (Bxviiin), as objects of the senses and of the understanding, on the one hand, and as objects that we merely think on the other. We see how much sense we can make of things from this two-fold standpoint, in particular—within the first Critique—with a view to the theoretical sciences. We are then supposed to compare this to considering things only "from one and the same point of view," and if this leads to "an unavoidable conflict of reason with itself" the experiment confirms the critical turn. Kant is here clearly alluding to the conflicts presented in the Dialectic, i.e., as to whether the world is finite or infinite or as to whether matter has ultimate constituents or not —antinomies which are to be dissolved or lose their real potency within his own critical philosophy.

Metaphysics is thereby framed as *the most general attempt to make sense of things*—very much in accordance with A. W. Moore's characterization thereof in his account of "modern metaphysics" (2012). "The transformation in the way of thinking" is even said to be "analogous" to the Copernican hypothesis (Bxxiin). To be sure, Kant goes on in the same footnote to state that it "will be proved, not hypothetically, but apodictically." Hence, it is the "first attempts at such a transformation" that are hypothetical. It does not follow that the end result of the critical enterprise is hypothetical as well. After all, metaphysics is ultimately to be developed in a systematic form that meets the highest evidential standard—that of exhibition of necessity, or of apodictic certainty, like in mathematics.

What is suggested in the preface to CPR, though, is that the crucial *thought experiment* of pure reason is a *critical test* of which of two alternative standpoints in metaphysics that is the best way of making sense of things—the revolutionary one he proposes or the traditional one. In the B-introduction to CPR, the general problem, or task ["Aufgabe"], of pure reason is said to be "How are synthetic judgments a priori possible?" (B19). Kant goes on to state that "On the solution of this problem, or on a satisfactory proof that the possibility that it demands to have explained does not in fact exist at all, *metaphysics now stands or falls*" (B19; my italics). This shows that Kant is here presenting his end as that of giving an *explanation* of the possibility of metaphysics—in the form of offering a way out of the antinomies of pure reason, as presented in the Transcendental Dialectic. The two-fold standpoint presented in his Transcendental Analytic, which is the framework for synthetic a priori judgments, is also the only acceptable solution to the problems of whether the world is infinite or finite in extension, infinitely divisible or of finite composition, etc. He thus reaches the framework for immanent metaphysics through abduction, i.e., in the form of an inference to the best explanation of paradoxical features of our natural world-picture and of the historically evolved conflicts to which these features have given rise.

Moreover, in the same place in the preface where Kant speaks of the "experiment of pure reason," he brings up both a "synthetic procedure" in the sciences as well as the "analysis of the metaphysician." In the Architectonic, he also speaks of the "analysis" of the chemists (A842/B870), which he likened to isolation in metaphysics, as we saw earlier, in Section II. It is clear, then, that his "experiment of pure reason" includes both analysis into resources for sense-making, i.e., the elements of cognition—through an "*analysis of the faculty of understanding* itself" (A65/B90)—as well as the very sense-making from these resources, i.e., the "synthesis," in the form of a system from such elements, or at least the outline of its plan (A13/B27). Thus, in very broad terms, the experiment of reason has affinity to the analytic-synthetic method with which he was familiar from the natural sciences, not only in chemistry but in a more abstract version in physics, from the works of Galileo, Descartes, and Newton. This method has two sides—one of resolution and one of composition. Even the axiomatic method in mathematics—as handed down from Euclid—has these two sides, i.e., one of systematization into axioms and one of inferring theorems from the axioms within the system.

Still, while Kant has suggested two major abductive entry points to his critical philosophy in CPR, he also holds that metaphysics as a science has its own "seed" in reason itself. There is a sense, then, in which its "growth" into a full-fledged, coherent, and unique system is rationally pre-determined, as we shall see. (Cf. also the preface to the B-edition, xxxvii-viii). This follows from a general view of what a proper rational science is. Metaphysics has remained at a very immature stage indeed, though, prior to Kant's critical philosophy.

IV. Is There a "Seed" in Reason for a Metaphysics of Nature?

According to the Architectonic in CPR (A833/B861), any proper science has to be a *system anchored in reason itself*. What requires systematization through the analytic-synthetic method for its *modo cognoscendi*, we might say, is in its *modo essendi* a rational system that unfolds immanently from pure reason itself. Science that is systematic in this strong sense has to contain a manifold of cognitions under a *unifying idea of reason*. The ordering of its parts is made possible not directly by the idea but by a *schema* for the idea that makes its execution ["Ausführung"] possible.

The schema for any science cannot merely be a technical one for ordering empirical cognitions. It has to provide *architectonic unity a priori*—at least if the system is to be a *proper* science, like physics. The ordering is according to an *end of reason*, which is also an *inner end of the system*. All parts of the system can be seen as "derived" from this end and are internally related to each other. The system can also grow *organically*, like an "animal body," without addition of members ["Glieder"]. The schema contains the outline ["Umriß"] and division of the whole into members, and thus also purports to set the boundaries for the system's growth. At an immature stage, the idea of the science is still just a *seed* in reason (A834/B862). What might be presented as its "schema," or even "definition," at this stage need not correspond very well to reason's idea of the science—an idea which is to bring the cognitions together in a *natural unity*. After all, there are no real genetic definitions in metaphysics.

Accordingly, if there is to be transcendental philosophy as a science, there must be origins in reason itself for its genesis, and the task of CPR is to present at least the schema for the idea of its system. As Kant puts it, "the critique of pure reason is to outline the entire plan [for the idea of transcendental philosophy as a science] architectonically, i.e., from principles" (B27). Philosophy itself as a science is the "system of all philosophical cognitions" (A838/B866). More specifically, Kant's concern is with pure rather than empirical philosophy (A840/B868), and thus with the philosophy of "pure reason," where "reason" is taken to comprise the entire higher faculty of cognition (A835/B863), and, hence, both the understanding and reason more narrowly construed.

So what precisely is the *science of metaphysics* here, according to Kant? In CPR, Kant distinguishes between pure philosophy in the sense of (1) critique; (2) the system of pure reason/pure philosophy; (3) the system of pure speculative reason/theoretical philosophy (A841/B869). It has already been noted in Section II how a critical methodology has to precede any science in the form of a system of cognitions even according to the pre-critical Kant. The critique of pure reason, however, offers a distinctive critical method. In accordance therewith, it brings out the elements for a system as these are rooted in the faculties of the human subject, i.e., in the pure understanding together with the pure sensibility. Hence, CPR has two main parts—The Doctrine of Elements and The Doctrine of Method.

Whereas metaphysics as science is one of the systems (2 or 3), it can be taken to comprise critique (1), or its outcome, in its base. Transcendental critique, on the other hand, thereby also includes metametaphysics in its reflexivity, in the form of a "metaphysics of metaphysics," as Kant states explicitly in the letter to Marcus Herz (of May 11, 1781). Furthermore, 2 is "metaphysics" in a wider sense, and 3, in a narrower sense. The former comprises both *metaphysics of nature*, or of pure reason in its speculative use, and *metaphysics of morals*, or of pure reason in its practical use. The latter brings us to Kant's take on the special metaphysics of the tradition. Indeed, Kant refers to the "first part" (Bxviii-xix) and the "second part" of metaphysics (Bxix) in the preface to the B-edition, clearly alluding to the traditional distinction between general and special metaphysics. However, whereas the general metaphysics, or ontology, of the tradition purports to be about things as such, and thus about all things whatsoever, including the soul and God, the transcendental analytic lays claim only to be concerned with objects of possible experience, and thus only with things as they appear to us, not with things as they are in themselves.

The metaphysics of nature is itself divided into transcendental philosophy and "rational physiology" (cf. A845/B873), corresponding to the two parts of traditional metaphysics, respectively. "Physiology" here deals with "physis," i.e., with nature, under some aspect or other. Transcendental philosophy, as the system of all principles of pure reason (cf. B27), is the transformed general metaphysics (cf. A247/B303). Unlike mathematics or empirical science, such a system can be *complete*—indeed, Kant states in the preface to the *Metaphysical Foundations of Natural Science* that any metaphysical system—be it general or special—has the table of categories as the "schema" for its completeness (Ak. 4: 473–474). The critique of pure reason remains a *propaedeutic* to a complete system for metaphysics, though, but at least it has to present a "complete enumeration of all ancestral concepts [Stammbegriffe]" (A13/B27) for it, i.e., the categories (A81/B107). From these, there is room for further conceptual developments, since they also have "derivative concepts," the *predicables*, which belong to a "complete system of transcendental philosophy" (A81–2/B107). In *Progress* (Ak. 20: 272), a further distinction is made between predicables which are pure and those which are not. In a letter to Beck, of January 20, 1792 (Ak. 11: 313–14), Kant speaks of a planned 'System of Metaphysics,' in which he would begin with the categories and there would also be exposition of all predicables "included under" each category, as well as examples of their use, i.e., their links to intuition or experience. He goes on: "From this there emerges a whole science of ontology as *immanent* thinking." He then refers to the "second part," which is clearly a move to transcendent and special metaphysics, where the latter is said to be dealt with in the "dialectic of pure reason."

This brings us to "rational physiology," which deals with specific domains of nature, and thus relies on specific existence claims, unlike transcendental philosophy. It can be either *transcendent*, which is here taken to be "rational cosmology" and "rational theology," or *immanent*, which is said to comprise "rational physics" and "rational psychology." The latter must deal with phenomena of inner sense without falling prey to "transcendental paralogisms." I shall not go more into what it could possibly comprise here.

It is rational physics of corporeal nature that is properly restrained by transcendental philosophy. After having characterized physics as "natural philosophy [philosophia naturalis]" (Ak. 21: 134) in the *Opus postumum*, Kant even goes on to call transcendental philosophy the "philosophy of philosophy [die Philosophie von der Philosophie]" (Ak. 21: 135). The ascension to a meta-level of clarification is unavailable to traditional ontology, which has universal validity and does not draw any line between what is immanent and what is transcendent. Indeed, according to the Transcendental Dialectic, transcendent physiology yields only illusory cognitions, but these are still included in the system, it would seem. After all, they do have a genuine regulative use in the sciences, e.g., "Always look for more distant non-empty regions!" is a norm for cosmological exploration suggested by the First Antinomy.

In the preface of *Metaphysical Foundations*, Kant distinguishes between the "*transcendental* part of the metaphysics of nature" (Ak. 4: 469) as well as of "a *special* metaphysical natural science (physics or psychology)" (Ak. 4: 470). "Special metaphysics" in this sense deals with existence, or with what is actual, or with real possibility, and not with mere constructibility, or thin formal possibility, as in mathematics. The laws of natural science—unlike the "laws" of mathematics—must therefore have a basis in more than the forms of pure intuition and the quantitative categories that underlie continuous quanta and discrete quantity. Furthermore, special metaphysics is concerned with "the special nature of this or that kind of things" (Ak. 4: 470). It therefore presupposes empirical concepts of natural kinds. A geometrical figure, by contrast, has an essence but no nature (Ak. 4: 467n), i.e., its essence is not real. (For a thorough discussion of the status of real essences in Kant, see Chapter 2 in the present volume.) The fundamental concept in geometry is a pure concept of space, and geometrical spaces are constructible as limitations of all-inclusive space. The fundamental concept in physics is an empirical concept of matter, which is not mathematically constructible as such.

Still, Kant emphasizes that *proper* natural science requires a *pure* part (Ak. 4: 469), which is synthetic a priori. The concept of phenomenal substance as the movable in space is empirical. The motion in question is not that of pure figurative synthesis executed by a constructing subject, as in geometry, but the objective motion of a portion of matter relative to other portions of matter. (Cf. Friedman 2013, 85–90, on "mathematical and empirical motion.") It therefore ultimately presupposes a distinction between real and apparent motion, which cannot be drawn without empirical input. However, relative motion can be mathematized—velocity and acceleration are intensive magnitudes—as are mass and force, and other physical magnitudes.

Now, the transcendental conservation principle for substance in CPR, in the form of the First Analogy, can be turned into a fundamental physical law of conservation of mass. The Second Analogy, into the law of inertia, i.e., no acceleration without a force, where force is a predicable derived from the category of causality, etc. In general, whatever belongs to the specific metaphysical framework for physics and can be represented mathematically can also be anticipated a priori of physical phenomena, which is why it constitutes a pure part of physics despite relying on an impure concept of matter.

Kant's approach to the foundations of physics remains transcendental, then, not naturalistic. The constitutivity thesis of critical philosophy is: "The conditions of the *possibility of experience* in general are at the same time conditions of the *possibility of the objects of experience*" (A158/B197). Hence, the methodological norm for conceptualization in metaphysics, as formulated in the Architectonic, becomes: "it is only possible for our reason to use the conditions of possible experience as conditions of the possibility of things" (A771/B799). Concepts that are freely fabricated independently of these conditions have no objective reality.

V. Concluding Remarks

Metaphysics that cannot account for the constitutive conditions and rational constraints of scientific discourse is not legitimate, according to Kant. Its methods and evidence are not adequate if they do not fulfil the critical task of clarifying and justifying the foundations for the exact sciences. The role of architectonic schemata shows the significance of intrinsic plausibility considerations in Kant, but we have seen that metaphysics also relies on a broadly abductive methodology, and that it is subjected to external pulls from mathematical inventions and empirical discoveries. The relationship between immanent metaphysics and science in its empirical parts is considered an intimate one and suggests that we should look for "natural" and "reasonable" extensions even beyond physics, e.g., in chemistry. The relationship between metaphysics, with its fixity of concepts, and mathematics—with its potential both for conceptual novelty and universal applicability—is also close but quite different. Subsequent developments within mathematics and physics add to the complexity in these relationships. They also indicate the need for a lower evidential standard than that of apodictic certainty to fulfil Kant's critical ambition—namely, a gradual one in terms of explanatory adequacy and unifying power in relation to the special sciences. Kant's general prescription for metaphysics in a theoretical context—that it can make sense of itself only by making sense of science—is not thereby undermined.

Acknowledgment

I am grateful to Houston Smit for comments on an earlier draft of this chapter.

References

Büchel, Gregor. *Geometrie und Philosophie: Zum Verhältnis beider Vernuftwissenschaften im Fortgang von der Kritik der reinen Vernunft zum Opus postumum*. Berlin: Walter de Gruyter, 1987.

Friedman, Michael. *Kant's Construction of Nature: A Reading of the 'Metaphysical Foundations of Natural Science'*. Cambridge: Cambridge University Press, 2013.

Lowe, E.J. *The Possibility of Metaphysics: Substance, Identity, and Time*. Oxford: Oxford University Press, 1998.

Menzel, Alfred. "Die Stellung der Mathematik in Kants vorkritscher Philosophie." *Kant-Studien* **16** (1911), 139–213.

Moore, A.W. *The Evolution of Modern Metaphysics: Making Sense of Things*. Cambridge: Cambridge University Press, 2012.

Parsons, Charles. *From Kant to Husserl*. Cambridge, MA: Harvard University Press, 2012.

Paton, H.J. *Kant's Metaphysic of Experience*. Vol. I. London: Allen & Unwin, 1936.

Sandmel, Thor. *Matematisk erkjennelse: Grunnriss av en transcendentalfilosofisk fundert matematikkfilosofi*. PhD-thesis. Oslo: Unipub, 2001.

Strawson, P.F. *Analysis and Metaphysics*. Oxford: Oxford University Press, 1992.

Sutherland, Daniel. "The Point of Kant's Axioms of Intuition." *Pacific Philosophical Quarterly* **86** (2005), 135–159.

Williamson, Timothy. "Abductive Philosophy." *Philosophical Forum* 47, No 3–4 (2016), 263–280.

2 Essence, Nature, and the Possibility of Metaphysics

Houston Smit

The critical Kant argues that the enterprise of ontology—when it is conceived, in the traditional way, as a science (and thus genuine rational cognition) of the nature of a thing in general—is impossible for us.[1] The traditional ontologist purports to achieve insight, through pure reason, into the nature of a thing in general. She purports thereby to put the categories—including crucially the concepts of substance and cause—to use to grasp what it is for something to be, as the subject of activity and power, a thing, and in such a way as to provide a real definition of a thing in general. To achieve this rational insight would be to generate purely a priori rational cognition of the real essence of a thing in general. And, Kant contends, we cannot achieve any cognition of real essences at all, even with the aid of experience, let alone through pure reason.

According to Kant, Hume helped to make evident the impossibility of ontology by attacking the legitimacy of our claim to have the idea of causation as an objectively necessary connection between distinct existences. Moreover, this attack puts us in a position to see that no cognition of things at all—even that which we take ourselves to have in taking ourselves to have experience—would be possible for us, unless we can put the categories to use to achieve *some sort of* insight into what belongs to the inner possibility of things. Since these concepts make up our fundamental concept of a thing in general, we can think a thing only in representing it through these concepts. Moreover, we can, in thinking a thing, cognize that thing only if we can prove its real possibility (Bxxvin), and in giving any such proof, even one that appeals to our experience, we presuppose that the categories have a priori objective validity in respect of that thing. Establishing the a priori objective validity of the categories for things that are to constitute possible objects of our cognition, insofar as they constitute such objects, is what, on Kant's account, requires us to achieve some sort of insight into how their conformity to the categories belongs to the inner possibility of these things themselves.[2]

Hume's skeptical treatment of causality, then, serves to make clear, not only that ontology is impossible, but also that it is a response, however misconceived, to a real need. For ontology purports to establish our right to use the categories to cognize things in a straightforward fashion: namely, by putting these concepts to use, in an act of purely a priori insight, to grasp how the determinations thought in the categories collectively suffice to constitute the real essence of a thing in general. Given the impossibility of ontology, and more generally of our achieving any cognition of real essences, there is a real puzzle about whether, and if so just how, we can put the categories to use to cognize things.

The Copernican Revolution Kant proposes for metaphysics sets out to meet this real need, but in a way entirely different from that pursued by traditional ontology. It aims to establish the a priori objective validity of the categories, if only in respect of the being things have, not in themselves, but merely as objects of our experience, by appeal to an entirely different sort of insight, a purely a priori rational insight that is not, itself, insight into the possibility of any thing. This insight is, rather, itself one merely of the possibility and necessity of "subsuming all appearances under" the categories, that is, of using these concepts "as principles of the possibility of experience" (Prolegomena, Section 28; Ak. 4: 311). But because "the conditions of the possibility of experience in general are at the same time conditions of the possibility of the objects of experience" (A158/B197), this insight is also an insight into the possibility and the necessity of the categories being, in this use, intellectual formal conditions of the possibility of objects of our experience and, thereby of things as they appear to us. And, as such, this insight is sufficient to establish the a priori objective validity of the categories, if only for things in respect of a being they have merely as objects of our experience.

The sketch I just gave of Kant's new, and distinctively critical, approach to general metaphysics, though familiar in some ways, is distinctive in others—perhaps most obviously in the way it highlights the central role that Kant's crucial, but generally neglected, notion of insight plays in his critical metametaphysics. Moreover, in doing so it employs a reading of this notion I proposed elsewhere, as part of a new reading of Kant's notion of the a priori. On this reading, Kant employs the term 'a priori' in the original, now archaic sense, one I dubbed the 'from-grounds' sense. To cognize something a priori (*etwas a priori erkennen*) in this sense—which in Kant's terminology is equivalent to having insight into it (*etwas einsehen*)—is to cognize it from the grounds that make it true. Moreover, in line with the rationalist tradition, Kant conceives of cognizing a priori in a robust fashion, as consisting in seeing *how* grounds make something true, i.e., of how, as objectively determining grounds, they objectively necessitate the consequences they determine. Kant tells us that "to cognize something a priori means to cognize it from its possibility" (MFNS; Ak. 4: 471), because, on his account, seeing how the grounds that determine a truth objectively necessitate that truth consists in cognizing how something follows from its mere possibility. Now what determines possibility, on Kant's account, are essences. Cognizing something a priori requires, then, representing whatever essence or essences make it necessary, in a way that is sufficient for seeing how that essence or essences exclude every alternative. So, for example, in constructing our concept of a triangle we have insight into this geometric figure (and indeed in such a way that we are in a position to give the real definition of this figure), because in this construction we cognize the essence of this figure as such, so as to see how it necessitates a triangle's possessing, as its essential properties, all and only the properties that enter into this essence.[3]

The present chapter begins to fill out the sketch I just provided of Kant's critical metametaphysics by drawing on, and developing further, this reading of his notion of the a priori. In particular, it explains the different sorts of essences that Kant distinguishes in the course of his critique of pure reason, and the sort of essence of which, on his account, he achieves cognition in executing this critique. Moreover, in doing so I clarify Kant's account of his new, critical transcendental philosophy, a science to which he at one point refers as "the metaphysics of metaphysics" (Correspondence; Ak. 10: 269). We will see that Kant distinguishes the sorts of essence that we can cognize, including the sort of essence that he claims to cognize in executing his critique, from real essences, where a real essence is the constitution of a thing (something that is real, as the subject of activity and power) itself, the constitution the possession of which makes a thing the thing that it is. Moreover, we will need to see that he distinguishes between two sorts of non-real essence. One sort of non-real essence is logical: the constitution of a concept as the general representation of something. Kant insists that the constitution of our concept of a thing must be strictly distinguished from that of the thing itself that we think in forming that concept. Another sort of a non-real essence is, like a logical essence, the constitution merely of a representation we can have, and thus what *itself* determines the inner possibility merely of that representation.[4] But this sort of non-real essence is one that, insofar as it can be realized in some suitable matter (consisting of the manifold of some given intuition), constitutes the sensible or intellectual form of an experience that is possible for us, given only the purely a priori nature of our capacity of cognition. Such an essence determines a priori the formal possibility of appearances that are given to us constituting an actual experience (and this, in judgments of experience that subsume these appearances under the schematized categories) and thereby the object (*Gegenstand*) of that experience itself. In this way, it determines the possibility, not of a concept of a thing as it appears to us, but rather of that (transcendentally ideal) thing itself, a thing that has its (empirical) reality only as the object of our experience. Let us call the second sort of non-logical essence, formal essence. ('Formal essence' is my term, not Kant's.) On Kant's account, the only synthetic a priori *rational* cognition of things of which we are capable is cognition we achieve in cognizing the a priori ground that the non-logical possibility of things as they appear to us has in some formal essence.

In proceeding, I will take as my point of focus the following intriguing passage from the Preface to his *Metaphysical Foundations of Natural Science* (MFNS):

> All true metaphysics is taken from the essence of the capacity of thinking itself and is in no way fictitiously invented on account of not being borrowed from experience. Rather it contains the pure actions of thought, and thus a priori concepts and principles, which first bring the manifold of empirical representations into the law governed connection through which it can become empirical cognition, that is, experience.
>
> (MFNS; Ak. 4: 472)

The sentence that immediately precedes this passage makes it clear that Kant is contrasting true metaphysics with the false, because dogmatic, metaphysics put forward by his predecessors: the latter, he contends, consists merely in "contriving possibilities merely at will and playing with concepts, which can perhaps not be presented in intuition at all, and have no other certification of their objective reality than that they merely do not contradict themselves" (ibid). A full explanation of just how true metaphysics is to be "taken from the essence of the capacity of thinking itself" is well beyond the scope of this chapter. Such an explanation would amount to a reading of the Transcendental Analytic of the first critique: it is over the entire course of the Analytic of Concepts (i.e., the Metaphysical and Transcendental Deductions of the Categories) and the Analytic of Principles that Kant takes true metaphysics from the essence of our capacity of thinking itself. Nonetheless, the MFNS's characterization of the Transcendental Analytic is highly illumining, because it brings to the fore a crucial, but not sufficiently appreciated, aspect of Kant's critical philosophy—namely, how the latter's line of argument consists, at bottom, in an appeal to rational cognition of this essence. Correctly interpreting Kant's account of this essence, and of how true metaphysics is to be taken from it, will require attending to his distinction between essences that are real, or natures, and essences that are not. In particular, since the critical Kant repeatedly denies that we can have any cognition of real essences and any true metaphysics, as a science, must be taken from the essence of our capacity of thinking itself in and through rational cognition of this essence, we can expect that this essence is a non-real one. Moreover, since metaphysical cognition proper is synthetic and logical essences can yield only analytic cognition, we can expect that this essence is not a logical essence, but rather a formal essence. Indeed, the force of 'itself' is, arguably, to distinguish the formal essence of the capacity of thinking itself from the logical essence of our concept of this capacity.

I. Essences, Real and Otherwise

In drawing the distinction between an essence in general and an essence that is real, and thus a nature, Kant employs the terms '*Wesen*' and '*Natur*' in their formal, or adjectival, as against their material, or substantival, senses. In the latter sense, nature is "the sum total of all things insofar as they can be objects of our senses and thus also of experience" (MFNS; Ak. 4:467). In the former sense, in contrast, a 'nature' means, roughly, the constitution (*Beschaffenheit*) of a thing. In a parallel fashion, '*Wesen*' in the material or substantival sense signifies a being (*ens*), whereas in its formal or adjectival sense, this term signifies an essence (*essentia*). (Hereafter, I will use 'nature' and 'essence' only their adjectival senses.)

Kant characterizes a nature more fully as "the first inner principle of all [*Alles*] that belongs to the existence of a thing [*Dasein eines Dinges*]" (MFNS; Ak. 4: 467–468). And he says that, if nature is understood in this, its adjectival sense, "there are as many natural sciences as there are specifically different things, each of which must contain its own peculiar inner principle of the determinations belonging to its existence" (ibid): each specifically different thing must contain its own distinctive nature.[5]

In order to understand Kant's characterization of a nature, then, we need first to understand what Kant means by "determinations that belong to the existence of a thing." Now a thing is "something real" (*etwas reales*): a subject of activity and power considered as the subject of activity and power that it is. And existence (*Sein*) is "the positing [*Position*] of a thing, or certain determinations in themselves" (A598/B624): what is to be asserted in using 'is,' not copulatively, but rather to posit a thing or "a certain determination" outside our representation of it. Kant adds the qualifier 'certain' to indicate that not all determinations are to be posited absolutely: as we will see shortly, a geometric figure is a determination, but it is nothing but a determination of the form of our outer sensibility, and as such it is a mere representation and is not to be posited absolutely. The determinations that belong to the existence of a thing, then, are those that are to be posited absolutely, in virtue of their deriving from the nature of that thing (perhaps only contingently, in virtue of real relations it bears to other things). The extension, or shape, of a body, unlike the determinations of a geometric figure, are to be posited absolutely, but only because they are determinations that belong to the existence of that body, in virtue of having a ground in its nature.

In characterizing a nature as the first inner principle of all that belongs to the existence of a thing, then, Kant is saying that a nature is the first inner principle of determinations that are to be posited absolutely, as determinations of a thing itself. The force of 'first' here is to distinguish the nature of a thing from just any inner principle from which a determination of a thing derives. For example, a body's extension is *an* inner principle that grounds its possession of its shape. But it is not the *first* inner principle that grounds its possession of its shape, along with all the other determinations that belong to that body's existence. *This* inner principle is its nature. Consider, for example, the piece of gold that is my wedding ring. The *first* inner principle that grounds its possession of its current shape (the one given to it by the artisan who made the ring), and indeed of all the other determinations, including its extension, that "belong to its existence," is the nature of gold.

Consider next Kant's characterization of essence, and how he introduces it in the course of explaining how a geometric figure can have only an essence, and not a nature:

> Essence is the first inner principle of all that belongs to the possibility of a thing. Therefore, one can attribute only an essence to geometrical figures, but not a nature (since in their concept nothing is thought that would express an existence [*da in ihrem Begriffe nichts, was ein Dasein ausdrückte, gedacht wird*]).
>
> (MFNS; Ak. 4: 467n)

We form the concept of a geometric figure in a certain act of pure productive imagination, one that delimits space, as we represent it in pure intuition. But space, on Kant's transcendental idealism, is nothing but the form of our outer sensibility. It "represents no property at all of any things in themselves, or any relation of them to each other" (A26/B42). Consequently, it is only insofar as space, as the form of our outer sensibility, conditions the possibility of any experience we can have, and so of any object of an experience that is possible for us, that it yields us any representation of a thing. It thus yields representation only of things as they appear to us. It is crucial to this view that the first inner principle of the delimitations of space we bring forth in doing pure geometry—and thus what we think in forming the concept of a geometric figure—is as such the first inner principle of everything that belongs to the non-real possibility of appearances being given to us that constitute a body with the corresponding shape. For what we think in constructing the concept of a geometric figure is itself the possibility merely of a possible way of delimiting space, the form of our outer sensibility, and so of a mere representation. But this form, together with our pure understanding and its concepts, determines, entirely a priori, all the possible ways in which it can be delimited in our cognition, and thereby the sensible formal possibility of any outer experience that is to be possible for us. The concept of a geometric figure thus constitutes synthetic a priori cognition of the formal possibility of appearances being given to us that constitute a body that has a shape that conforms to the conditions thought in that geometric concept. At the same time, because the content of the geometric concept that we form in this mathematical construction is exhausted by what we think in this act of delimiting our pure intuition of space, in this concept "nothing is thought that would express an existence," that is, a thing, or a determination that is to be posited absolutely. And an essence can be a nature—the first principle of everything that belongs to the existence of a thing—only if it is the content of a concept that "would express an existence."

For our purposes, it will prove crucial to appreciate a point implicit in the previous paragraph: namely, that in distinguishing between the notions of essence and nature, Kant does not mean to imply that a nature is not an essence. Indeed, on his account a nature is a species of essence, as he indicates in repeatedly treating the notion of a nature and that of a real essence as intersubstitutable. See for example, this passage from his letter to Reinhold of May 12, 1789:

> For I can easily find the logical essence of a given concept—namely what constitutes [*ausmacht*] its first [*ersten*] *constitutiva*—as well as the attributes, as *rationata logica* of this essence, by means of the analysis of my concepts into all that I think under them. But the real essence (the nature), that is, the first inner ground of all that necessarily belongs to a given thing, this is impossible for man to cognize with regard to any object [*von gar keinem Object*]. For example, extension and impenetrability constitute the whole logical essence of the concept of matter, that is, they are all that is necessarily and primitively contained in my, and every man's concept of matter. But to cognize the real essence of matter, the first inner sufficient ground of all that necessarily belongs to matter, this far exceeds all human capacities. We cannot discover the essence of water, of earth, or the essence of any other empirical object [*Object*]; but leaving that aside, even the real essence of space and time and the basic reason why the former has three dimensions, the latter only one, are unknowable [*unerforschlich*]. And the reason for this is precisely that since the logical essence is to be cognized analytically and the real essence must be cognized synthetically and a priori, there must be a ground of the synthesis of the latter, at which we, at least, must remain [*wobey wir wenigstens stehen bleiben müssen*].
>
> (Correspondence; Ak. 11: 36–37)

That we cannot cognize any real essences is a crucial point, to which we will return. For the present, the key point to see is that Kant conceives of a nature, that is, the first inner *principle* of *all* that belongs *to the existence of a thing*, as such a principle in virtue of its being, in the first place, the first inner *ground* of all that *necessarily* belongs to that thing. Indeed, a real essence is such a ground, that is, the first inner *sufficient* ground of all that belongs necessarily to a thing, in virtue of being the first inner sufficient ground of the *real* inner possibility of a thing: the necessary accidents (attributes) of a thing are all and only those of its accidental (and so non-essential inner) determinations that are entailed by its *essentialia*. The logical essence of a given concept, however, is also an essence, in the sense Kant specifies at Ak. 4: 467n. It is, after all, the first inner principle of everything that belongs to the non-real, because merely logical, possibility of a thing—a possibility that itself consists simply in the possibility of forming a concept of that thing, and so of this representation.[6]

Now recall that the essence that, at Ak. 4: 467n, Kant says we can ascribe to a geometric figure is an essence, despite not being a real essence, because it is the first inner principle of all that belongs to a possibility of a thing – which possibility is not the real possibility of this thing. The thing in question, I proposed, is a body insofar as it belongs to a class of material bodies with shapes that conform to the geometric figure in question. I also proposed that the relevant possibility of the so-shaped body—one that is to be distinguished from the possibility of delimiting space, which as such is the possibility merely of a certain sensible representation—consists in appearances being given to us that constitute a body that conforms to the conditions thought in the geometric concept of that figure. Now, in Kant's view, our pure understanding has the authority, under the principle of the synthetic unity of apperception (B135–136), to prescribe a priori to given appearances their conformity to the categories and the principles of pure understanding, as a conformity that is objectively necessary to their constituting possible experiences. As the principles of this prescription, the categories and the principles of pure understanding collectively constitute the intellectual formal conditions of the possibility of our experience in general. In the case of our pure geometric concepts, it is in virtue of *our sensibility*'s conformity to the axioms of intuition, and of the pure understanding's authority to prescribe to given appearances their conformity to these axioms as the condition of the possibility of their constituting *intuitions*, that these concepts are possible. But these concepts themselves represent only the inner possibility of delimiting space in a certain way. We represent the formal possibility of *the things* that fall under these concepts—bodies with shapes that conform to the conditions specified in these concepts—only in using the dynamical categories to think of the conditions, specified in the Analogies of Experience and the Postulates of Empirical Thought, that appearances that are to be given to us must meet if they are to constitute an outer experience that is possible for us, given only the nature of our pure understanding and the forms of our sensibility.

Consider now, if only in outline, Kant's conception of the non-real and not-merely-logical possibility of a human experience in general. In his account, this possibility consists in the objectively necessary conformity of appearances that can be given in our sensibility, given only the forms that its nature specifies, to the categories, a conformity that our capacity of understanding itself, as "the legislation for nature," prescribes to them a priori. And he terms this possibility—one that these appearances have in relation to our pure understanding—"their formal possibility as possible experiences" (A127; cf. A250). The categories and the forms of our sensibility serve, together and under the principle of the synthetic unity of apperception, to determine this formal possibility. The formal possibility of experience in general also constitutes the formal possibility of the things as they appear to us that are the objects (*Gegenstände*) of a human experience in general: Kant is talking about the conditions of formal possibility when we writes "The conditions of the possibility of experience in general are at the same time conditions of the possibility of the objects [*Gegenstände*] of experience" (A158/B197). Nonetheless, the formal possibility of the objects (*Gegenstände*) of our experience—which as such are things as they appear to us—must be distinguished from their real possibility. Their real possibility consists, not just in the formal possibility of a human experience in general (which as such is purely a priori), but in this possibility together with the material possibility of the experience of which these things are essentially the objects (*Gegenstände*). The real possibility of things as they appear to us, then, is one the conditions of which are determined a priori by the natures of our pure understanding and our pure sensibility, and a posteriori by the determined empirical character of the appearances that are given to us through sensation. The formal possibility of appearances as experiences—be it considered insofar as it is determined solely by conditions that our pure understanding itself sets, or insofar as it is determined further by the addition of particular conditions set by the forms of our sensibility—is, itself, the possibility, not of things, but only of mere representation. This possibility is not, any more than the possibility into which we have insight in constructing the concept of a geometric figure, to be confused with real possibility.[7]

As I specified earlier, I will call essences that are first inner principles of all that belong to the non-real possibility of a thing 'formal essences.' These essences include those of geometric figures, which as such have their determining ground in the nature of our sensibility. But they also include the essence of our capacity of thinking itself, an essence that determines the principles of our thinking purely a priori and thereby the formal possibility of acts of thinking that determine manifolds of given representations under these principles. One such act of thinking is that of the understanding, the legislation for nature, which as such grounds the possibility of acts of thinking in which one relates given intuitions to possible objects of our experience in cognition of the same. The essence of our capacity of thinking itself has its ground solely in the nature of our pure understanding. Indeed, most generally, formal essences determine non-real and not-merely-logical possibilities in virtue of having their ground solely in the purely a priori subjective constitution of our capacity of cognition.

Let us call all non-real essences—which thus include logical, as well formal essences—"mere essences." What distinguishes mere essences from real essences is that they are first inner principles of all that belong to the possibility of a thing that is, itself, merely the possibility of our representing that thing, and not of that (or any) thing itself. Any thing, as such, even a thing that is merely empirically real (e.g., matter), must—considered in itself (and so not merely as what meets the conditions of falling under a concept, or determined in a formal essence)—have a real essence.

Notice that, in explaining why we cannot attribute a nature to geometric figures, Kant implies that something in the concept of which is thought something that would express an existence can have a formal essence, as well as a nature. Now Kant's concept of a capacity is that of an active potency, a potency the operation of which is an activity, and as such a potency the inner possibility of a power.[8] A capacity, then, is at least a candidate for having a nature, or real essence, since our thought of it contains something (namely, activity) "that would express an existence." Indeed, as Kant's repeatedly speaking of the nature of our understanding (Prolegomena; Ak. 4: 308 and 328; B110; A648/B676) confirms, our capacity of understanding, on his account, has a nature—that is, a first inner principle of its real possibility as the subject of an activity, namely, of a distinctive sort of activity that, when exercised in certain circumstances, realizes a power to understand. This suggests that we ought, in interpreting Kant's account of our capacity of understanding, and the role this account plays in his critical philosophy, to distinguish between appeals to the nature, or real essence, of this capacity, and those to its formal essence. With this point in mind, we turn now to applying these distinctions between logical, real, and formal essences to Kant's account of the possibility of a true metaphysics.

II. The Essence of the Capacity of Thinking in General and Its Role in Transcendental Philosophy

What, then, is "the essence of the capacity of thinking itself" from which Kant says all true metaphysics is taken? If this essence is one from which a true metaphysics is to be taken, one might think it must be one of which we can have cognition, and thus not a real essence. But if this essence is not a real essence, how can rational cognition of it yield a true metaphysics? After all, metaphysics purports to be a science of the real (where, again, to be real is to be the subject of activity and power). Before pursuing these questions, however, we need to take care of two preliminaries. The first is to distinguish rational insight—which is the sort of insight that yields science—from a *sui generis* insight Kant posits in providing his distinctively critical account of our understanding: this insight is what constitutes the form of our understanding's characteristic act—namely, that of prescribing to given intuitions their objectively necessary conformity to the categories as the condition of their constituting objects of our cognition. The second is to see, if only briefly, how Kant conceives of metaphysics as a science that goes beyond physics in virtue of enjoying a certain independence from experience not had by the latter.

As I explained at the outset, to have insight into something is, on Kant's account, to cognize it a priori, which, in turn, is to see how the essence or essences that determine its possibility render it necessary. We now need to see that rational insight is *a species of* insight. An insight is rational, only if and to the extent that it consists in seeing not only *how*, but also *why*, the essence or essences that determine something's possibility necessitate it. Rational insight, moreover, requires one to have, not just any representation, but *cognition* of the essence from which one cognizes that something follows from its mere possibility. It is this rational perception of why grounds necessitate their consequences that constitutes what Kant calls 'apodictic certainty', i.e., the certainty characteristic of demonstration. The example of insight I provided previously—the insight into a geometric figure that one has in constructing one's pure sensible concept of that figure—is one of *rational* insight: it consists of the rational cognition of this figure's essence—a formal essence—as the first inner principle that necessitates its possessing, as its essential properties, all and only the properties that enter into this essence.

However, not all insight is, on Kant's account, rational insight: in particular, he maintains that it is essential to our capacity of understanding itself that its objective use consist, if only in its form, in an act of cognizing an object of our cognition a priori that is not itself an instance of rational insight. This insight—which Kant terms 'insight of the understanding' (*Verstandeseinsicht*) (A134/B173; cf. Correspondence; Ak. 10: 135; and CPJ; Ak. 5: 180)—consists in the capacity of understanding itself, or "the legislation for nature" (A127), insofar as it is directed to the manifold of a particular given intuition in cognizing how it exhibits, in one's apprehension of it, the distinctive objectively necessary conformity to the categories in virtue of which it constitutes an object (Object) of our cognition. What makes this act a species of cognizing a priori (again, in the from-grounds sense of 'a priori') is that it is the act in virtue of the possibility of which the manifold of a given intuition constitutes an object of our cognition.

What subjects all manifolds of given intuition that one is to encounter in oneself (so insofar as one can be conscious of them in the operation of one's inner sense) to the categories, and thereby to all laws of nature, is one's capacity of understanding itself, the legislation *for* nature. This legislation is what first renders the conformity to the categories that these manifolds themselves have, as representations of one's sensibility—a conformity that is itself merely contingent—objectively necessary, one in virtue of which they constitute objects of our cognition. And it does so, in the first instance, collectively, in determining a priori in all manifold of given intuition the relation they bear to each other in a single universal human experience. But the legislation for nature subjects these manifolds to the categories in this way only on the condition that one can, in and through the insight of the understanding, subsume each of these manifolds under the concept of some determined object, and this in prescribing to that manifold, as one is conscious of it in apprehending it, a distinctive objectively necessary conformity to the categories, as the condition of its constituting an object of our cognition.

Now all given intuitions, including the pure intuitions of Space and Time (which are given by the nature our sensibility), are to be contrasted with intuitions that are made—including those the geometer constructs in the pure intuition of Space. The latter, unlike the former, are products of voluntary (*willkürliche*) syntheses in and through which we have rational insight into how Space can be delimited. The pure intuition of Space is what, together with the mathematical categories and principles, determines the formal possibility of appearances that are given in our outer sense constituting bodies with determined shapes. The laws to which our pure understanding subjects the manifold contained in the pure intuition of Space are the laws of nature under which appearances given to us in Space are to constitute objects of our outer experience as bodies that have some determined shape or other. Kant thus is also speaking of the lawfulness appearances are to have under the axioms of intuition when he claims that the categories are "the highest" of the laws that "come from the understanding itself a priori" and that "must provide the appearances with their lawfulness and by that very means make experience possible" (A126).

In the insight of the understanding, one sees only *how*, and not *why*, the manifold of a given intuition has the distinctive objectively necessary conformity to the categories in virtue of which it constitutes an object of our cognition. Indeed, we cannot, in principle, see *why* the manifold of a *given* intuition has this necessary relation. That would require us to cognize the transcendental object (Gegenstand) of our cognition, as such, and so as what originally affects our outer sensibility in such a way as to determine the matter of our experience in general: all appearances that belong to a single possible human experience. For it is only in virtue of having this object as its transcendental ground that the manifold of a given intuition exhibits, in our apprehension of it, a conformity to the categories that our understanding, in and through the insight of the understanding, can subsume under the categories as one that is objectively necessary. Seeing *why* the manifold of a given intuition, as a manifold of appearances that belongs to the content of a single possible human experience, has this necessary relation to an object of that experience would be to cognize the real essence of this object. And the critical Kant, with Locke, holds that we cannot cognize any real essence.

We can now see that the fundamental problem Kant takes Hume to have raised for our understanding's claim to have a right to put the categories to use to cognize things is one that concerns, specifically, its right to prescribe to manifolds of *given* intuition their objectively necessary conformity to these concepts. And since we cannot establish this right by achieving the sort of rational insight into things at which ontologists aim, we must look, rather, to a different sort of rational insight, one into the possibility of the categories' constituting first principles a priori of our cognition to which we have the right, in and through the insight of the understanding, to prescribe to the manifolds of given intuition that exhibit conformity to these concepts in our apprehension of them, their objectively necessary conformity. As we will see, this rational insight is, on Kant's account, one into the essence of the capacity of thinking itself. This insight essentially employs the concept of the capacity of understanding as the original synthetic unity of apperception. Indeed, it is rational cognition of the principle of this unity of apperception as the essence of the capacity of thinking itself. Nonetheless, this insight is not one into the original synthetic unity of apperception itself. It is, rather, a rational insight in which we see how and why our capacity of thinking itself, conceived as the inner possibility of our power to prescribe a priori to representations given in our intuition their conformity to intellectual principles of our thought and cognition, *would have to* consist in the original synthetic unity of apperception.

I turn now to our second preliminary task, that of sketching Kant's conception of metaphysics as a science that includes, among the principles of its rational explanations, only purely a priori ones. Kant conceives of science (*Wissenschaft*), again in a traditional vein, as *scientia*—namely, as "a system, that is, a whole of cognitions ordered according to [*nach*] principles" (MFNS; Ak. 4: 467). Reason orders cognitions in this way, using principles to exhibit ground-consequence relations in them, where a principle (*Princip*), in turn, is "a cognition that is a ground of a ground according to [*nach*] a certain rule" (M. Mrongovius; Ak. 29: 747). Now Kant distinguishes between proper [*eigentliche*] natural science, which is science in the strict sense [*in strengem Sinn*], and a natural science in a looser sense of the term, and that is better called "a systematic art" (MFNS; Ak. 4: 468). The former "handles its object [*Gegenstand*] completely according to [*nach*] a priori principles" and has "apodictic certainty," in virtue of their principles being ones that carry with them some "consciousness of their necessity" (ibid). It is crucial to see that the mark of a proper science, on Kant's account, is that its principles are not *merely* empirical, but are a priori—namely, they are cognitions in which the subject sees how and why they, as principles, necessitate a ground's determining its consequence according to a certain rule. Even an empirical science (such as physics) constitutes a proper science only in virtue of handling its object completely according to principles that are a priori, even though these principles are only impurely a priori.[9] In this way, Kant's conception of proper science retains a highly demanding conception of rational explanation, one that requires rational cognition of some essence or essences in respect of how and why they determine certain universal and necessary truths.

On Kant's account, any metaphysics worthy of the name would have to be a proper science. Now philosophy, more generally, is rational cognition from concepts, and so to be contrasted with mathematics (A713/B741). And what distinguishes metaphysics from other material philosophy (as against general logic, which is merely formal)—in particular, from physics and ethics—is that it is pure: the a priori principles that it employs as a proper science and so as principles of the ground-consequence relations it exhibits, are *purely* a priori. In the *Groundwork* he expresses this point by characterizing metaphysics as a species of pure philosophy, one that, unlike physics and ethics, lacks any empirical part:

> All philosophy insofar as it is based on grounds of experience can be called empirical; but insofar as it sets forth its teachings simply from a priori principles it can be called pure philosophy. When the latter is merely formal, it is called logic; but if it is limited to determined objects [*Gegenstände*] of the understanding, it is called metaphysics.
> (GW; Ak. 4: 388)

And in MFNS he writes that a metaphysics of nature "must always contain solely principles that are not empirical (for precisely this reason it bears the name of a metaphysics)" (Ak. 4: 469). Physics and ethics, in contrast, are systematic bodies of cognition that reason generates in exhibiting ground-consequence relations under empirical principles. As I emphasized in the previous paragraph, because physics and ethics are proper sciences, these principles are not *merely* empirical, but also a priori. But because they are empirical, the rational explanations physics and ethics provide, unlike those of metaphysics, have an a posteriori, as well as an a priori, aspect.

It is crucial to see that the purity characteristic of metaphysics is one only of the principles under which, as a proper science, it exhibits ground-consequence relations a priori in rational cognition. This point is crucial to understanding how Kant distinguishes the transcendental part of a metaphysics of nature, which as such prescinds from "the nature of this or that thing in the sensible world," and a part of this metaphysics that is concerned "with a particular nature of this or that kind of thing, for which an empirical concept is given" (Ak. 4: 469–470). The latter is still a part of metaphysics, despite employing this concept—and indeed empirical principles contained in this concept—because the principles under which it organizes this concept, among others, into a system are purely a priori. As Kant puts it, a special, as against a general, metaphysical natural science is

> concerned with the particular nature of this or that kind of thing, for which an empirical concept is given, but still in such a manner that, outside of what lies in this concept, no other empirical principle is used for its cognition (for example, it takes the empirical concept of matter or of a thinking thing as its basis, and it seeks that sphere of cognition of which reason is capable a priori concerning these objects), and hence such a science must still always be called a metaphysics of nature, namely, of corporeal or of thinking nature. However, it is then not a general, but a *special* metaphysical natural science (physics or psychology), in which the above transcendental principles are applied to the two species of objects [*Gegenstände*] of our senses.
>
> (MFNS; Ak. 4: 470)

Indeed, in the body of MFNS Kant goes on to develop a special metaphysics of physics, one that employs empirical concepts of matter in its cognition. Now a principle is "a cognition that is a ground of a ground according to [*nach*] a certain rule" (M. Mrongovius; Ak. 29: 747). The empirical principles contained in the empirical concept of "the particular nature of this or that kind of thing" are principles as cognitions conformity to which our empirical understanding has the authority to prescribe, a posteriori, to the appearances of that thing that are actually given to us, in respect of their determined matter: in this prescription, the cognition is a ground of these appearances constituting grounds according to this empirical concept (rule) and thereby an experience of this thing. Moreover, in the course of executing the critique of pure reason, we determine that our empirical understanding has this authority under the dynamical principles of pure understanding, and this given the legitimacy of its presupposing that *these* appearances are ones that, in its act of prescription (judgment of experience) in fact constitute an experience. But an empirical concept enters into an impurely a priori *principle* of special metaphysics in being used, in this metaphysics, to apply the principles of pure understanding to its object. Consider here Kant's account of how the practitioner of the special metaphysics of physics achieves *rational* insight into matter as it constitutes the object of given empirical concepts. She does so, for example, when she achieves cognition of how *and why* the matter that is actually given to her, in experience, as the subject of repulsive power must as such also be the subject of attractive power—and vice versa.[10] This rational cognition is cognition of *why* matter, as it is given to us in our experience, must be the subject both of attractive and repulsive power: it must, on pain of the given appearances that we take to constitute our outer experience failing to conform to the transcendental conditions of the formal possibility of outer experience specified in the Transcendental Analytic of the first critique, which conditions are as such pure a priori conditions of any appearances constituting objects of any outer experience in general that is to be possible for us. This use of empirical concepts is what generates the impurely a priori principles under which the practitioner of this special metaphysics achieves an impurely a priori rational insight into matter: principles under which she exhibits ground-consequence relations a priori in rational cognition, and so in such a way as to cognize how *and why* the ground determines its consequence.

Let us return now to Kant's claim that "all true metaphysics is taken from the essence of the capacity of thinking itself." Recall that a capacity just is the inner possibility of a power, which as such consists in an active potency, that is, a potency the operation of which is an activity. The capacity of thinking itself, then, is the active potency that, as such, constitutes the inner possibility of any power to think that the subject of this capacity is, through its operation, to realize. And this capacity, in us, is our capacity of discursive understanding, where 'understanding' is taken in the broad sense of "the capacity for bringing forth representations itself, or the spontaneity of cognition" (A51/B75). Now the outer possibility of a power consists in the obtaining of conditions in which the operation of a capacity must take place if this activity is to actuate it as that power. In the case of the capacity of thinking itself, this occasion consists in some particular manifold of representations being given to the subject in its sensible intuition in such a way that she can relate them to an object in an act of thinking. And for a manifold to be given in this way is for it to meet the conditions for the unification of these manifolds in thinking that are determined in the intellectual form of representations that the operation of the spontaneity of cognition brings forth itself so as to realize in us the power to think an object through these given representations. Moreover, as we have just seen, the cognition of this essence in and through which all true metaphysics is to be taken from this essence, must be a purely a priori synthetic and rational cognition of this essence as the first and highest principle of all true metaphysics: for it is only such cognition that can constitute the principle under which all genuinely scientific, and so true, metaphysics is to exhibit real ground-consequence relations. Is this essence from which true metaphysics is to be taken logical, real, or formal?

This essence cannot be logical, i.e., the essence of our concept of the capacity of thinking. For, on Kant's account, cognition of this essence can yield only analytic judgments, and properly metaphysical judgments are "one and all synthetic" (Prolegomena; Ak. 4: 273): it belongs to metaphysics to analyze concepts, but all such analysis, and the analytic judgments they yield, are directed to the proper aim of this science, which is to generate synthetic a priori rational cognition of the objects of the concepts that are analyzed (ibid).[11] Indeed, the force of 'itself' in 'the capacity of thinking itself' is, plausibly, to specify the essence of the capacity itself, as against the essence of our concept of this capacity. Moreover, in introducing the project of the Analytic of Concepts, Kant stresses that this analytic is not the analysis *of concepts* at all, but rather the

> analysis of the capacity of understanding itself, in order to research the possibility of a priori concepts by seeking them only in the understanding as their birthplace and analyzing their pure use in general; for this is the proper business of transcendental philosophy.
>
> (A65–66/B90)

The essence from which a true metaphysics is to be taken cannot, as such, be the *real* essence of the capacity of thinking itself either. To be sure, this capacity, as an active potency, does have a real essence. Indeed, on Kant's account, this capacity is our capacity of understanding in the broad sense of 'understanding', that of the spontaneity of cognition, insofar as its operation, on the occasion of having representations given to us in our sensibility, brings forth itself intellectual representations to realize in us the power to think objects through these representations. The nature of our capacity of thinking itself, then, is the first inner principle of all that belongs to our existence merely as the subject of the operation of the spontaneity of our cognition in which we realize this capacity as our power to think. But, in Kant's view, we cannot cognize any real essence, or nature. So, a fortiori, it cannot be in and through any cognition we can have of the real essence of this capacity that we are to produce a true metaphysics. The essence of our capacity of thinking itself from which we are to take any true metaphysics, then, cannot *as such* be a real essence.

Assuming that all essences are either logical, real, or formal, it follows that the essence of our capacity of thinking itself from which we take true metaphysics is, as such, a formal essence: the first inner principle of everything that belongs, not to the real possibility of this capacity, but rather to some sort of non-real and not-merely-logical possibility of this capacity. This possibility, I propose, is the purely intellectual possibility of our capacity to think; one determined, collectively, by all the intellectual principles inherent in our capacity for the spontaneity of cognition itself. On this proposal, the essence of our capacity of thinking itself is the first inner principle under which these intellectual principles constitute the principles conformity to which we are, as thinkers, to prescribe to given manifolds of representations: namely, the principle Kant dubs "the principle of the synthetic unity of apperception" (B136). By way of explaining and motivating this proposal I can here only sketch the outlines of a reading of Section 16 of the B-edition Transcendental Deduction of the Categories, on which this section presents the rational cognition we can, in doing transcendental philosophy, achieve of this principle as the essence of our capacity of thinking itself.[12]

The original synthetic unity of apperception is the intrinsic unity of an individual thinker's pure apperception throughout all representations that are to be anything to her, a unity essential to her pure apperception. Now pure apperception is, in Kant's terminology, a thinker's purely intellectual *de se* consciousness (itself obscure) of a certain spontaneity that first makes concepts possible, an inner activity he dubs 'reflection' (Anthropology; Ak. 7: 135n). And this reflection, in turn, is her a priori consciousness of how the sources of her cognition (understanding or sensibility) objectively necessitate how she is to relate given representations to an object in her thinking (A260/B316). Pure apperception, then, is the purely intellectual *de se* consciousness that a thinker has of itself insofar as she is subject to principles, inherent a priori in her capacity of cognition, in such a way that she is capable of acts of thinking that are imputable to her, and thus as she is to constitute a psychological person. Now, in the opening paragraph of Section 16 of the B-edition Deduction, Kant tells us that pure apperception is the "act [*actus*] of spontaneity" that "brings forth that thinker's representation <I think>" (B132). Indeed, pure apperception is essentially reflective, in that it brings forth the representation <I think> in representing itself *de se* as the token act that constitutes the I as the individual transcendental subject that it is. A central thesis of Section 16 is that the <I think>, as the representation that an individual thinker brings forth in its pure apperception, constitutes the most general principle that is always actually operative in any and all of her thinking. All other intellectual principles of her thinking, including those, such as the logical functions of judgment or the categories, that are to constitute the intellectual form of an act in which she determines her own thinking out of her consciousness of how she ought to, derive from her representation <I think> as its specifications. For principles of thinking are, as such, principles conformity to which a thinker has, merely as a psychological person, the purely a priori warrant to prescribe to her representations a priori, on pain of their not having any use in her thinking, and so not being *her* representations. And what makes them such principles is their, one and all, issuing from that subject's pure apperception as such principles, that is, as principles conformity to which is a condition of a manifold, and the representations that make it up, given to her in her inner sense being taken up atemporally into the original synthetic unity of its apperception, and thereby constituting a manifold that has a use in *her* determined thinking.

Consider now how later in Section 16, Kant introduces a principle that he will immediately go on to refer to as "this principle of the necessary unity of apperception":

> Synthetic unity of the manifold of intuition, as given a priori, is thus the ground of the identity of apperception itself, which precedes a priori all **my** determined thinking. Combination does not lie in the objects [*Gegenstände*], however, and cannot as it were be borrowed from them through perception and by that means first taken up into

> understanding, but is rather only an operation of the understanding, which is itself nothing further than the capacity of combining a priori and bringing the manifold of given representations under unity of apperception, which principle is the supreme one in the whole of human cognition.
>
> (B134–135)

Moreover, in the heading of the next section, Kant names this principle 'the principle of the synthetic unity of apperception', claiming that it is "the supreme principle of all use of the understanding" (B136). The identity of apperception itself has its ground in a certain synthetic unity specifically of the manifold of intuition, as it is given a priori, and this, as what provides the matter for pure apperception's realizing a relation (*Beziehung*) that "all manifold of intuition that is to be encountered" in the subject of that pure apperception bears to that subject (B132). For without this matter, given to this subject in an operation of the spontaneity of her cognition he earlier dubbed 'combination in general' (B129f.), the original synthetic unity of apperception that constitutes her as transcendental subject would not be possible. One is a transcendental subject only as a subject of the capacity for acts of thinking that are imputable to one, under principles of thinking that govern these acts, and that is, as the subject of this capacity, numerically identical throughout all representations that are to be anything to one (as representations that have a use in one's determined thinking). Since this synthetic unity of the manifold of intuition must, as this ground, itself be a product of the operation of the understanding (as spontaneity), combination in general (the operation Kant here refers to as "combining a priori"), the capacity of understanding is itself "nothing further" than the capacity not just for combining a given manifold of representations a priori (in combination in general), but for accompanying this combination with the action that brings this given manifold under unity of apperception (pure apperception). The principle of the synthetic unity of apperception just is this rational cognition of what the capacity of understanding itself is.

This cognition, the principle of the synthetic unity of apperception, is not just any a priori principle, but the highest principle of the whole of human cognition, and this as rational cognition of the essence of the capacity of thinking itself. This cognition is, to be sure, cognition we achieve only in carrying out the critique of pure reason, and so in the service of establishing the purely a priori warrant each of us has, as a subject of pure understanding, to prescribe a priori to representations that are given to her their conformity to principles of her thinking. But it is, nonetheless, as a rational cognition that we are all (in principle, if not in practice) able to achieve, the first principle in which, and under which, each of us, as a thinker, has the authority to prescribe to manifolds of representations given in her intuition, as she does in any of her own determined thinking, the conformity of these representations to the purely intellectual conditions that intellectual principles of her thinking specify of their having any use in her thinking. Being taken up into a thinker's pure apperception to realize original synthetic unity of her apperception is what, on Kant's account, subjects a manifold of representations given in her sensible intuition to the intellectual principles of her thinking and thereby first gives those representations the use in her thinking that makes them *her* representations. In this way, the original synthetic unity of a thinker's apperception is what constitutes her capacity of understanding itself. Moreover, pure apperception realizes the original synthetic unity of apperception in all representations that are to have any use in a subject's thinking only in virtue of this unity's constituting that thinker's numerical identity as transcendental subject throughout all these representations. This last point captures the sense, and true force, of Kant's characterization of the understanding as "the capacity for bringing forth representations itself, the spontaneity of cognition" (A51/B75): this spontaneity is the self-activity in which our higher capacity of cognition brings forth the intellectual principles of our thinking in realizing itself as the capacity of understanding itself (Smit 2009).

In one respect, it is the understanding's authority to legislate of all intuitions that are to be encountered in one their objectively necessary conformity to the categories as the intellectual condition of their belonging to an experience in general that grounds its a priori warrant to prescribe to all other manifolds of representations that can be given to it in its intuition their conformity to principles of its thinking. But, in another, and even more fundamental, respect, it is the possibility of our achieving complete insight (as we do in the Transcendental Analytic as a whole) into how and why all of the intellectual principles of our thinking collectively, and under the principle of the original synthetic unity of apperception, constitute the formal possibility of our capacity of thinking itself that is fundamental. For it is in this insight that we establish our a priori entitlement to prescribe to all manifolds of representations that are to be given in our intuition their conformity to intellectual principles, including that for prescribing specifically to all manifold of intuition that is to be encountered in us their objectively necessary conformity to the categories.

The transcendental philosopher's rational cognition of the principle of the synthetic unity of apperception is not rational cognition of pure apperception itself as the noumenal ground of the original synthetic unity of apperception, nor of this unity itself. Indeed, such cognition is not possible for us, for our thought of this ground constitutes cognition *of this ground* only if we can prove that this ground is really possible, and we can provide no such proof. It is, rather, rational cognition merely of this principle as the first a priori principle under which we each have, as persons, the a priori warrant to prescribe a priori to all manifolds of representations that are to be given in her intuition their conformity to all the principles of thinking inherent in our pure understanding. This insight, in other words, is one into the inner possibility of these principles as mere representations, and not an insight into (which as such is a rational cognition of) the inner possibility of a thing. The transcendental philosopher's insight employs, and need employ, only a mere representation, as against a cognition, of combination in general and pure apperception as the actus of spontaneity that would, as its noumenal ground, have to originate this unity and thereby our capacity of understanding itself, if any intellectual principles at all are to be possible for us. For all that this insight needs to be is one into how and why our originally acquiring the categories in and through pure apperception, as the actus that brings forth original synthetic unity of apperception, *would* make these concepts "self-thought a priori first principles of our cognition," and this as concepts that "contain the ground of the possibility of all experience in general from the side of the understanding" (B167). They are such *intellectual principles* as ones that, in determining for given intuitions a use they have in our thinking that constitutes cognition of their objects, first make possible all other intellectual principles of our thinking, and thus our very capacity of thinking itself. This insight suffices, thereby, to establish a *de se* a priori entitlement[13] that each of us has, merely as thinkers, to *presuppose* that the transcendental conditions of the legislation for nature, which is our capacity of understanding, actually obtain. This insight into the formal possibility of our capacity of thinking itself, then, *concerns* the original synthetic unity of apperception, and moreover in respect of the noumenal grounds that would have to constitute it, without being cognition *of* these grounds themselves, or indeed *of* any *thing* at all: that, again, would require that we be able, in having insight into these grounds as its *constitutiva*, to prove that the thing that thinks is really possible. It is, itself, rational cognition only *of* the formal possibility of our capacity of thinking itself—the possibility of certain purely intellectual representations, merely as representations. In this way, the essence of our capacity of thinking itself is as such the first ground of the inner possibility only of these principles, the purely a priori intellectual representations that constitute the intellectual form of our possible thinking and cognizing of an object in general. This essence, in other words, is the first inner principle of everything that belongs to the non-real, and merely formal, possibility of the thing that thinks, as against the first inner principle of everything that belongs to the being of this thing (cf. Section I). And the rational insight we have into this principle, then, is itself one only into the possibility of a mere representation, much as the possibility into which we have insight when, in doing pure geometry, we construct the pure sensible concept of a triangle is, of itself, the possibility merely of a purely a priori sensible representation in and through which we are to delimit our pure intuition of space.

This brings us to the second horn of the dilemma we noted at the outset of the present section: how can a true metaphysics be taken from an essence that is not a real essence? Kant's answer, schematically stated, is that, among the intellectual principles the possibility of which we achieve insight into in achieving rational cognition of the essence of our capacity thinking itself are the categories and the principles of pure understanding. And these principles, unlike those of pure formal logic, are principles objectively necessary conformity to which we prescribe of given appearances a priori, in legislating this conformity as what constitutes their formal possibility as experience. This is why Kant tells us that true metaphysics, as rational cognition, is to be drawn from the essence of our capacity of thinking itself because true metaphysics "contains the pure actions of thought, and thus a priori concepts and principles, *which first bring the manifold of empirical representations into law-governed connection through which it can become empirical cognition, that is, experience*" (Ak. 4: 472; my italics). Notice that Kant here infers ('thus') that true metaphysics contains the categories and the principles of pure understanding from its containing "the pure actions of thought": these thoughts include, not just the categories, but even more fundamentally, the *actus* of combination in general and pure apperception, the inner actions that true metaphysics represents as what together realize a thinker's capacity of understanding itself, and this in constituting it as the numerically identical transcendental subject of all its determined thinking. But, again, true metaphysics contains these actions in what are mere representations *of these actions*, representations that constitute cognitions—more specifically, comprehension (Logic; Ak. 9: 65), and so the content of rational insight that is sufficient for the purposes of true metaphysics—only of the principles of thinking that the understanding itself brings forth in these actions, as cognitions of the sort they purport to be. This cognition is one of these principles as ones that we have the a priori warrant to prescribe that they purport to have in virtue of their having this origin, one in which the thinker first realizes herself as transcendental subject in originally acquiring them. Kant is laying claim to this sort of insight, specifically with respect to the categories, when he claims "I have complete insight into not only the possibility but the necessity of subsuming all appearances under these concepts [categories]—i.e., of using them as principles of experience" (Prolegomena; Ak. 4: 311).

We saw earlier that we cannot, on Kant's account, achieve insight into the real possibility of the objectively necessary conformity to the categories that our understanding demands in the legislation for nature in which it consists. But we need not in order for the understanding to have the authority to legislate it. All we need be able to do is to achieve an insight that allows us to establish our a priori entitlement to presuppose that these intellectual principles are representations of the sort they purport to be, as we do in making judgments of experience: principles to which the appearances that we relate to objects in these judgments in fact conform, and principles conformity to which is sufficient for these appearances to have the relation to objects of our experience that we thereby judge them to have. This insight, moreover, is as such one that allows us to establish our pure understanding's authority, in its real use, to prescribe this conformity to appearances that are actually given to us as manifesting to us regularities in kind that are subsumable under the principle of the Second Analogy. This insight, however, is itself one merely into the formal possibility of our capacity *of thinking* itself. Each of us has a *de se* a priori entitlement, merely as a thinker, to presuppose that the transcendental conditions of the possibility of this legislation in fact obtain in herself. These conditions are, first, that one's sensibility is originally affected by the transcendental object (*Gegenstand*) of our cognition, an affection that provides the matter of a single universal human experience, and, second, that the supersensible operation of one's capacity of cognition timelessly realizes original synthetic unity of apperception in the matter that is given to one in the original (noumenal) affection of one's sensibility. Each of us has a *de se* a priori entitlement to presuppose that these conditions in fact obtain, because we can in principle determine them, in this insight, as the conditions that *would, if actual, constitute, and so be necessary and sufficient for*, our being subjects of the capacity to think. And actually achieving this insight into our capacity of thinking itself also serves, thereby, to establish that each of us has a purely a priori entitlement to presuppose as actually obtaining all the conditions that we can determine, in this way, to be both necessary and sufficient for our being subjects of the capacity to think.

In short, Kant holds that our understanding has the authority to prescribe to given appearances their conformity to the categories only in virtue of our being able, in principle, to establish, in their transcendental deduction, that this conformity is a condition of the inner possibility of our having any power to think at all. In establishing this last claim, Kant appeals to an insight into the formal essence of the capacity of thinking itself. This insight employs, constitutively, a mere representation (as against cognition) of our capacity of understanding itself as the original synthetic unity of apperception—a mere representation of the real essence of this capacity. This, in outline, is how, on Kant's metametaphysics, we can, and over the course of the Transcendental Analytic of the first critique do, take from this mere essence a true metaphysics. Pure reason achieves the needed insight in the Transcendental Analytic only in virtue of pure understanding, under pure reason's guidance, "isolating itself completely not only from everything empirical, but even from all sensibility" over the course of the Transcendental Analytic (A64–65/B89). And in virtue of pure understanding's doing so—so as to isolate the most general, and purely intellectual, conditions on our thinking an object in general, with which traditional ontologists were concerned—the Transcendental Analytic constitutes a mere analytic of the understanding, and does not deserve the traditional title 'ontology' (A246–247/B303). The Transcendental Analytic does, however, in the service of providing a transcendental deduction of the categories, establish the principle of the synthetic unity of apperception as the highest principle of all our cognition (B135–136). And this principle, when taken in relation to all manifolds of appearances that are to be given to us as what constitutes the highest transcendental condition of the possibility of their constituting experience, as it is in transcendental philosophy, constitutes the highest principle of any true metaphysics of which we are capable. On Kant's account, it is in this way, and only in this way, that we can provide a true general metaphysics—albeit one that constitutes cognition of things only in respect of the being they are to have as objects of our experience and so only in appearing to us. And any other true metaphysics, which as such must be a special metaphysics, must derive from this general metaphysics—in the case of a special metaphysics of nature, through an impurely a priori rational insight that employs certain empirical concepts of matter.

On Kant's account, Leibniz was right to hold that our cognition of phenomena depends, for its possibility, on the nature of our understanding. But he was wrong to hold that we can ground the former on cognition of the latter: we cannot have cognition of any nature, including the nature of our understanding. And because we cannot cognize the nature of our understanding, all true metaphysics must, following the critical method, be taken rather from the formal essence of our capacity of thinking itself, by way of a rational cognition of this essence that puts us in a position to establish our right, by appeal to the supersensible nature of our understanding, to apply the categories to objects of our experience.[14]

Notes

1. In what follows, I will use 'ontology' and its cognates to refer only to what purports, falsely on Kant's view, to be such a science. In doing so, I follow Kant, who famously rejects the title 'ontology' for the enterprise that the tradition misconceived as a science of the nature of a thing in general, and proposes in its stead the modest title of "a mere analytic of the pure understanding" (A246–247/B303).
2. To be sure, Kant holds that we can prove the real possibility of a particular thing of which we have experience "by the testimony of experience from its actuality" (Bxxvin). But experience—in the relevant technical sense in which Kant employs the term 'experience' (*Erfahrung*)—is itself a species of empirical cognition (*Erkenntnis*). In appealing to the testimony of experience, then, we already presuppose the a priori objective validity of the categories. Hume's skeptical attacks on our ideas of causality, substance, distinct and continued existence, etc., served to call into question, not only our right to presuppose the objective validity of experience, but even the very possibility of experience, in this technical sense of 'experience.' These attacks showed that we need to establish both our claim to have the categories (including our idea of causality as an objectively necessary connection of distinct existences), and our right to put them to use to cognize things. Indeed, these attacks showed that we need to establish this claim and this right prior to and independently of all our experience.

3. I first developed this reading of Kant's notion of the a priori in Smit (2009). See especially, 206–207. However, in this piece I did not distinguish, as I ought to have, between rational insight and a distinctive sort of insight Kant refers to as insight of the understanding' ('*Verstandeseinsicht*'). As we will see in the following, Kant holds that the latter, unlike the former, does not itself consist in an apodictic consciousness, one in which we see *why* the ground determines its consequence to the exclusion of any other. Moreover, I do not, any longer, hold that, on Kant's account, the only way in which we can cognize a thing a priori, even in rational insight, is to cognize it from its *rationes essendi* as they constitute its nature. As I explain in Section II, Kant maintains that in the form of a special metaphysics of body, we achieve a degree of rational insight into matter, even though we cannot have any cognition of natures, or real essences, including those of things as they appear to us.
4. Here, and throughout, I will use 'representation' to denote what is represent*ed* in an act of representing, i.e., the representational content of a representing, as against the act of representing. A representation, in this sense, is what is to be related, in an operation of a capacity of representation, to some object.
5. Kant here uses 'natural science' in his most permissive sense, on which even a "historical doctrine of nature, which contains nothing but systematically ordered facts about natural things" (MFNS; Ak. 4: 468) counts as a natural science. But this is not even a rational science, let alone a science in the proper sense of 'science,' which designates "that whose certainty is apodictic" (ibid). In this most permissive sense of 'science,' any system, or "a whole of cognitions ordered according to [*nach*] principles"—even principles that are merely ones of "natural description," "a system of classifications for natural things according to [*nach*] their similarities" and that as such would better be called "a historical doctrine of nature"—counts as a natural science (ibid). Thanks to Frode Kjosavik for prompting me to provide this clarification.
6. The "first *constitutiva*" of a given concept are the marks that I think as constituting that concept (its essentialia) in grasping its essence: on Kant's account, to form the concept <bachelor> just is actually to think (perhaps only obscurely) the concepts <unmarried> and <male> as what belong, under this essence, to the non-real, because merely logical, possibility of the thing that is thought in forming this concept, which possibility consists merely in the possibility of thinking it. These marks—which are constitutiva merely of the concept, and not of the thing itself that we think in forming this concept—are to be distinguished from the "*rationata logica*," namely, the marks that derive from the essence of this concept, a logical essence.
7. This reading of what Kant means by 'formal possibility' is at odds with that which Nicholas Stang develops in his excellent recent book *Kant's Modal Metaphysics* (2016: especially chapters 7 and 8). On Stang's reading, formal possibility is a species of real possibility—namely, real possibility with grounds that are "immanent to experience" and thus "subjective conditions of phenomena" as against transcendental grounds, or noumena (2016, 200). On my reading, by contrast, formal possibility is itself the possibility of a mere representation, and as such to be contrasted with real possibility. Moreover, I maintain *pace* Stang that, on Kant's account, formal possibility also has a transcendental ground: namely, in the noumenal constitution or nature of our capacity of cognition itself. In this, and other, respects, the readings of Kant's notions of essence and nature, as well as of the use to which Kant puts them in his critical philosophy, differ from those Stang offers.
8. I develop a detailed reading of Kant's notion of a capacity (*Vermögen*), and its attendant notions of action, activity, and power, in Section iii of Smit (2009). I owe my recognition of the importance of reading Kant's account of our cognitive capacity in light of these metaphysical characterizations of a

capacity to Béatrice Longuenesse (1998). See, especially, 7–8 of the Introduction of this ground-breaking work.

9. Kant holds that the chemistry of his day is, unlike Newtonian physics, not a proper science because it does not meet this demand. To be sure this chemistry (unlike a "historical doctrine of nature," cf. fn 5) is "a *rational* science," because "the connection of cognition in this system is an interconnection of grounds and consequences" (MFNS; Ak. 4: 468). Nonetheless, since "the grounds and principles themselves" of this chemistry are "still in the end merely empirical," the "whole of cognition" had in chemistry "does not deserve the name of a science in the strict [*in strengem*] sense" (ibid).
10. Kant presents the content of this rational insight in the second chapter of MFNS (Ak. 4: 498f.) in the form of what has come to be called 'the balancing argument.'
11. Some commentators have read the Discipline's characterization of philosophical cognition as "rational cognition from concepts" as specifying that philosophical cognition is analytic. I argue that this reading is mistaken, and provide an alternative reading, in Smit (2013).
12. I develop central aspects of this reading, though not under this description, in some detail in Smit (Forthcoming).
13. I use the term 'entitlement' here in a sense akin to that Tyler Burge specifies in his influential Burge (2003). Entitlement is warrant that is externalist to the extent that the subject that has this warrant need not, in fact, be able to understand it. In the case of our a priori entitlement to presuppose that the conditions of the possibility of our having any power to think actually obtain, what we cannot understand (or indeed cognize in any way), even in principle, is that these conditions actually obtain, or indeed that they are really possible. Nonetheless, we have this entitlement, on Kant's view, only in virtue of our being able, in principle, to understand these conditions as ones that are necessary, and that would (were they to obtain) be sufficient, for our having this power. To this extent, our a priori entitlement to presuppose that the conditions of the possibility of our having the power to think actually obtain is internalist.
14. I first developed many of the ideas in this chapter in 2015–16, during a fruitful year spent at the Centre for Advanced Study at the Norwegian Academy of Science and Letters, as a member of the research project "The Possibility of Metaphysics in the Age of Science." I thank the Centre, and the Academy, for their support. I am also grateful to other project members for many helpful conversations. This chapter benefitted especially from conversations with Christian Beyer, Dagfinn Føllesdal Michael Friedman, Olli Koistinen, Charles Parsons, and the project leaders, Frode Kjosavik and Camilla Serck-Hanssen. Indeed, Frode Kjosavik first suggested to me that I pull together my ideas in a paper that focuses on Kant's characterization of true metaphysics in the Preface to MFNS, at 4: 472. I presented a version of this chapter at the 2018 North American Kant Society Biennial in Vancouver, and I thank that audience for helpful discussion. Finally, I am grateful to Jonas Indregard, Frode Kjosavik, Santiago Sanchez, Camilla Serck-Hanssen, and Mark Timmons for helpful comments on a penultimate draft.

References

Burge, Tyler. "Perceptual Entitlement." *Philosophy and Phenomenological Research* 67 (2003), 503–548.

Kant, Immanuel. *Grundlegung zur Metaphysik der Sitten* [*Groundwork to a Metaphysics of Morals*]. Berlin: Academy Edition. Vol. 4, 387–463. Ed. by P. Mentzer. Berlin: Georg Reimer, 1911a. (Cited as [GW; Ak. 4]).

Kant, Immanuel. *Kritik der reinen Vernunft: Erste Auflage 1781* [*Critique of Pure Reason, A Edition*]. Berlin: Academy Edition. Vol. 4, 5–252. Ed. by B. Erdmann. Berlin: Georg Reimer, 1911b. (Cited, in standard A/B edition pagination, as 'A', with page number following.)
Kant, Immanuel. *Kritik der reinen Vernunft: Zweite Auflage 1787* [*Critique of Pure Reason, B Edition*]. Berlin: Academy Edition. Vol. 3, 2–552. Ed. by B. Erdmann. Berlin: Georg Reimer, 1911c. (Cited, in standard A/B edition pagination, as 'B', with page number following.)
Kant, Immanuel. *Metaphysische Anfangsgründe der Naturwissenschaft* [*Metaphysical Foundations of Natural Science*]. Berlin: Academy Edition. Vol. 4, 467–565. Ed. by A. Höfler. Berlin: Georg Reimer, 1911d. (Cited as [MFNS; Ak. 4]).
Kant, Immanuel. *Prolegomena zu einer jeden künftigen Metaphysik, die als Wissenschaft wird auftreten können* [*Prolegomena to any Future Metaphysics that Will Be Able to Present Itself as a Science*]. Berlin: Academy Edition. Vol. 4, 255–383. Ed. by B. Erdmann. Berlin: Georg Reimer, 1911e. (Cited as [Prolegomena; Ak. 4]).
Kant, Immanuel. *Kritik der Urteilskraft* [*Critique of the Power of Judgment*]. Berlin: Academy Edition. Vol. 5, 167–485. Ed. by Wilhelm Windelband. Berlin: Georg Reimer, 1911f. (Cited as [CPJ; Ak. 5]).
Kant, Immanuel. *Anthropologie in Pragmatischer Hinsicht* [*Anthropology from a Pragmatic Point of View*]. Berlin: Academy Edition. Vol. 7, 119–333. Ed. by Oswald Külpe. Berlin: Georg Reimer, 1911g. (Cited as [Anthropology; Ak. 7]).
Kant, Immanuel. *Logik* [*Logic*]. Berlin: Academy Edition. Vol. 9, 1–150. Ed. Max Heinze. Berlin: Georg Rimer, 1911h. (Cited as [Logic; Ak. 9]).
Kant, Immanuel. *Briefwechsel. Band I, 1747–1788* [*Correspondence*]. Berlin: Academy Edition. Vol. 10. Ed. by R. Reicke. Berlin and Leipzig: Walter de Gruyter, 1922a. (Cited as [Correspondence; Ak. 10]).
Kant, Immanuel. *Briefwechsel. Band II, 1789–1794* [*Correspondence*]. Berlin: Academy Edition. Vol. 11. Ed. by R. Reicke. Berlin and Leipzig: Walter de Gruyter, 1922b. (Cited as [Correspondence; Ak. 11]).
Kant, Immanuel. *Metaphysik Mrongovius*. Berlin: Academy Edition. Vol. 29, 747–940. Ed. by G. Lehmann. Berlin and Leipzig: Walter de Gruyter, 1983. Cited as (M. Mrongovius; Ak 29).
Longuenesse, Béatrice. *Kant and the Capacity to Judge*. Princeton: Princeton University Press, 1998.
Smit, Houston. "Kant on Apriority and the Spontaneity of Cognition." In S. Newlands and L.M. Jorgensen, eds., *Metaphysics and the Good: Themes from the Philosophy of Robert Merrihew Adams*. Oxford: Oxford University Press, 2009, 188–251.
Smit, Houston. "Kant on the Apriority and Discursivity of Philosophy." In C. La Rocca, A. Ferrarin and S. Bacin, eds., *Kant und die Philosophie in weltbürgerlicher Absicht: Akten des XI. Kant Kongresses 2010*. Berlin: Walter de Gruyter, 2013, 739–750.
Smit, Houston. (Forthcoming). "Kant's 'I think' and the Agential Approach to Self-Knowledge." *Canadian Journal of Philosophy*. Published online, May 28, 2019. https://doi.org/10.1080/00455091.2019.1615682
Stang, Nicholas. *Kant's Modal Metaphysics*. Oxford: Oxford University Press, 2016.

3 Toward a Husserlian (Meta)Metaphysics

Christian Beyer

Husserl's conception of the constitution of reality has it that there are layers of reality which are more basic than others, as far as their constitution in consciousness is concerned. The view that reality is structured in such a way that some of its elements depend on other, more fundamental ones is also held by proponents of more recent analytic (meta)metaphysics, but in connection with a rather dogmatic naturalism. I will begin by explaining Husserl's conception of constitution. This requires us to go into the motives that led him to "transcendental idealism" to some extent, a position which it may be helpful to compare to (some aspects of) what Hilary Putnam calls internal realism (section II). It will become clear that as it stands Husserl's position is not well-argued and less than convincing, but that it can be modified so as to yield a more persuasive view, which can also be used to give substance to (the more plausible aspects of) internal realism. Finally, I will indicate how the claim that Husserl's transcendental phenomenological analyses provide a non-naïve alternative to today's naturalistic mainstream metametaphysics (to be sketched in section I) can be substantiated further. To this end, I shall briefly sketch his analysis of the constitution of nature as conceived of in physical (as well as cognitive) science, as developed in the second volume of his work *Ideas* (*Ideas* II) (section III).

I. Analytic (Meta)Metaphysics and Referential Magnetism

A useful way to draw a terminological distinction between "metaphysics" and "ontology" is to adopt a Quinean notion of ontology, according to which *ontology* concerns the question "What is there?," where the ontology a person, or group of persons, subscribes to comprises those objects they commit themselves to by the use of their quantifiers (cf. Quine 1948). The term "*metaphysics*" can then be used to refer to the study of the *way* or mode in which existing things exist—whether they are basic or dependent on more basic entities (Schaffer 2009), and what their ontological status is, i.e., to which category they belong or (to use a Husserlian term) which general essence they instantiate. Many recent contributors to so-called analytic metametaphysics (e.g., Sider, Fine, Schaffer) assume that ontological claims made in a philosophical tone of voice are substantial, rather than trivial or "internal" (Carnap 1956), and that if some of these claims are true, then there are *basic* objects or properties—"referential magnets" as Lewis called them—whose existence helps make these claims true. These fundamental truth-makers, or the corresponding ontological claims, are supposed to "carve out nature at its joints" (Lewis). They "ground" all other existing entities and the corresponding true existence statements. The notion of grounding is held to be an undefinable basic notion, a view already to be found in Bolzano. Thus, Schaffer writes:

> Grounding should . . . be taken as *primitive*, as per the neo-Aristotelian approach (cf. Fine 2001, 1). Grounding is an unanalyzable but needed notion—it is *the primitive structuring conception of metaphysics*. It is the notion the physicalist needs to explicate such plausible claims as 'the fundamental properties and facts are physical and everything else obtains *in virtue of* them' (Loewer 2001, 39). It is the notion the truthmaker theorist needs to explicate such plausible claims as: 'Must there not be something about the world that makes it the case, that serves as an ontological ground, for this truth?'
>
> (Schaffer 2009, 364f.)

As a consequence of what Schaffer says, philosophical statements regarding ontology are made true by physical properties and facts, and metaphysics investigates which of these are fundamental—which of the corresponding entities are "substances," as Neo-Aristotelians put it. Metaphysics goes hand in hand with science, and it is nature as investigated in science that metaphysical statements "carve out at its joints," thereby teaching us *how* existing things exist.

From the viewpoint of Husserl's transcendental phenomenology, this metametaphysical picture is naïve. Here he agrees with Putnam, who criticizes the natural properties or "elite classes" postulated by Lewis (as those "classes of things 'out there,'" which are such that "it is a 'natural constraint' on reference . . . that as many of our terms as possible should refer to these 'elite classes'") as "something 'spooky'" (Putnam 1990, 38). (I will return to the relationship between Husserl and Putnam in section II.) To be sure, Husserl also distinguishes between substances, which he refers to as "substrates" (he uses the term "substance" in a more special sense, see section III), and other entities, but he has a complex notion of substrate embedded in his epistemological theory of the pre-predicative, or pre-judgmental, constitution of empirical reality. Husserl would agree that there are referential magnets, in the sense of fundamental truth-makers or their underlying basic referents, but deny that in the final analysis this is due to natural properties as conceived of in (present or future) physical science, as mainstream analytic (meta)metaphysics has it. Rather, referential magnetism is a "constitutive" achievement; it is due to the essential and epistemic structure of perception and, more generally, pre-predicative experience, to be studied in transcendental phenomenology, and to the structure of empirical reality thus "constituted." It is to the relevant notion of constitution that I now turn.

II. Husserl's Notion of Constitution and His "Transcendental Idealism"

Husserl is a realist, in the sense that he is convinced of the existence of objects transcending what is given to our (present) consciousness. But he is not a *naïve* realist. Rather, he investigates into what he calls the *constitution* of objects—a notion that will be elucidated further in what follows. Nor does he adhere to what he calls the "image theory" of representation, the view that we experience objective reality only indirectly, by representing subjective appearances or sense-data. (A special case of this view is indirect realism or representationalism about perception.) Against this view he objects, among other things, that it leads to a false duplication of represented objects and merely shifts the problem of intentionality it is supposed to solve:

> The same Berlin which I represent also exists, and the same would no longer exist if judgment fell upon it as upon Sodom and Gomorrah.
> (Husserl 1994, 347; Hua XXII, 305f)

> It is overlooked that the phantasy content first must become the [r]epresenting image *of* something or other, and that this 'pointing-beyond-itself' in the image—which first makes it into *an image* . . .—is a 'more [*plus*; CB]' that is essential to consider.
> (Husserl 1994, 348; Hua XXII, 306)

But then, what is it that enables and entitles us to represent things transcending our present consciousness, if not mental images or sense-data that we could draw upon to infer the existence of transcendent objects causing those images, as the image theory of representation has it? This is the (meta-)epistemological "*problem of transcendence*" (Hua X, 345). In this context, Husserl is interested in a justification, regarding the assumption of entities transcending the "self-givenness" of one's own consciousness, that is "immune to all reasonable doubt" (Hua X, 344). All presuppositions or premises based on this realistic assumption are to be systematically bracketed in a "phenomenological epoché," so that we have to start from a methodological solipsism (see Hua VIII, 65f).

Given that Husserl is a realist, and only a *methodological* solipsist, what is the point of his "transcendental idealism"? During the years in which his transcendental phenomenology took shape, he developed a number of "proofs" of this position, most of which are based upon his conception of a "real possibility" regarding cognition or the acquisition of knowledge. By a "real possibility," Husserl understands a possibility that is such that "something—more or less—'speaks in favour of it' [*'für die etwas spricht' und bald mehr, bald weniger spricht*]." (Hua XX/1, 178) Real possibilities are, in other words, conceived of as more or less (rationally) *motivated* (epistemic or practical) possibilities; and Husserl understands motivation in such a way that it is always *someone* who is motivated a certain way (see Hua IV, 222). This is why Husserl subscribes to the following *dependency thesis*:

> The real possibility to acquire (empirical) knowledge regarding a contingent object A [possible world, individual thing, state of affairs involving such thing; cf. Hua XXXVI, 139f] "requires" an "epistemic subject [*Erkenntnissubjekt*]," which "either experiences A, or acquires knowledge regarding A on the basis of experience, or else has the practical possibility (or the practical ability) to experience A and acquire knowledge regarding it."
>
> (Hua XXXVI, 139)

Husserl also adheres to the following *correlation thesis* with regard to empirical reality and real epistemic possibility:

> If a contingent object A is real (really exists), then the *real* (as opposed to the merely logical) possibility obtains to acquire knowledge regarding A [*rechtmäßige Erkenntnis von A*].
>
> (cf. Hua XXXVI, 138, l. 35f)

From these two propositions—the dependency and the correlation thesis—he derives the conclusion that the existence of a contingent object A requires "the necessary co-existence of a subject either acquiring knowledge [*eines erkennenden*]" regarding A "or having the ability to do so [*erkenntnisbefähigten*]." (Hua XXXVI, 139f) This is nothing but the thesis of *transcendental idealism*:

> The thesis of transcendental idealism runs: A nature without co-existing subjects of possible experience [*Erfahrung*] regarding it is unthinkable; possible subjects of experience are not enough.
>
> (Hua XXXVI, 156)

Why are *actual* subjects of experience supposed to be necessary? Husserl's answer refers to the notion of *full* epistemic justification (or *full degree* of real possibility)—a notion he takes to be applicable (as far as empirical consciousness is concerned) in case of *truth* only:

> In order for [a thing of nature] to really exist, and thus in order for the assumption that it exists to be reasonable or justified [*berechtigt*] not merely in a restricted but rather in an unrestricted way, i.e. to the fullest extent [*voll und ganz*], there must be an actual ego in [*an*] whose experiences [. . .] the being of the thing manifests itself [. . .] in such a way that nothing in the consciousness of this ego stands in the way of this being by scoring it out [*durchstreichend*], and also in such way that the course of experience [*Bewusstseinslauf*] of this ego does not leave the being of this thing open.
>
> (Hua XXXVI, 76f)

Behind this lies a conception of full epistemic justification that is at the same time *internalist* and *non-fallibilistic*—a combination of views I regard as problematic, because only an omniscient subject would seem to be in a position to possess such justification. In any case, full justification should not be *identified* with truth, as will become clear in a moment. Some passages in Husserl regarding truth indicate that he was attracted to such a false identification. Thus, in §16 of *Formale und transzendentale Logik* he first equates "truth" with "genuine knowledge" which is "repeatable by any . . . rational being" (Hua XVII, 46) and then makes the (somewhat weaker) claim that truth and such knowledge are "correlated" (Hua XVII, 47).

However, as Künne demonstrates in his critique of the "epistemic conception of truth" once held by Putnam, and associated with his "internal realism" (see Putnam 1990), there are truths (although not exactly truths corresponding to "outer experience") regarding which one cannot *even in principle* have a justified, let alone fully justified (or "idealized[ly] rational[ly] acceptable;" cf. Putnam 1990, 41) belief. The truth which results if one specifies the number *n* such that (i) and (ii) is a case in point: (i) the number of blossoms of the cherry tree in Bolzano's garden on May 15, 1830, is *n*, and (ii) no one is ever justified in the belief that (i). If someone were justified in believing the conjunction of (i) and (ii), he or she would also be justified in believing (i), which would falsify (ii) (Künne 1992, 240). Yet, the conjunction of (i) and (ii) is true (given that there was a blooming cherry tree in Bolzano's garden on May 15, 1830, and that nobody counted its number of blossoms—or might God be busy counting everything all of the time?). So the notions of truth and (fully) justified assertibility are not even co-extensional, and the epistemic conception of truth is false (except that you postulate, *ad hoc*, a special kind of philosopher's God, or deny the truth of (ii) for some other reason). It is not the case that that what is true is always "*warrantable* on the basis of experience and intelligence for creatures with a 'rational and sensible nature'." (Putnam 1990, 41) Near the end of section II, I shall return to the relationship between Husserl and Putnam once again, in order to compare their respective views concerning another, less problematic feature of "internal realism."

There is something wrong with Husserl's "proof of transcendental idealism." The dependency thesis underlying it is too strong. There may be real epistemic possibilities without co-existing epistemic subjects. As Husserl himself stresses, "surely no human being and no animal" *must* exist in the actual world (adding that their non-existence would, however, already result in a "change of the world") (Hua XXXVI, 121).[1] Husserl nevertheless holds that real possibilities require a "substrate," and "a merely logically possible subject" cannot function as a substrate (Hua XXXVI, 139). Here Husserl relies on a metaphysical assumption; he seems to regard real epistemic possibilities as epistemic dispositions, or habitualities. But does the real existence of a world or object really depend on a co-existing subject's actual dispositions, as the correlation thesis would state under this reading? After all, there is a sense in which real epistemic possibilities and corresponding epistemic justification regarding the actual existence of A may obtain without there actually being a subject co-existing with A that possesses such justification. This sense may be spelled out in terms of what could be called *real higher-order possibilities*, i.e., possibilities for acquiring mental dispositions to acquire knowledge in cases where epistemic subjects *would* be co-existing with the world or object whose existence is in question. I thus propose to replace the dependency thesis by the following *modified dependency thesis*, which is compatible with a possible world without co-existing epistemic subjects:

> The real possibility to acquire knowledge regarding a contingent object A requires the real higher-order possibility of an "epistemic subject [*Erkenntnissubjekt*]," which would "either experience . . . A, or acquire . . . knowledge regarding A on the basis of experience, or else [have] the practical possibility (or the practical ability) to experience A and acquire knowledge regarding it."

Accordingly, the thesis of transcendental idealism should be replaced by the following claim, which could be labelled as *weak transcendental idealism* and follows from the correlation and the modified dependency thesis:

> The existence of a contingent object A requires the real higher-order possibility of "a subject either acquiring knowledge [*eines erkennenden*]" regarding A "or having the ability to do so [*erkenntnisbefähigten*]."

On this modified view, there do not have to be co-existing subjects in order for environmental stimuli, perceptual and judgmental possibilities, etc., to be there, and to be *ready to motivate* epistemic subjects. However, what would *count* as a real possibility of the required sort is indeed dependent on epistemic subjects (and their context of epistemic assessment). It is dependent on *us*, the phenomenological subjects reflecting about such counterfactual situations in the methodological context of our analysis. In this sense, the real higher-order possibilities entailed by the actual existence of A are a function of the epistemic standards we—the phenomenologists—recognize as sufficiently justified for a counterfactual subject, in the light of our analysis (under the guideline of both the epoché and the problem of transcendence), with these possibilities resulting in a certain way A is structured. It is this idea that I take to be the true kernel of Husserl's transcendental idealism.

For instance, the phenomenological analysis of the "constitution" of perceptual objects makes it clear that we regard these objects as possessing various properties, such as color and shape, and that we are entitled to do so due to the holistic justification that can be (meta-)justified with regard to the pre-predicative and predicative experience motivating the "individual concepts" we form with respect to such objects. Thus, for example, if you look at a flat circular object from above, it looks round. If you move away from it and keep looking at it, it looks more and more elliptical, but in such a way that these various visual appearances continuously represent one and the same circular shape, displayed by (what consistently appears to be) one and the same object, throughout the whole process of observation. It will then be justified to keep the entry "it is circular" in the concept associated with the object in question; it has been constituted as circular in shape. By "constitution" Husserl means an "activity [*Tätigkeit*]" (not to be confused with an intentional action)[2] of the experiencing subject in a process of experience aiming at knowledge; where knowledge, or truth, defines the relevant *interest* of the subject leading it to pursue this activity.[3] So it occurs in a context of epistemic justification, which is why the phenomenological *analysis* of constitution, or "elucidation of the origin [*Ursprungsklärung*]," takes the form of a *meta*-justification.[4] As described in the following quotation, this activity consists, at the level of *predicative* experience (empirical judgment), in the formation and processing of *individual concepts*. At the *pre*-predicative level, it consists in the experiences *motivating* such (actual or potential) processings, such as, in Husserl's paradigmatic case, the experiences occurring in the course of the continuous observation of an object and its behavior across a period of time.

> I have experienced the object multifariously, I have made 'multifarious' judgments about it and have gained multifarious [pieces of] knowledge about it, at various times, all of which I have connected. Thanks to this connection I now possess a 'concept' of the object, an individual concept. . . . [W]hat is posited in memory under a certain sense [*mit einem gewissen Sinn*] gains an epistemic enrichment of sense, i.e., the x of the sense is determined further in an empirical way [*erfahrungsmäßig*].
>
> (Hua XX/2, 358)

These individual concepts may be looked upon as variable or dynamic systems of belief about a particular object, what Perry refers to as "mental files." Perry introduces this notion in order to explain what enables us to keep in mind a particular object across time and to collect information about that object (cf. Perry 1980). On Husserl's view, such files (i.e., their respective current version) influence the further course of predicative experience, i.e., subsequent empirical judgments:

> We speak of a *sedimentation of sense regarding the object* [*Sinnesniederschlag am Gegenstand*]. That is to say: Just like any step of receptive [i.e. pre-predicative; CB] experience, every step of predicative judging has its lasting result. It generates [*stiftet*] habitualities which influence the further course of actual judging in a great variety of ways.
>
> (Husserl 1999, 250)

The identification of a "concept" in the present sense of the term with a mental file à la Perry is supported by the fact that Husserl characterizes such individual concepts as being infinitely "open" and "in flux" (Hua XX/2, 359).[5] Note that the elements of these concepts or files (or the experiences motivating them) involve a *sense of identity through time*, which holds them together—they belong to the same *determinable* X or "*x* of the sense," as Husserl puts it in the penultimate quotation.

The determinable X holding together the elements of an individual concept is apt to lead us back through time towards the original situation where the reference of the relevant concept was fixed, like for instance the occasion of the subject's first perceptual encounter with a particular object in space, which Husserl (in *Experience and Judgment*) describes as the constitution of an "original [*ursprüngliches*]" or "ultimate substrate [*letztes Substrat*]" (or as the "original doxa [*Urdoxa*]") for potential empirical judgments about this "individual object."

Qua ultimate substrate, this object is merely constituted as "on hand [*vorhanden*]" (cf. Hua IV, 186f; Husserl 1999, 159f). However, that ultimate substrate would be "categorically formed" by (attributes deriving from) subsequent judgments regarding it,[6] or by their pre-predicative foundations,[7] thus gaining an "epistemic enrichment" of the "sense" (Hua XX/2, 358) under which it can henceforth be experienced.

Part of the reason for the original perceptual object's being the referent of the relevant file is the fact that the file's function is precisely to accumulate and adapt information about this object, because the relevant subject has become epistemically *interested* in this very object—the subject *wants to cognize it*, hence the "activity" of file processing, i.e., (predicative) object-constitution. So it depends on the subject's interests *which* objects are constituted, and it depends on the sense emerging in the process of object-constitution *how* they are constituted.

There are remarkable parallels between this view of object-determination and a later version of Putnam's "internal realism," in which the above-criticized epistemic conception of truth plays no decisive role anymore. The following quotations from the paper "A Defense of Internal Realism," in which Putnam still sticks to the latter conception, express the core ideas of his later, purified version:

> In my picture, objects are theory-dependent in the sense that theories with incompatible ontologies can both be right.
>
> (Putnam 1990, 40)
>
> [V]arious representations, various languages, various theories, are equally good in certain contexts . . . sentences [have] this dependence on a theory for their truth-value . . . . [O]bjects are, at least when you get small enough, or large enough, or theoretical enough, theory-dependent.
>
> (Putnam 1990, 41)
>
> That we do not, in practice, actually construct a unique version of the world, but only a vast number of versions . . . is something that 'realism' hides from us.
>
> (Putnam 1990, 42)

To be sure, what Putnam has in mind are primarily theoretical entities, rather than perceptual objects. However, in his book *The Many Faces of Realism* (Putnam 1987) he explicitly refers to Husserl's critique of the naive sort of "scientific metaphysics" sketched in section I (which Husserl and Putnam refer to as "Objectivism") and states that perceptual objects, including colors, are just as real as their "underlying" microphysical structure, if any. (Regarding color, Putnam points out that there actually is no single underlying physical structure it could be "reduced to" in the naturalistic attitude of what Sellars calls the "scientific image," as opposed to the "manifest image," which does not provide for basic theoretical entities; cf. Putnam 1990, 5f.) What is real is both a matter of the conceptual scheme underlying our corresponding existential judgments ("conceptual relativity") and the way the world is ("realism") (cf. Putnam 1987, 17–20). One of the examples used by Putnam to illustrate this brings out that, at least in some cases, the resulting notion of reality (in the case at hand: causal reality) is a function of the *epistemic interests* leading to the choice of a particular conceptual scheme: If the escape valve on a pressure cooker is malfunctioning and the cooker explodes, we will not say that "an arbitrary irregular shaped piece of the surface of the cooker, 0.1 cm in area" caused the explosion, even if the presence of that tiny piece may provide a physical explanation. The "explanation space" determined by our interests only comprises two alternatives: explosion and proper functioning of the pressure cooker; it does not include the alternative that that 0.1 cm piece was missing. Rather, the fact that the piece is in place is considered as included in the background conditions (Putnam 1987, 37f.). While Husserl usually reserves the term "causation" for explanations given within the "naturalistic attitude" (see section III), rather than the attitude we take when dealing with use-objects like pressure cookers, he agrees with Putnam that the choice of concepts (including "naturalistic" concepts like causation) determining truth-conditions (via object- and property-constitution) is relative to epistemic interests, without those truth-conditions thereby ceasing to be objective; where "concept" here refers to sense. Recall the previous explanation of Husserl's notion of individual concept or mental file processing in terms of the sedimentation of sense in contexts of epistemic justification.

Another example Putnam draws upon in order to illustrate the point of (the present, less problematic version of) his internal realism is the following (cf. Putnam 1987, 18ff.). We can describe "a world with three individuals" as containing no more than three objects x_{1-3}. However, if we introduce the concept of a (non-empty) mereological sum, we will say that it contains seven objects: the three individuals x_{1-3} as well as the four sums $x_1 + x_2$, $x_1 + x_3$, $x_2 + x_3$, $x_1 + x_2 + x_3$. Depending on our choice of concepts, we will either regard three or seven corresponding existential sentences as true. However, once we have chosen a conceptual scheme, there *is* a true answer to the question "How many objects are there?," the (correct) decision about the truth of which is not up to us, the users of the conceptual scheme. (It is in this sense that the corresponding existential statement can be said to have *objective* truth-conditions.) Husserl would agree. Once mereological sums ("collections [*Kollektiva*]," "multitudes [*Mengen*]") are "constituted," reality contains more objects than before, on his view, notably both collections and their parts. The constitution of collections (and other "categorical objectualities") is analyzed both in his first work, *Philosophie der Arithmetik*, and in later writings (such as *Erfahrung und Urteil*):

> The specific interest in the formal already arose in me by my *Philosophie der Arithmetik* (1891), which, as a first book, was still immature yet represented a first attempt to achieve clarity with regard to the original . . . sense of the basic concepts of the theory of multitudes [*Mengen*] and numbers by making recourse to the spontaneous activities of collecting [*des Kolligierens*] and counting in which collections . . . and numbers are given in the original mode in which they are created [*in ursprünglich erzeugender Weise*]. So it was already, to put it in my later terminology, a phenomenological-constitutional investigation [regarding first- and higher-order 'categorical objectualities'].
>
> (Hua XVII, 90f)

These constitution analyses, which are not to be found in Putnam, help to make sense of the two versions of reality described by Putnam that contain different numbers of objects but are "deeply related" (Putnam 1987, 20). The conceptual scheme underlying the second version can be explained in terms of the first along with the newly introduced concept of a sum. Husserl's analysis of the constitution of the extension of the latter highlights the "polythetic" "pre-constitution" of collections on which their constitution as higher-order objectualities (i.e., as objects ontologically dependent on, or "founded in," the objects they consist of) depends. This pre-constitution simply consists in connecting a number of any type of (already constituted) objects in thought. As Husserl puts it in *Philosophie der Arithmetik*:

> If we ask what constitutes the connection between a multitude of things as disparate as redness, the Moon and Napoleon that we think of [as a multitude], then the answer runs that this connection merely consists in the fact that we connect these contents in thought [*daß wir diese Inhalte zusammendenken*], that we think of them in one act.
>
> (Hua XII, 74)

If you think of the number seven, the property of being red, and the person Napoleon and connect these objects in thought, say by thinking of them one after another while still keeping the others in mind (cf. Husserl 1999, 296), then you are in a position to "constitute" the *collection* of these objects by reflecting on that polythetic thought and unifying the series of objects represented in the course of it into *one* object (thus representing them in a *mono*thetic act), namely a particular second-order collection of three objects. (I say "second-order" because I assume that none of these objects has itself been constituted as a collection—otherwise we would be dealing with a collection of at least third order, such as the collection $(x_1 + x_2) + (x_2 + x_3)$; cf. Hua XII, 74). On Husserl's view there does obtain, contrary to what the preceding quotation may suggest, a *non*-psychological connection between the objects thus collected, but it is a purely formal one—a "syntactical form of unity [*syntaktische Verbindungsform*]":

> Collecting does not serve to constitute a sensual whole; the elements of the manifold (where it is assumed that it is sensual objects that are collected) are not related to it in the manner in which the parts of a sensual whole are related to that whole. . . . [T]here is a sense in which elements of the manifold remain 'apart from each other' ['außereinander']. Their form of unity is not a sensual one but a syntactical one, precisely their '*being collected*.' And since we can collect each and every object of any type whatsoever, this means that this form of unity is completely independent of the conditions of homogeneity, or at least of the relations of similarity and dissimilarity that

> apply in the case of sensual-intuitive unification. We are dealing with a *syntactical form of unity*.
>
> (Husserl 1999, 296f)

Once collections have been constituted in this way (so that the concept of mereological sum is applicable), the formal-ontological apparatus of mereology can be applied to yield corresponding true existential statements, such as the statement that reality contains the mereological sum $x_1 + x_2 + x_3$. This illustrates how Husserlian constitution analysis may be used to give substance to Putnam's internal realism about formal-ontological (and other) facts. In the final section, I shall briefly sketch how it can be used to make sense of the notion of nature as conceived of in physical and cognitive science.

III. Prospect: Natural and Mental Properties Conceived From a Husserlian Viewpoint

I propose to make recourse to Husserl's method of constitution analysis in order to develop an alternative to the metametaphysical picture previously sketched in section I, which invokes the idea of "carving out nature at its joints." Husserl's own analyses provide a suitable starting point for this project, as I will try to make clear in what follows. In particular, they yield a non-naive conception of natural properties, and also of supervenience. The latter notion is rejected by Schaffer, who associates it with the attempt "to fake ordering structure within a flat ontology" he ascribes to Quine (Schaffer 2009, 363) and who approvingly cites Kim:

> [Supervenience] is a 'surface' relation that reports a pattern of property covariation, suggesting the presence of an interesting dependency relation [i.e., grounding; CB] that might explain it.
>
> (Kim 1993, 167)

Unlike these more recent authors, Husserl thinks that in order to naturalize the mind, scientists *need* to look upon the stream of consciousness as supervenient on physical reality, because what he calls "physical substances" are essentially extended in space. As the following passage from *Ideas* II makes clear, he conceives of physical substances as units of causal dispositions ("real properties") as studied in physics; where these dispositions are regarded as multi-track ones (cf. Hua IV, 39: "It is *the same objective property* that manifests itself both in the radiance [*Glanz*] and the smoothness [*Glätte*] [of a physical body]").

> *Reality or, what amounts to the same thing here, substantiality, on the one hand, and causality, on the other, inseparably belong together.* Real properties are eo ipso causal ones. Therefore, knowing a thing

> means knowing in an empirical manner how it behaves [*sich benimmt*] under pressure and impact [*Stoß*], bending and breaking, calefaction and cooling, i.e., how it behaves [*sich verhält*] in connection with its causalities [*im Zusammenhang seiner Kausalitäten*], into which states [*Zuständlichkeiten*] it gets and how it remains the same throughout these [varying states]. It is the task of physics (in an extended sense of the term) to investigate into these connections and to determine the real properties in a scientific manner [*wissenschaftlich denkend*].
>
> (Hua IV, 45)

By contrast, Husserl holds that what he calls "*spiritual* substances," i.e., units of *mental* dispositions as studied in naturalistic *psychology*, per se lack spatial extension (because they cannot be separated into "pieces," i.e., independent parts displaying the same general essence; cf. Hua IV, 33). This implies that the manifestations ("Verhaltungsweisen;" Hua IV, 44) of physical and mental dispositions, respectively, cannot be identical (*contra* Davidson) (Hua IV, 29).

> *Spiritual nature, understood as animal nature, is a complex composed of a* lower stratum of material nature, whose essential feature is extension, and an inseparable upper stratum which is of a fundamentally different essence and which, above all, excludes extension.
>
> (Husserl 1989, 32; Hua IV, 29; also cf. Hua IV, 281, keyword: "conditional dependence on nature")

Thanks to the supervenience of the mental on the physical, mental events can be said to be located spatiotemporally in a secondary sense (Hua IV, 138), but their essence prevents them from being themselves spatially extended in the sense in which material objects are. Their peculiar essence cannot be studied in the naturalistic attitude (see Hua IV, 124). Rather, it is to be studied in the phenomenological attitude, notably in the context of the analysis of the (self-)constitution of consciousness, which does not take any assumptions concerning the relationship between conscious experiences and physical events for granted.

Husserl also distinguishes the naturalistic attitude from what he calls the "personalistic" attitude. In the third main section of *Ideas* II, titled "The Constitution of the Spiritual World," Husserl analyzes the world as we are conscious of it from the personalistic attitude, which he characterizes as follows:

> the attitude we are always in when we live with one another, talk to one another, shake hands with one another in greeting, or are related to one another in love and aversion, in disposition [*Gesinnung*] and action, in discourse and discussion. Likewise we are in this attitude

> when we consider the things surrounding us precisely as our surroundings and not as '[o]bjective' nature, the way it is for natural science.
>
> (Hua IV, 183; Husserl 1989, 192)

Husserl claims that the world as represented in the naturalistic attitude, i.e., "objective' nature," is constituted on the basis of the prior constitution of the personalistic world. It is only due to an "abstraction," going along with "a kind of self-forgetfulness of the personal Ego," that it (falsely) appears to us as if the objectively determinable nature were completely independent of our personal environment (cf. Hua IV, 183ff.). The personalistic world is constituted as relative to persons who are able to "communicate" with one another, i.e., to "determine one another" by performing actions in the intention to motivate the other to display "certain personal modes of behavior" on the basis of his or her grasping that very communicative intention (Hua IV, 192). If an attempted piece of communication such as this, also called a "social act" (Hua IV, 194), is successful, then certain "relations of mutual understanding [*Beziehungen des Einverständnisses*]" are formed (Hua IV, 193), which in turn help determine the common environment for a group of persons; the world as constituted this way (i.e., with recourse to one's own experience of such mutual agreement) is called a "communicative environment [*kommunikative Umwelt*]." This constitutive achievement then forms the basis for the constitution of the "'objective'" world of "nature," which is why Husserl refers to the latter as "*secondary* environment":

> the whole physical nature is of no concern for . . . the primary environment. For, we can also say that it is the *secondary environment*. Just as the evaluated thing becomes an object of value and at the same time . . . an element of the [primary] environment, physical nature, which gets determined theoretically on the basis of 'appearances' [i.e., observation; CB], is a secondary environmental object, whose primary environmental object is the appearance [i.e., that which appears in the personalistic attitude; CB].
>
> (Hua IV, 285)

For example, the color of a perceptual object belongs to a differently constituted environment than its corresponding (micro-)physical properties, but the resulting two versions of reality are "deeply related," as Putnam would formulate it (Putnam 1987, 20), thanks to the constitutional relationships between primary and secondary environment. It seems plausible to assume that in the case of color this "deep relation" is similar to the one between consciousness and physical events, i.e., that color supervenes on certain physical properties. However that may be, it is necessary, on Husserl's view, to study the constitutional relationship between color attributes and physical properties in order to naturalize the former, at least if this naturalization project is supposed to be a philosophical one.

Unlike mainstream analytic (meta)metaphysics, the Husserlian approach is epistemological. It investigates into both the way objects of all sorts constitute themselves in epistemic consciousness and the sense (or mode of presentation) they acquire in this way. Husserl's own approach is associated with a problematic and poorly justified version of (what he calls) transcendental idealism that can, however, be modified to yield a more plausible position, which is similar to Putnam's internal realism but may possess even more philosophical substance. Moreover, this approach promises to offer an alternative to the dogmatic naturalism underlying much of more recent (meta)metaphysics.[8]

Notes

1. It should be mentioned that Husserl has later titled the paragraph containing this remark with a big question mark (cf. Hua XXXVI, 121, fn. 2).
2. See Husserl 1999, 89ff.—Note that constitution *can* take the form of intentional action, but in the sphere of perception and empirical thought this is the exception rather than the rule. So clearly, "constitution" does not mean *creation*.
3. To be more precise, it is only at the level of *predicative* experience that the subject's epistemic interests take the form "of the *will to knowledge* [*des Willens zur Erkenntnis*];" cf. Husserl 1999, 92.
4. See Husserl 1999, 44; keyword: "justification of the doxa [*Rechtfertigung der Doxa*]."
5. For further textual evidence supporting this identification, see the following quotation.
6. See Husserl 1999, 18ff, 60.
7. See Husserl 1999, 132f.
8. This chapter was written at the Centre for Advanced Study (Oslo), within the framework of the research group 'Disclosing the Fabric of Reality—The Possibility of Metaphysics in the Age of Science,' led by Frode Kjosavik and Camilla Serck-Hanssen, who also organized a workshop at which an earlier version of the article was read. I gratefully acknowledge the support. I would also like to thank the participants of the workshop for their helpful comments and particularly Frode Kjosavik, whose suggestions led to substantial improvements of the article.

References

Carnap, Rudolf. "Empiricism, Semantics, and Ontology." In *Rudolf Carnap, Meaning and Necessity*, 2nd ed. Chicago: University of Chicago Press, 1956, 205–221.

Fine, Kit. "The Question of Realism." *Philosophers Imprint* 1 (2001), 1–30.

Husserl, Edmund. *Husserliana: Edmund Husserl—Gesammelte Werke* [Hua]. Den Haag/Dordrecht: Nijhoff/Kluwer/Springer, 1950ff.

Husserl, Edmund. *Ideas Pertaining to a Pure Phenomenology and to a Phenomenological Philosophy, Second Book*. Dordrecht: Kluwer, 1989. [Hua IV]

Husserl, Edmund. *Early Writings in the Philosophy of Logic and Mathematics*. Dordrecht: Kluwer, 1994. [From Hua XII, XXI, XXII and XXIV. The treatise "Intentionale Gegenstände" is contained in Hua XXII.]

Husserl, Edmund. *Erfahrung und Urteil*, 7th ed. Hamburg: Meiner, 1999.

Kim, Jaegwon. "Postscripts on Supervenience." In *Jaegwon Kim, Supervenience and Mind—Selected Essays*. Cambridge: Cambridge University Press, 1993, 161–174.

Künne, Wolfgang. "Bolzanos blühender Baum—Plädoyer für eine nicht-epistemische Wahrheitsauffassung." In Forum für Philosophie, ed., *Realismus und Anti-Realismus*. Frankfurt/M: Suhrkamp, 1992, 224–244.

Loewer, Barry. "From Physics to Physicalism." In Carl Gillet and Barry Loewer, eds., *Physicalism and Its Discontents*. Cambridge: Cambridge University Press, 2001, 37–56.

Perry, John. "A Problem about Continued Belief." *Pacific Philosophical Quarterly* **61** (1980), 317–332.

Putnam, Hilary. *The Many Faces of Realism*. La Salle: Open Court, 1987.

Putnam, Hilary. "A Defense of Internal Realism." In *Hilary Putnam, Realism with a Human Face*. Cambridge, MA: Harvard University Press, 1990, 30–42.

Quine, Willard Van Orman. "On What There Is." *Review of Metaphysics* 2 (1948), 21–38.

Schaffer, Jonathan. "On What Grounds What." In David J. Chalmers, David Manley and Ryan Wasserman, eds., *Metametaphysics—New Essays on the Foundations of Ontology*. Oxford: Clarendon Press, 2009, 347–383.

4 Frege on "Es gibt," Being in a Realm and (Meta)Ontology*

Leila Haaparanta

I. Introduction

Gottlob Frege did not call any of his philosophical views by the name "ontology," let alone "metaontology." When applied to Frege's thought, the term "metaontology" is clearly anachronistic. Ontology is often understood as the study of the categories of being, hence, as general metaphysics, and ontological views are construed as answers to the question "What is there?", "What are the entities that exist?", and "What is there fundamentally?". The editors of *The Routledge Companion to Metaphysics* tell us that, besides general metaphysics as the study of the most fundamental concepts and principles, "ontology" may mean the list of basic kinds of entities; the basic kinds of entities assumed by a philosopher; the study of all possible general arrangements of the world; and, in computer science, a set of categories for programming and data representation (Poidevin et al. 2009, 591). How one understands the terms "metaphysics" and "ontology" naturally depends on which philosophical tradition or school one is working in. For Heidegger, for example, the most general categories discussed in contemporary analytic metaphysics would not be the basic categories of his fundamental ontology; such distinctions as that between being-there [*Dasein*], being present-at-hand [*vorhanden sein*], and being ready-to-hand [*zuhanden sein*] are what matters in his *Sein und Zeit*. In the "Introduction" of *Metametaphysics: New Essays on the Foundations of Ontology*, David Manley states that metaphysics is concerned with the foundations of reality, while metametaphysics is concerned with the foundations of metaphysics (Manley 2009, 1). Peter van Inwagen, for his part, calls the question "What are we asking when we ask 'What is there?' the meta-ontological question," the attempt to answer it "meta-ontology" and any proposed answer to it "*a* meta-ontology" (van Inwagen 2001, 13). Among other things, metaontology is interested in the conceptual tools and methods of ontology construed on the model of scientific theory; hence, metaontology is the study of the foundations of that very theory. The term "metaontology" also covers an old subfield of the philosophy of language and logic. That subfield is interested in the words "to be" and "to exist" of natural language and the possible epistemological and ontological commitments that lie behind the ways the words are used in natural languages and analyzed by means of formula languages.

Gottlob Frege was interested in what the words "is" and "exists" mean. He is a prime example of a philosopher who partly motivated his choice of a formula language by its ability to give a correct analysis for the concept of being. I have discussed that topic in my earlier studies, whose theses the present chapter elaborates.[1] I discuss two philosophical positions that Frege held and that have metaontological relevance. They are the doctrine of the ambiguity of the word "is" and the doctrine of three realms. Frege's semantics and his relation to Kant and to the so-called ontological argument for God's existence have been analyzed a great deal. Likewise, his distinction between three realms has been studied in detail. The two doctrines, one about the ontological argument and the other about three realms, are usually considered separate themes that occur in Frege's writings. It is true that Frege does not combine the two topics. He does not raise the question whether they are related, and if they are, how that relation ought to be understood. To a contemporary reader, it may also seem that the two doctrines are not compatible. On the one hand, Frege argues that being or existence is not a real first-order property; on the other, he makes a distinction between three realms whose denizens seem to have the property of being in one way or other. This chapter seeks to show that the doctrines can be connected, and this is precisely what it seeks to add to Frege scholarship, including my earlier studies on Frege. It argues that the concept of existence that is expressed as a second-order concept in the formula language plays an important logical role, while the doctrine of three realms is a means to express a division between three modes of being or three ways to exist. It seeks to show that the modes of being are predicated of objects, but they have a special, constitutive role in relation to the entities of which they are predicated; that very role makes them differ from normal predicates. If we use the terminology of medieval philosophers, we may say that the verb "to exist" is an analogous rather than a completely ambiguous word.

I will proceed as follows: First, I outline Frege's distinctions between the meanings of the word "is," which are written down in his formula language. I also make a few remarks on the view on existence that Frege puts forward in his discussion with Pünjer; that view illuminates the way he treats existence in his formula language. I then introduce his doctrine of three realms, discuss its possible ontological and metaontological relevance, and compare it with the medieval theory of the analogy of names. In section IV, I will discuss Frege's view on existence in the context of his distinction between sense and reference. In section V, I will combine the doctrine of the ambiguity of "is" and the doctrine of three realms. A.W. Moore (2012) has argued that Frege neither has metaphysics nor metametaphysics; still, he admits that what he calls Frege's theory of sense was important in view of the quest for sense in analytic philosophy, including later analytic metaphysics. I will return to that argument in section V and discuss Frege's thought in relation to Kant's and Aristotle's views on being and existence. Section VI concludes the article and suggests that Frege's views on being contain elements both of Aristotelian and of Kantian approaches to metaphysics.

II. Frege's Struggle With Natural Language and the Word "Is"

Frege's great achievement was his conceptual notation, "Begriffsschrift," which was meant to be a genuine universal language of Leibnizian spirit that speaks about all that there is. Frege states in the preface of his *Begriffsschrift* (1879) that Leibniz's idea was too grand to be reached, but he seeks to realize it by giving it a more modest form. Natural language leads us astray, Frege argues, and therefore a new language, a new tool for philosophers, is needed (BS, XI–XII). Frege can be interpreted as saying that the world has a structure independently of languages, and that structure can be shown by an ideal language. That, however, is something that he does not formulate as an explicit philosophical thesis. The claim he makes is precisely that natural language leads us astray, hence, that it ought to be revised or even replaced by a better language.

Since the 1960s, there have been studies on Frege's metaphysics and on the similarities and differences between Aristotle and Frege. Frege scholars like Ignacio Angelelli (1967) and Eike-Henner W. Kluge (1980) already considered Frege in relation to the tradition of metaphysical thought. They compared Frege with Aristotle, medieval philosophers, Leibniz, and others, and argued that Frege tried to present a proper metaphysical system. Kluge sought to show that Bolzano was a transitional figure between Leibniz and Frege (Kluge 1980, 280). Reinhard Grossmann claimed that if we take as our paradigm of an object a spatiotemporal individual thing and as our paradigm of a function a property or relation, then Frege's ontology seems to be a modification of Aristotelian ontology. However, as Grossmann notes, Frege's ontology includes all kinds of entities, such as value-ranges and truth-values, thus differing radically from that of Aristotle (Grossmann 1976, 31–32).

The very word "is" receives various interpretations in Frege's conceptual notation. "There is" is construed as a second-order concept and expressed by means of the symbol for generality, that is, the universal quantifier, and two negation signs. Existence is attached to objects only as an empty predicate of self-identity. Non-empty predication is expressed by means of function-names and parentheses, and class-inclusion or subordination between concepts is captured by Frege's symbol for generality, together with a symbol for implication or conditionality, plus function-names and parentheses. The symbol for identity is a natural choice for expressing the "is" of identity. In his vocabulary Frege also has what he calls the judgment stroke or the assertion sign, which can be read as "it is the case that . . . " or "it is true that . . . ." That Frege regards the word "is" as ambiguous in the described ways is shown by the very fact that he makes the mentioned distinctions in the ideal language, which is meant to be free from ambiguities typical of natural language. Besides the "is" expressed by the assertion sign, the ambiguity that Frege finds in natural language is thus between the following expressions:

1. "being" that expresses a property of a concept (the "is" of existence, existence as a second-order concept, e.g., There *is* at least one philosopher; $(\exists x)\ (P(x))$.
2. "being" that expresses the existence of an object (an empty concept "to be identical with oneself," e.g., Socrates *is*; $(\exists x)\ (s = x))$.
3. "being" as a part of predication (the "is" of predication, e.g., Socrates *is* a philosopher; $P(s))$.
4. "being" that expresses identity (the "is" of identity, e.g., The Morning Star *is* the Evening Star; $a = b)$.
5. "being" that expresses class-inclusion or generic implication (the "is" of class-inclusion, e.g., Man *is* an animal; $(x)\ (M(x) \rightarrow A(x)))$.

Frege argues that his language is better than natural language in several respects, one of them being that it distinguishes between objects, on the one hand, and concepts and relations, on the other.[2] One important argument is presented in the critical remarks against Boole.[3] Frege notes that Boole does not recognize the distinction between predication and subordination, such as the distinction between "Socrates is a man" and "Man is an animal," because he does not keep individuals [*Einzeldinge*] and concepts apart. In his article "Über Begriff und Gegenstand" (1892) Frege points out that the "is" of identity and the "is" of predication must be distinguished from each other in order to show clearly that objects and concepts differ from each other (KS, 167–168). Hence, the "is" in "The Morning Star is the Evening Star" is the "is" of identity, and the sentence is about an object, while the sentence "The Evening Star is bright" contains the "is" of predication, and it relates an object to a concept under which the object falls.

In his *Begriffsschrift* of 1879, Frege distinguishes between a thought [*Gedanke*] and a judgment [*Urteil*], which is an acknowledgment of the truth of a thought, a thought as judged. In his conceptual notation the distinction is shown by two strokes, one being the content stroke or the horizontal stroke and the other being the judgment stroke or the vertical stroke (BS, § 2). In "Einleitung in die Logik" (1906), he remarks:

> In fact at bottom the sentence 'it is true that 2 is prime' says no more than the sentence '2 is prime.' If in the first case we express a

> judgement, this is not because of the word 'true,' but because of the assertoric force we give the word 'is.'
>
> (NS, 211; "Introduction to Logic," PW, 194)

In his "Meine grundlegenden logischen Einsichten" (1915) he writes: "In language assertoric force is bound up with the predicate" (NS, 272; "My basic logical insights," PW, 252). As noted earlier, for Frege, the judgment stroke or the assertion sign is the sign for "it is the case that . . ." or "it is true that . . . ." Charles Kahn calls this meaning of "is" the veridical "is" (Kahn 1973, 331–370).[4]

In his dialogue with Pünjer on existence (before 1884), Frege assumes for the sake of argument that being is a real, non-empty property of an object. His argument goes as follows. If being is a real property of an object, it is a characteristic [*Merkmal*] of a first-order concept and the negation of the sentence "A is" can be true; hence, it is possible that there are entities which do not have the property of being. But then the expression "there are" [*es gibt*] cannot be replaced with "being," that is, "There are Bs" is not equivalent in meaning to "Something that has being [*einiges Seiende*] falls under the concept B." For if the sentence "There are entities which do not have the property of being" means the same as "Something that has being falls under the concept of not-being [*der Begriff des Nichtseienden*]," it is a contradictory sentence. Frege concludes that if the sentence "There are B's" is equivalent in meaning to the sentence "Something that has being is B," then the concept of being does not distinguish between any two objects and being is not a real property of an object.[5] The other obvious alternative is that "There are Bs" and "Something that has being is B" are *not* equivalent in meaning. That is also what natural languages, or at least some of them, for example German, suggest. Frege concludes that what is expressed by "there is" [*es gibt*] is a property [*Eigenschaft*] of a concept, not its characteristic [*Merkmal*], and the word "exist" cannot be regarded as the characteristic mark of a concept, either, because it adds nothing to the individual objects that are discussed (NS, 83–85). He then adds:

> We can see from all this how easily we can be led by language to see things in the wrong perspective, and what value it must therefore have for philosophy to free ourselves from the dominion of language. If one makes the attempt to construct a system of signs on quite other foundations and with quite other means, as I have tried to do in my concept-script, we shall have, so to speak, our very noses rubbed into the false analogies in language.
>
> (NS, 85; PW, 67)

It seems, however, that natural language gives hints at the right perspective on things. That is precisely the case with the German distinction between "es gibt" on the one hand, and "ist" or "existiert," on the other. However, as will be shown in section IV, "is" or "exists" can be analyzed in such a way that it turns out to be a second-order predicate.

III. Frege's Doctrine of Three Realms and "Being" as an Analogous Term

At first glance, the doctrine of three realms appears to be an answer to a direct ontological question concerning what there is. A similar doctrine to that of Frege's can be found in Hermann Lotze, who was one of Frege's teachers. According to Lotze, the actuality [*Wirklichkeit*] of abstract objects, such as thoughts, is not like the actuality of concrete objects. Abstract objects are valid [*geltend*]. Lotze distinguishes between what is valid and what is (Lotze 1874, 16 and 507). Gottlob Frege made a sharp distinction between subjective psychological acts, such as thinking and judging, and the contents of those acts, namely, thoughts. It is not clear what kind of being he was ready to ascribe to those contents. One could argue that Frege's *Inhalte*, marked by the content stroke in the conceptual notation, that is, *Gedanken*, thoughts, are ontologically real.[6] It is also thinkable that Frege merely supported epistemological realism, which presupposes the distinction between the act of knowing and the object that is known. Frege himself did not use the term "realism." It is well known that Frege was not alone in nineteenth-century philosophy with his distinction between psychological acts and their objective contents. Bernard Bolzano, for example, had argued along similar lines as early as 1837, in his *Wissenschaftslehre*. Frege distinguished between the thought, the judgment [*das Urteil*], and the assertion [*die Behauptung*], which is the linguistic expression of the judgment.[7] Bolzano distinguished between propositions in themselves [*Sätze an sich*], which neither exist in space and time nor depend on mental acts, and the thinkings or assertions of those propositions (Bolzano 1929, §19). He argued that since the proposition is different from the subjective idea [*Vorstellung*] which happens to be present in the consciousness of a thinking being and also different from the judgment, or affirmation, of the proposition, we cannot attribute existence (or actuality [*Wirklichkeit*]) to propositions in themselves. Bolzano maintains that "it is only the proposition asserted or thought, i.e., only the thought of a proposition, likewise the judgment containing a certain proposition that has existence in the mind of the being who thinks the thought or makes the judgment" (ibid., § 19). For Bolzano, propositions are matter that thinking beings grasp in mental acts and they exist only in those acts in which they are thought of or held to be true (ibid., § 122). Göran Sundholm (2000) has paid special attention to connections between Bolzano and Frege, while Wolfgang Künne has argued against the thesis that there is a historical link between the two philosophers (Künne 2010, 769).

Whether Frege received his doctrine of three realms from any particular source, it naturally follows from and is motivated by his antipsychologism. Frege was an antipsychologist in various ways, one of them being his view that the foundations of logic do not lie in psychology. In the *Grundlagen* (1884) he formulates his basic principle: "always to separate sharply the psychological from the logical, the subjective from the objective" (GLA, X). By the doctrine of the third realm, Frege expresses his view on abstract objects, such as thoughts and the categories and laws that constitute thoughts, hence, logical objects. In the first volume of his *Grundgesetze der Arithmetik* (1893) and in his article "Der Gedanke" (1918) he makes a distinction between the realm of objective and actual [*wirklich*] objects, the subjective realm of ideas [*Vorstellungen*], and the realm of objects which do not act on our senses but which are objective, that is, the realm of such abstract objects as numbers and thoughts (GGA I, XVIII–XXIV; KS, 353). That distinction can be interpreted as a straightforward ontological view, or it can be construed as a doctrine proposed by a transcendental philosopher, whose ontology is ontology within the limits of knowledge or language.

It is usually assumed that Frege's acknowledgment of the realm of thoughts and other abstract objects is a Platonic doctrine. Some interpreters have challenged the received view, but others, for example Tyler Burge (1992), have argued for the interpretation that Frege held a Platonic ontology; Burge notes, though, that Frege did not seek to defend his position, except for showing problems in competing views, and that he did not develop a sophisticated version of his ontology. Other interpretations are also possible. When Frege discusses his third realm in his "Der Gedanke," he remarks that he must use metaphorical language. In other words, such expressions as "the content of consciousness" and "grasping the thought" must not be understood literally (KS, 359, n. 6). Frege does think that the objects of the third realm are epistemically independent of subjective minds, but their being is not like the being of the denizens of the objective and actual realm. Thomas Seebohm argued as early as 1989 that on Frege's view, the existence of mathematical objects and logical categories is a necessary condition of the meaningfulness of mathematical and logical practice; Seebohm claimed that Frege's view did not amount to arguing directly for the thesis that those objects exist (Seebohm 1989, 348).

In the Preface of the *Grundgesetze der Arithmetik I* Frege writes:

> That the logical laws should be guiding principles for thought in the attainment of truth is generally admitted at the outset; but it is only too easily forgotten. The ambiguity [*Doppelsinn*] of the word 'law' is fatal here. In one sense it states what is, in the other it prescribes what should be. Only in the latter sense can the logical laws be called laws of thought, in laying down how one should think. Any law that

> states what is can be conceived as prescribing that one should think in accordance with it, and is therefore in that sense a law of thought. This holds for geometrical and physical laws no less than for logical laws. The latter then only deserve the name 'law of thought' with more right if it should be meant by this that they are the most general laws, which prescribe universally how one should think if one is to think at all. But the expression 'law of thought' tempts us into viewing these laws as governing thinking in the same way as the laws of nature govern events in the external world. They can be nothing other than psychological laws, since thinking is a mental process . . . authoritative only with qualifications (like fashion etc.) . . . I understand by logical laws not psychological laws of holding as true, but laws of being true.
>
> (Frege, "Vorwort," GGA I, in Künne 2010, 68–69; "Preface," GGA I, in Beaney 1997, 202–203)

In "Der Gedanke" (1918) Frege writes:

> So the result seems to be: thoughts are neither things in the external world nor ideas. A third realm must be recognized. Anything belonging to this realm has it in common with ideas that it cannot be perceived by the senses, but has in common with things that it does not need an owner so as to belong to the contents of his consciousness.
>
> ("Der Gedanke," in Künne 2010, 101; "Thought," in Beaney 1997, 336–337)

The three realms are thus constituted by different kinds of laws.

For Frege, the word "is" mirrors the acknowledgment of the truth of a thought. We, as empirical subjects, are the ones who acknowledge and assert. But what is it to acknowledge or recognize [*anerkennen*] the third realm? A natural reading of Frege's text is that it means acknowledging its constitutive laws. However, it does not follow that one who acknowledges those laws always obeys them. In that sense the laws of the third realm resemble the constitution of a state; they express the essence of the realm, but an individual as a thinking subject may acknowledge and still sometimes violate those laws.[8]

As we saw previously, for Frege "being" as a first-order predicate is empty. In the context of the doctrine of three realms being is treated as a first-order property, but it is also a property that belongs to all entities, as it is the property of belonging to one of the three realms. However, in his doctrine of three realms Frege distinguishes between what we could call *modes of being*, that is, being as being objective and actual, being as being subjective and unactual, and being as being objective and unactual. These are real in contrast to empty predications, although real in the sense of actuality [*Wirklichkeit*] only in one alternative, and each entity has one and only one of the three alternatives. If we think of the three realms on the model of categories and use the terminology of the Scholastic tradition, we may say that for Frege being as a first-order concept is transcategorial. It means that it overcomes the limits of categories. That view is on line with what Aristotle states about being, although Aristotle's categories are clearly not those of Frege's. Aristotle assumes that to be is always to be either a substance of a certain sort, a quality of a certain sort, etc. (An. Post. I 22, 83b13–17). The medieval analogy theory was based on Aristotle's view on being.

The distinction between the order of knowing, or the terminological order (*ordo nominis*), and the order of being (*ordo rerum*) was important for theologians. In *Summa contra gentiles* Thomas Aquinas writes:

> When a description is used in a related sense (by analogy), priority in terminology may diverge from factual priority. For terminological order follows the sequence in which we get to know the things. . . . Sometimes it happens that what comes first in fact is not known to us first.
> (Aquinas 1974, Cap. XXXIV)

Medieval philosophers construed nouns, verbs, and adjectives as names or terms. They distinguished between different types of analogy, based on the location of similarities. In the analogy theory, similarities could be found in the notions expressed by the name, in the things named or in both. If they were found in both, the type of analogy was *analogia secundum intentionem et secundum esse*. Names, or terms, as well as things could be called analogous (*analogia nominum*, *analogia entis*).

Using the vocabulary of the Scholastic tradition, especially of Cardinal Cajetan (1498, 1953), we could say that as a transcendental term, as a term that transcends the limits of Frege's three realms, the term "being" is an analogous rather than an ambiguous or a totally multivocal term. The medieval analogy theory was a theory for positive statements about God. On that theory, we can tell what God is like, hence, ascribe positive predicates like "being good" or "being wise" to God meaningfully, if we think that those predicates apply to God on the basis of analogy. For example, Socrates's wisdom serves as a model which tells us something about God's wisdom, and the predicate "being wise" is thus analogous. That is not to say that we know all about God's wisdom if we know what a human being's wisdom is like. In the terminological order, Socrates's wisdom is prior to God's wisdom, while in the order of being, God's wisdom comes first. In the philosophy of religion, the analogy theory is meant to show that the predicates that are applied to creatures in the world are meaningfully applicable to God without the danger of anthropomorphism. Univocity in theology would lead to anthropomorphism, whereas total equivocity or multivocity would make nonsense of positive theological language. The analogy theory was proposed as the third alternative. In the preceding example the predicate "being wise" is thus used transcategorially. The terms "being" and "existence" are also analogous, for example, in the expressions "Socrates's being" and "God's being," and "Socrates's existence" and "God's existence."[9] If we think of Frege's realms along these lines, we may say that the word "being" is analogous. We may also suggest that the word "being" is primarily applied to the first realm, and the being of the denizens of the second and the third realm is understood on the model of the being of the denizens of the first realm. That is something that Frege does not propose himself, even if he tells that he uses metaphorical language when he talks about abstract objects like thoughts.

If we now combine Frege's thesis that existence as "es gibt" is a second-order concept with my construal of the three modes of being, we may say that the sentence "There are entities which are objective and unactual" expresses that the concepts of objectivity and unactuality are both instantiated in at least one entity. In that sentence we have both being as a second-order concept and being which is clarified in one of the ways Frege's doctrine of three realms allows. Hence, in Frege's formula language there is the second-order concept which captures the instantiation of a first-order concept, while in the non-formal vocabulary of ontology there is the property of being which is attached to objects, but which reduces to the three alternative modes of being. I suggested earlier that we could think of the being of abstract objects, the being of subjective ideas, and the being of actual objects in the world to be analogous. The three modes of being have a special, constitutive role among the properties of objects, even if they can be distinguished in the formula language only by means of different symbols for concept-words and they thus seem to be similar to common predicates. In order to understand their status as special predicates, we need the vocabulary of ontology of our natural language, that is, such words as "objectivity" and the like. In that vocabulary the objects of the outer world and their being are primarily known to us and the being of the processes and states of our minds as well as the being of abstract objects is known but not reducible to the being of the objects of the first realm. By allowing this kind of ontological vocabulary, Frege partly gives up the idea that all predicates are on a par.

IV. Senses, References, and Existence

I argued in Section II that Frege makes a distinction between "There are Bs" and "Something that has being is B" in his ideal formula language. Frege's concept of sense reveals new features in the way Frege understands the phrase "es gibt" in contrast to being as a first-order concept. In his dialogue with Pünjer, Frege notes that if we talk about Leo Sachse, such statements as "Leo Sachse is" and "Leo Sachse exists" are self-evident, because we presuppose that the words that we use designate something, that they are not empty (NS, 78). Pünjer, who opposes Frege's view that existence is not a real predicate of objects, makes a Kantian suggestion that the word "is" carries the same meaning as "is something that can be experienced" [*ist erfahrbahr*]. He considers the set of objects of experience [*Gegenstände der Erfahrung*] to be a subset of the set of objects of ideas [*Gegenstände von Vorstellungen*]. Frege shows that Pünjer's view results in a contradiction: If "A is not" means that A is not an object of experience, what we say here is that there is something that is not an object of experience. Frege argues that if we accept Pünjer's way of defining the concept of existence, this statement means that there is an object of experience which is not an object of experience, which is a contradiction (NS, 71–72). Pünjer could object that "A is not" means that there is an object of ideas which is not an object of experience; however, that move in the discussion would mean giving up the thesis that there is one non-empty concept of existence, which can be predicated of individual objects. Again, this example suggests that "there is" behaves in the context in a different manner from the "is" that follows the proper name. Relying on Frege's later distinction between the three realms, we could also translate the expression as "There are entities that are not denizens of the first realm."

If we apply Frege's later distinction between senses [*Sinne*] and references [*Bedeutungen*] to the example "Leo Sachse exists," we get one more analysis. For Frege, the senses of sentences are thoughts and the references of sentences are truth-values, the True and the False. Sentences are compounded out of proper names, which refer to objects, and function-names, which refer to functions. The senses of function-names are simply parts of thoughts.[10] In "Über Sinn und Bedeutung"(1892) Frege remarks that the sense of a proper name is a way in which the object to which this expression refers is presented, or a way of "looking at" the object. He gives a few well-known examples of senses, such as 'the Evening Star' and 'the Morning Star' as senses of Venus, and 'the teacher of Alexander the Great' and 'the pupil of Plato' as senses of Aristotle, and states that senses belong to objects to which proper names refer (KS, 144).

Frege thinks that objects are given to us in a variety of ways. His examples suggest that he takes it to be our knowledge or at least our true beliefs concerning an object that determines what sense, or what senses, the name of the object expresses to us. He does not consider cases where we believe, but not truly, that an object falls under a particular concept or has a particular relation to another object. His concept of *Sinn* expresses our epistemic access to objects, but our knowledge of an object is always one-sided [*einseitig*]. Frege argues:

> Complete knowledge [*allseitige Erkenntnis*] of the reference would require us to be able to say immediately whether any given sense belongs to it. To such knowledge we never attain.
>
> ("Über Sinn und Bedeutung," KS, 144; Geach and Black, 58.)[11]

Frege does not argue that knowing an object as falling under a particular concept constitutes knowing the object completely or knowing its essence. Nor does he distinguish between essential and accidental predicates.[12]

By introducing the concepts of sense and reference, Frege tries to give a natural reading to identity statements and to solve the problems of what we call intensional contexts. In his "Über Sinn und Bedeutung" he argues that in those contexts, expressions refer to their normal senses. By this move, he preserves the principle, later called the principle of compositionality, which demands that the reference of a complex expression is determined by the references of its components, and there is no need to take senses into account. However, there is more to the distinction between senses and references, and it is also relevant if we wish to understand Frege's way of thinking about being. For his *Sinne* are something that we cannot avoid when we try to reach the world by means of our language. His belief in the inescapability of *Sinne* is a special form of the broadly Kantian belief that we must always consider objects through our conceptual systems. In "Ausführungen über Sinn und Bedeutung" (1892-1895) he remarks that it is via a sense and only via a sense that a proper name is related to an object (NS, 135).[13]

If we start with the epistemically loaded semantic views described above, Frege's doctrine of existence in case of sentences like "Leo Sachse exists" or "Leo Sachse is" can be construed as follows. If we say that Leo Sachse exists and if someone asks us who is Leo Sachse, we give an answer by listing some of Leo Sachse's properties, hence by listing concepts under which the individual called "Leo Sachse" falls. We can say that the sentence "Leo Sachse exists" means that there is an object, a person, who has certain properties. Here "x exists" reduces to "there is an x such that . . .," and existence is thus understood as a second-order concept, which means instantiation of a bundle of properties. It is via those properties that we know the object, and it is precisely existence as a second-order concept that draws our attention to those properties. However, we cannot say what the object or the person is as abstracted from the bundle of properties.[14] Frege does not discuss that kind of reduction. His texts do not help us to decide whether he would be a bundle theorist or whether his objects would contain non-describable bearers of properties.

V. Frege, the Metaphysician, the Metametaphysician or Neither?

I mentioned previously that in Frege's view, a proper name is always related to an object via sense. Still, his universal language is meant to mirror directly those very objects and functions to which it refers. When Frege introduces his conceptual notation, he mentions Adolf Trendelenburg's article "Über Leibnizens Entwurf einer allgemeinen Charakteristik" (1867). Trendelenburg proposes that philosophers ought to construct a Leibnizian universal language by taking into account Kant's ideas. On his view, Kant continued Leibniz's project by studying the conceptual component of thought after distinguishing it from the empirical component. Frege regards himself as a follower of Leibniz, even if he emphasizes, like Trendelenburg, that Leibniz's aim was somewhat unrealistic. For Trendelenburg, it was Kant's focus on the conceptual component in his transcendental logic that made the project realizable.

In Frege's formula language there is a symbol formed by means of the symbol for generality and two negation signs, which is taken to be the symbol for existence. The symbol is thus compounded out of three parts, which are a name of a second-order function of one argument, the argument being a name of a first-order function, and two instances of a name of a first-order function of one argument, which are the negation signs. This is how we can characterize what is called existence as a second-order concept if we use the terminology and the list of primitive logical function-names which Frege gives in § 31 of his *Grundgesetze*. In the conceptual notation, or in the formula language of pure thought, as Frege calls his notation in the *Begriffsschrift*, existence appears as a compound logical concept. We may say that the compound logical function-name described previously can be translated into German by the expression "es gibt" and into English by the expression "there is" or "there are."

If we consider Frege's doctrine of three realms, we get another story, which has to do with the empty predicate of being or existence, but which becomes meaningful when it is understood as the predicate of belonging to one of the three realms. That predicate applies to all objects, as the three realms exhaust all that there is. In this construal of Frege's view, being is an empty first-order predicate, but entities instantiate various modes of being, depending on which realm they belong to. Their mode of being, hence, which belongingness they instantiate, is not shown by logical function-names of the conceptual notation. It is a topic of ontology to distinguish between subjectivity and objectivity, on the one hand, and actuality and unactuality, on the other. In the vocabulary of the conceptual notation those predicates would be represented as names of non-logical first-order functions. In Frege's semantic framework, a particular object that is named is considered via a bundle of properties which it instantiates. Seen this way, existence is a property of concepts. Simultaneously, according to the doctrine of three realms, the object has one of the three alternatives as its properties.

But is that all metaphysics or metametaphysics? A.W. Moore considers metaphysics to be making sense of things. Throughout the history of the field, Moore finds three questions under discussion that concern its scope and limits, first about "transcendent" things, second about the novelty of sense-making, and third about our creativity in the sense-making (Moore 2012, 9). On Moore's view, Frege is not himself a metaphysician, because his logic, even if it seeks to be maximally general, is an attempt to make sense of sense, not of things (ibid., 215). Moore does not deny that Frege's project has relevance to what metaphysics is today in the analytic tradition; on the contrary, he argues that the project of making sense paves the way for contemporary metaphysics and metametaphysics. However, there is more to be said of Frege's contribution.

Frege's conceptual notation is intended to be a mirror of all that there is, and his distinction between objects and functions is meant to show how all that there is is structured. Certain distinctions are therefore made in his ideal formula language, such as that between objects and functions. As a universal framework, Frege's conceptual notation is transcendental. Hence, there is no making sense outside that framework. If we emphasize this feature of Frege's project, we may say that he is a metaphysician if Kant is. Second, Frege's distinction between objects and functions and their senses is peculiar in that one cannot talk about functions, including properties, *as* functions or senses *as* senses, hence, without making them into objects. For Frege, they turn into objects once we begin to consider and speak about them, simply because in our sentences we refer to them. In that sense, semantic theorizing is not possible in Frege's philosophy. Therefore, on Frege's view, metaphysical theory about properties and relations is not possible, either. That does not prevent him from suggesting several things in his writings about metaphysical matters. We can conclude from his doctrine of senses and references, even if Frege himself does not, that every time we think of an object, we think of it under some description which captures at least one of its properties, hence, there is no direct access to what the object is in itself. Therefore, epistemically, objects are for Frege like bundles of properties.

Moore is certainly right that Frege's project is not metaphysics if we demand that it is making sense of things, and if Frege should be answering the question concerning what there is explicitly, arguing for his answer, and criticizing its alternatives. However, even if his project were considered to be merely that of making sense of sense, as Moore claims, it can be seen to have more content in view of metaphysics or metametaphysics than Moore thinks. Frege has the quest for being and existence, because he gives the guidelines for how to make sense of "being" and "existence." That is, if not metaphysics, at least discussing the topics that the philosophers before Frege discussed and paving the way for future studies of the concepts of being and existence. I have sought to show that in his preliminary work Frege followed Kant in that for him as well as for Kant being or existence was not a real property of an object, but Frege also thought that more analysis is required. In particular, Frege argued that as a real property, existence is a property of a concept, or more generally of a function. Frege can also be seen as a follower of Aristotle, if we consider his view that an entity has a specific mode of being depending on which realm it resides in. That view, which Frege does not elaborate in detail, comes close to the idea of the analogy of names and being discussed and supported by Aristotle and the Scholastic tradition.

VI. Conclusion

We saw in this chapter that Frege finds a variety of meanings for the words "being" and "existence" in what we might now call his tentative metaontology or metametaphysics. He does not tell us what being as a non-empty first-order predicate and as detached from all realms is, as each entity belongs to some realm. Still, he regards it as important to specify the mode in which an entity is, besides instantiating a bundle of properties through which we have access to it. The concept of existence, namely, existence as being an object of possible experience, which Pünjer suggests, is not far from what Frege says about the denizens of his first realm. However, the essential feature in Frege's view is that Frege makes a distinction between the logical concept of existence as a second-order concept and the concept of being or existence which is given content when the specific mode of being of an object is specified. The idea suggests itself that for Frege the first-order predicate "being" or "existence" is analogous in the sense of the medieval analogy of terms. It can be primarily applied to the denizens of the first realm and then transferred and applied to the denizens of the second and the third realm. In this chapter I have no suggestions concerning the ontological order, which would naturally be important if Frege were a Platonist. Frege's view combines, although not without problems, features from what we would now call Aristotelian and Kantian metametaphysics.

Notes

* I am very grateful to Frode Kjosavik for useful comments.

1. See, e.g., Haaparanta (1985), (1986), (2001), and (2012).
2. Frege emphasizes this distinction in several writings. See "Boole's rechnende Logik und die Begriffsschrift"(1880/1881), in NS, 9-52, "Dialog mit Pünjer über Existenz" (before 1884), in NS, 60-75, GLA (1884), X, "Über Begriff und Gegenstand" (1892), and a letter to Hilbert (6.1.1900), BW, 70-76.
3. See "Boole's rechnende Logik und die Begriffschrift" (1880/1881), NS, 19.
4. See also Haaparanta (2012).
5. See Haaparanta (1985, 129), where it is argued that Pünjer combines a Kantian with an anti-Kantian view. Also see Haaparanta (1986). Cf. Rosefeldt (2011), where the same idea is elaborated. Stuhlmann-Laeisz (1975, 126) already argued that Frege came up with regarding existence as a logical first-order concept.
6. See, e.g., BS, § 2, and "Der Gedanke" (1918), KS, 346.
7. See, e.g., BS, § 2, and GGA I, § 5.
8. See also MacFarlane (2002).
9. See, e.g., Aquinas (974) and Cajetan (1953). Also see Haaparanta (1992). Cf. Goris and Aertsen (2013).
10. See "Über Sinn und Bedeutung," KS, 144–147, GGA I, X, and § 26, and "Über die Grundlagen der Geometrie I–III" (1906), KS, 285. Also see "Einleitung in die Logik" (1906), NS, 203.
11. Geach and Black use the translation "comprehensive" for "*allseitig*."
12. See "Der Gedanke," KS, 350, "Über den Begriff der Zahl" (1891/92), NS, 95, and "Über Sinn und Bedeutung," KS, 144.

13. This distinction is similar to Edmund Husserl's idea of aspect-wise grasping through different perspectives. See, e.g., Hua VI, § 45. For Frege, unlike for Husserl, the aspects are conceptual or something that can be expressed by means of concepts.
14. This point is also made in Haaparanta (1985, 142).

References

Angelelli, Ignacio. *Studies on Gottlob Frege and Traditional Philosophy*. Dordrecht: Reidel, 1967.

Aquinas, Thomas. *Thomae Aquinatis Summae contra gentiles libri quattuor*. Edited, translated and annotated by K. Albert and P. Engelhardt. Darmstadt: Wissenschaftliche Buchgesellschaft, 1974.

Aristotle. *Analytica Posteriora*. In *The Works of Aristotle*, Vol. I. Ed. by W.D. Ross. London: Oxford University Press, 1928. (Cited as An. Post.)

Beaney, Michael (Ed.) *The Frege Reader*. Oxford: Blackwell, 1997.

Bolzano, Bernard. *Wissenschaftslehre in vier Bänden* (1837), hrsg., von W. Schultz. Leipzig: Verlag von Felix Meiner, 1929–1931.

Burge, Tyler. "Frege on Knowing the Third Realm." *Mind* **101** (1992), 633–650.

Cajetan (Tommaso de Vio). *The Analogy of Names and the Concept of Being*. Translated and annotated by E.A. Bushinski in coll. with H.S. Koren. Pittsburgh: Duquesne University Press, 1953 (originally published in 1498).

Frege, Gottlob. "Begriffsschrift, eine der arithmetischen nachgebildete Formelsprache des reinen Denkens (1879)." In *Begriffsschrift und andere Aufsätze*, hrsg., von I. Angelelli. Hildesheim: Georg Olms, 1964, 1–88. (Cited as BS.)

Frege, Gottlob. *Die Grundlagen der Arithmetik: eine logisch mathematische Untersuchung über den Begriff der Zahl* (1884). Breslau: Verlag von W. Koebner; Repr. and translated by J.L. Austin in *The Foundations of Arithmetic/ Die Grundlagen der Arithmetik*. Oxford: Basil Blackwell, 1968. (Cited as GLA.)

Frege, Gottlob. "Über Sinn und Bedeutung." *Zeitschrift für Philosophie und philosophische Kritik* **100** (1892a), 25–50; repr. in KS, 143-162.

Frege, Gottlob. "Über Begriff und Gegenstand." *Vierteljahrschrift für wissenschaftliche Philosophie* **16** (1892b), 192-205; repr. in KS, 167-178.

Frege, Gottlob. *Grundgesetze der Arithmetik, begriffsschriftlich abgeleitet, I. Band*. Jena: Verlag von H. Pohle, 1893. (Cited as GGA I.)

Frege, Gottlob. "Der Gedanke, eine logische Untersuchung." *Beiträge zur Philosophie des deutschen Idealismus* **I** (1918), 58-77; repr. in KS, 342-362.

Frege, Gottlob. *Kleine Schriften*, hrsg., von I. Angelelli. Darmstadt: Wissenschaftliche Buchgesellschaft, Hildesheim: Georg Olms, 1967. (Cited as KS.)

Frege, Gottlob. *Nachgelassene Schriften*, hrsg., von H. Hermes, F. Kambartel, und F. Kaulbach. Hamburg: Felix Meiner Verlag, 1969. (Cited as NS.)

Frege, Gottlob. *Wissenschaftliche Briefwechsel*, hrsg., von G. Gabriel, H. Hermes, F. Kambartel, C. Thiel, und A. Veraart. Hamburg: Felix Meiner Verlag, 1976. (Cited as BW.)

Frege, Gottlob. *Posthumous Writings*. Translated by P. Long and R. White. Oxford: Basil Blackwell, 1979. (Cited as PW.)

Goris, Wouter and Aertsen Jan. "Medieval Theories of Transcendentals." In *Stanford Encyclopedia of Philosophy*, 2013. https://plato.stanford.edu/entries/transcendentals-medieval/

Grossmann, Reinhardt. "Structures, Functions and Forms." In M. Schirn, eds., *Studien zu Frege/Studies on Frege, Band II*. Stuttgart-Bad Cannstatt: Fromman-Holzboog, 1976, 11–32.

Haaparanta, Leila. "Frege's Doctrine of Being." *Acta Philosophica Fennica* 39 (1985).

Haaparanta, Leila. "Frege's Concept of Being." In S. Knuuttila and J. Hintikka, eds., *The Logic of Being*. Dordrecht: Reidel, 1986, 269-289.

Haaparanta, Leila. "The Analogy Theory of Thinking." *Dialectica* 46 (1992), 169–183.

Haaparanta, Leila. "Existence and Propositional Attitudes: A Fregean Analysis." *Philosophiegeschichte und logische Analyse/Logical Analysis and History of Philosophy* 4 (2001), 75–86.

Haaparanta, Leila. "On "Being and Being: Frege between Carnap and Heidegger." In L. Haaparanta and H.J. Koskinen, eds., *Categories of Being: Essays on Metaphysics and Logic*. New York: Oxford University Press, 2012, 319–337.

Husserl, Edmund. *Die Krisis der europäischen Wissenschaften und die transzendentale Phänomenologie: Eine Einleitung in die phänomenologische Philosophie*. Husserliana VI. Ed. by W. Biemel. The Hague: Martinus Nijhoff, 1954. (Cited as Hua VI.)

Inwagen, Peter van. *Ontology, Identity, and Modality: Essays in Metaphysics*. Cambridge: Cambridge University Press, 2001.

Kahn, Charles H. *The Verb 'Be' in Ancient Greek*. Reidel: Dordrecht, 1973; new edition: Indianapolis/Cambridge: Hacket Publishing Company, 2003.

Kluge, Eike-Henner W. *The Metaphysics of Gottlob Frege: An Essay in Ontological Reconstruction*. The Hague/Boston/London: Martinus Nijhoff, 1980.

Künne, Wolfgang. *Die Philosophische Logik Gottlob Freges, Ein Kommentar*. Frankfurt am Main: Vittorio Klostermann, 2010.

Le Poidevin, Robin, Simons, Peter, McGonigal, Andrew, and Cameron, Ross P. (Eds.) *The Routledge Companion to Metaphysics*. London and New York: Routledge, 2009.

Lotze, Hermann. *System der Philosophie, Erster Teil: Drei Bücher der Logik*. Leipzig: Verlag von G. Hirzel, 1874.

MacFarlane, John. "Frege, Kant, and the Logic in Logicism." *The Philosophical Review* 111 (2002), 25–65.

Manley, David, "Introduction: A Guided Tour of Metametaphysics." In D.J. Chalmers, D.J. Manley and R. Wasserman, eds., *Metametaphysics: New Essays on the Foundations of Ontology*. Oxford: Oxford University Press, 2009, 1–37.

Moore, A.W. *The Evolution of Modern Metaphysics*. New York: Cambridge University Press, 2012.

Rosefeldt, Tobias. "Frege, Pünjer, and Kant on Existence." *Grazer Philosophische Studien* 82 (2011), 329-351.

Seebohm, Thomas. "Transcendental Phenomenology." In J.N. Mohanty and W.R. McKenna, eds., *Husserl's Phenomenology: A Textbook*. Washington: Center for Advanced Research in Phenomenology & University Press of America, 1989, 345–385.

Stuhlmann-Laeisz, Rainer. "Frege's Auseinandersetzung mit der Auffassung von 'Existenz' als ein Prädikat der ersten Stufe und Kants Argumentation gegen ontologischen Gottesbeweis." In C. Thiel, hrsg., *Frege und die moderne Grundlagenforschung*. Meisenheim am Glan: Anton Hain, 1975, 119-133.
Sundholm, Göran. "When, and Why, Did Frege Read Bolzano?" In T. Childers, ed., *The Logica Yearbook 1999*. Prague: Filosofia Publishing House of the Institute of Philosophy of the Czech Academy of Sciences, 2000, 164–174.
Trendelenburg, Adolf. *Historische Beiträge zur Philosophie, Dritter Band: Vermischte Abhandlungen*. Berlin: Verlag von G. Bethge, 1867.

Part II

Critical Metaphysics

The Scope and Limits of Metaphysics

5 Thinking-the-World

Joseph Almog and Olli Koistinen

I. Science, Philosophy and Religion's Threefold Quest for the One Infinitary Ur-Being

Our quest is for unification of three ways of *thinking-the-world*, to transliterate Descartes's phrase *penser-le-monde*. The three ways were once, e.g., in his time, symbiotic, but for a while now, they are sharply segregated. We urge that in spite of modernity's professional segregationism between (A) science, (B) philosophical metaphysics, and (C) religion, the trio not only *could* be unified but *must* be.

To redeploy a phrase Descartes used for the mind-body union, *L'union vecu*, we submit in what follows that we all—cosmic scientists, metaphysical philosophers, and religious practitioners—are each bound by a *lived-union* with the world. Ferdinand Alquié, Descartes's insightful French editor, which we meet in this chapter, calls this "experience ontologique" ("ontological experience"), and we might well use his phrase, or allude to the variant "mundane experience." The latter conveys the two-edged (and interrelated duo of) meanings: (1) the experience is *of* the *mundus*; and (2) the *mundanity-commonality* of the experience.

Descartes himself says to Princess Elizabeth, about the "lived experience" of the mind-body union (in his June 28, 1643 letter):

> and finally, it is by using only life and ordinary conversations, and abstaining from meditation and studying things that exercise the imagination, that we learn to conceive the union of the soul and the body.
> (Descartes 1991, vol. 3, 227, translation modified)[1]

Substitute for mind-body our duo of man (thinker) and the world, and you get what we submit in the chapter about our mundane-ontological experience of the world. Just as we each have, by life and conversation, and with no delving into dissective analysis and other such intellectual abstractions, lifelong ongoing experience of this one local *being*—me—we have by life and conversation, and no intellectual abstractions, lifelong direct experience of this ur-Being, The World. The two existential mundane experiences are inseparable—the first, me-experience, inherently carries *inter alia* the second, World-experience.

One final allusion to Descartes as launching the mundane-ontological experience theme. Our mundus-whisperer Descartes says in a key paragraph near the end of Meditation III:

> when I reflect on myself I not only know that I am something [imperfect], incomplete and dependent on another, which incessantly aspires after something which is better and greater than myself, but I also know that He on whom I depend possesses in Himself all the great things towards which I aspire [and the ideas of which I find within myself], and that not indefinitely or potentially alone, but really, actually and infinitely; and that thus He is God.
>
> (Descartes 1991, vol. 2, 35, translation modified)

A little bit later, in meditation VI, he calls this unlimited-infinitary Being *Nature in its most general respects*. We submit that what has been described by many as purely *local*, the inherently subjective-island-like and inward-looking "I think" and "I exist," is enveloped from the outset by the global infinitary Being or Nature in its most general respects. To think, which I do, and to exist, which I do, to *exist-thinking*(-ly), is not to bask in the glory of anything *monadic* (in both the logical and (Leibniz-)metaphysical senses of the adjective), an all by myself thinking-cum-existing. Quite the contrary—to (and read this with the dashes) *exist-thinking* is to be submerged all the way down to one's toes by the mundus—to exist and think is by the very *logic* of these *action* verbs to be relationally bound, to be pervaded by the mundus; *Cogito ergo Mundus Est*.

So, inspired by this-Descartes, we say—beyond this or that abstractive stopgap, one could not *ultimately* engage in the modernist-purist projects of the famed justice-theory construction "*Fiat iustitia et pereat mundus*" or, substituting freely whatever modernist-purist project one cares for in "*Fiat*. . . .," e.g., *Fiat Cognitionis* or *Fiat Ratio* or *Fiat Pulchrum or* (recalling the medievalists' sought-after "transcedentals") *Fiat unum or Fiat Verum or Fiat Bonum* . . . et *pereat mundus*. For the mundus is that without-which-not; not anything, nothing at all, and in particular, not my existing-thinkingly and striving for all the aforementioned. The mundus is the beginning of it all and the end of it all; I am just a flash in the pan, desperately hanging on to it and lo, rather ambitiously, while hanging by a thread, struggling to *understand* this *mundus*, not least my own local existing-thinkingly in it included.

For all those understanding projects, we must—and we borrow here Dostoyevsky's fertile phrase—first and foremost "save the world." Whether it is his own candidate saviour, Beauty, that will bring the world back from the brink or some other kind of mundane experience, is the question we undertake in a moment. But saving the world, bringing our experience of *it* to ground our experience of ourselves and human others as more such children of the world, is the first step. With the world lost from my radar and thus my experience, I am like a blindsided orphaned solitary child inevitably bent on inventing a tell-tale about himself as an independent, self-made and self-sufficient, island being. With the world regained on my radar, I benefit from a dose of reality, this phrase now read literally, a dose of Reality. My seeming island-reality turns out to be a local dose of Reality. Even detached islands—e.g., the Galapagos—can develop no coherent sense of their own individual islandhood without experiencing, cognizing, and re-cognizing the surrounding ocean that is one's maker and sustainer.

A Path to Saving-the-World

We discuss in this chapter the connection between two issues. We track a relation between (1) the professional modernist segregation—and allegedly imposed of *necessity*—between science, philosophical metaphysics and religion; and (2), again of supposed necessity, the exiling of the world/universe as *entium non gratum*. The disqualification dismantles the world both ontologically as a real unitary being, indeed The Mother Being, and as one we humans (again, of necessity!) bear fundamental cognitive-moral-aesthetic relations to. Instead of the one and only global unus mundus, the modernist segregators engage in locally controllable reverse engineering, each in his separate professional box, where they construct simulacra *models* or *aufbaus* that locally simulate the old universe, up to a host of *propositions (dicta)* the local-box professionals are interested in.[2]

In contrast to the professional segregators and the universe-reducing skeptics, we submit: (1) That in their different *ways*, science, metaphysics, and religion pursue this one and only underlying ultimate-unity, the mundus; and (2) This unus mundus is indeed (2.1) the ur-being of which all local beings are unfoldings, and (2.2) we humans relate to this mundus at the most fundamental existential level by experiencing it, thinking about it, and (morally, aesthetically, etc.) caring for it.

What is more, in our pursuit of this lost cosmic unity, we submit in a historical vein that: (3) A unitary approach to the segregated ABC domains was common among thinkers in between the renaissance and the end of the enlightenment; and (4) we further suggest that even the one intellectual figure who is standardly classed as the renaissance mold-breaker, the mundus dis-integrator, the champion segregationist pulling in 1781, at the end of early modernity, A apart from B apart from C, Immanuel Kant, is, in spite of the iterated press headlines, a role-model *unifier* of ABC. Throughout his life, and moved very much by the unitary-spirit of early modernity, Kant was driven by a quest for such cosmic unity that he understood, better than anybody, was the *necessary* (1) *source* and (2) *ground* of our thinking-cum-being.

The World Lost

The foregoing somewhat urgent sounding discourse about *saving* the world presupposes the prior *loss* as a brute fact. And so, to start this journey to save the world, we must first ponder—when and how was the world lost?

Such departures and the loss induced are hard to date precisely. But sometime in the late enlightenment, we lost The World. Philosophers heed the message, when the 1780s are mentioned and Kant's three epoch-making—indeed more than epoch, *modernity*-making—*Critiques* come to mind. Those critiques of pure reason, practical reason, and (aesthetic) judgment lead to the deconstruction of an absolutely-real universal objectual entity, the space-time universe of natural science, and in tow, the realms of God-given absolute morality and objective beauty.

As collateral damage of the dismantling of this absolute universality of The World, we see in the 1780s also gone what the ancients (and even the logical-construction-minded medievals) took as "the transcendentals," viz. what applied *universally* throughout space and time to any category of Being—what we may call "Fundamentals of Being," or, in short, "Fundamenta." Those consisted of *Unum* (Unity), *Verum* (Truth), *Bonum* (Goodness), and *Pulchrum* (Beauty). As uncritical reason's absolute Universe was being dis-unified and dismantled in the 1780s, disqualified with it were the ideas of a Universe-made *fundamenta*. The Universe's four *fundamenta* were now deconstructed by limitative critical reason, only to be re-formed—as I shall say later, *re-modeled*—by mere bottom-up constructed *limited* models of the unlimited original, capitalized Reality.

Modernity's intellectual reformation transmutes the very meaning of the now stigmatized uncritical "transcendental." It is now purified of the idea of a universal recurring throughout space and time and prior to any human construction. Instead, our sophisticated self-critical Reason now posits, prior to the old transcendentals, the new purifier and new God in town, the (rational-)Self, *whose* constructions (in the strongly possessive-genitive form) the old four fundamenta are now made to be. *Transcendental* has now been made to mean *Self-Made* meaning both (1) made by The Self and (2) purely auto-made, viz. not received from some self-prior primitively existing World or God.

More generally and beyond the exclusionary concerns of strict professional philosophy, a sense—an intellectual atmosphere or *zeitgeist*—was taking hold in those late years of the 1780s that if we are to understand the world, then *the world itself*—that ultimate object here—must . . . *not* be an object of this understanding. We can put this late enlightenment limitative *motif* paradoxically, if we use the right verbal emphasis in stating this directive that I will call *Kant's critical understanding axiom*:

> (**K-understanding**) To understand the world, make sure not to try to understand *the world*.

We can use also use at this juncture the German evocative term *weltanschauung* to fashion an analogue paradoxical directive when it comes to envisioning—intuiting or picturing—the world we live in:

> (**K-intuition**) To form a *weltanschauung* of the world you live in, make sure not to have the *welt* itself in the *anschauung* envisioned.

The world itself remains inaccessible, exiled as a beyonder—it transcends any human *anschauung* and any human understanding. By the way, notice that we speak here of different types of transcendings—what we humans can *envision* and what we humans can *conceptualize*. The world escapes them both.

This modernity-made paradoxical first commandment, viz. *to understand (envision) the world, don't try to understand (envision)* ***the world***, is all the more confounding because for the 200 preceding years, from the (late) Renaissance to the Critical enlightenment of Kant, and throughout natural science, as throughout philosophy and theology, the world qualified not merely as one among many other *entia grata* for investigation but starred as the *ens entium*, the Mother Entity, the ur-entity that is prior to and is the unlimited maker of all other merely limited (by This Maker) local objects.

So ran the seventeenth and early eighteenth century (and we re-run this period's story later, in Section II) investigative frameworks of Descartes and Galileo, Spinoza and Malebranche, Leibniz and Newton, and all the way to the young Kant. As late as 1762–3, in a series of brilliant lectures and essays on cosmology and metaphysics that we linger on in this chapter, The World, this ur-*totality* of Being, was not merely for Kant still alive and kicking but the *ens originarium*, the ur-*ground* of all (A) being, (B) thought and (C) possibility. In the mid-eighteenth century, philosophers and scientists alike, very much typified by Kant's person (who dabbled in both in his early career), can still assert in earnest, *The world rules, ok?*

Then, in the blink of an eye, as with many such all-powerful absolutistic rulers, The World's demise struck, swift and brutal. The one absolute global original was gone, replaced by a spectrum of relative-local constructions. This natural original, (i) the absolutely *infinite* and (ii) absolutely *unique* World, was replaced by artificial surrogates, *models*, each (I) of *limited* infinity and (II) part of a *spectrum* of constructed surrogates, world-s or models-s in the plural. Our original absolute "experience ontologique" or absolute 24/7 living by mundane experience of a one and-only world to which we are attached by a umbilical chord has been replaced by intellectual-conceptual manipulation of spectra of *narratives* or *texts*, virtual *re*-presentations of the one world we live in by being *present in* it.

In turn, Modernity's turn, we find that the four absolute Reality-made fundamenta have been de-absolutized and relativized to Self-made constructions. So it goes for Unity—only, say the modern apostles of critical-reason Kant, within our *synthesis power* of forming unities out of manifolds. So it also goes for Truth (relativized to a constructed model/world) and for Good(-ness), only within the bounds of rationality strictures and formal universalizability constraints. Last but not least of the fundamenta, Beauty is brought down, but not so much to earth as to within the bounds of the constructive imaginative flights of fancy of human artists (beatified in modernity as Nature-free *de-novo*-creative "geniuses"). Beauty is thus not only made to *live* in the eye of the beholder, it is *made* tout court; it is *created* by someone who is not merely a beholder, but one—a (genius) Self—who is its demi-god maker-creator.

Aware of its disappearance from view, we now turn to this world-salvaging operation. We conduct it simultaneously on two levels. There is first the historical vein, whereby we examine the period (roughly between 1600–1780) before The Fall and not least the common suggestion that it was Kant who brought about The Fall. We also point out, beyond nostalgia for that before–The Fall period, a tangible blueprint for us right now to "have" the world as our prime object of experience. And it is this very prime object of our individual and shared experience with which we circle back to our original trio of science, metaphysics, and religious attunement. The trio shares this one world, the one and only, as the object we live by, experience every moment of our existence, and try to understand.

II. The Quest for Nature, Reality, and God in Early Modern Philosophy 1600–1800

Infinity as God's Essence

There are two ways to think about God, Descartes tells in the second replies. The first starts from the attributes of God through which God is traditionally characterized, whereas the second commences from his infinite essence and proceeds to his attributes. It is the former way that one sees time and time again in the textbooks on philosophy of religion. God is first thought to be a perfect being and then his perfection is seen to be captured by a complex concept which has omniscience, omnipotence, etc. as its conjuncts. But it is the latter way that is more interesting.

This latter way is through infinity. According to Alexander Koyré (1922, 11–13), Descartes was among the initiators of the movement of placing infinity not only as something that is in God because he is perfect, but, rather, God is perfect because he is infinite. Infinity in Descartes's hands ceases to be a negative notion (not finite) but something that is prior to the finite. The infinite for Descartes is something that is possible to understand, even though it is not possible to comprehend. Only finite things can be comprehended; i.e., we can comprehend a house in being able to see it as a finite entity that has its limits. However, the absolute infinite cannot be comprehended and being in intellectual touch with it is not an act of comprehension but more like an intuition. My own existence as a finite thinker I cannot doubt—I can, as it were, singularize me, and this depends on my limiting myself to something that is infinite. But if God is incomprehensible, i.e., beyond thinking of him as a finite quantity, and granting that through progressive thinking we do not reach an infinite quantity, how does it come about that we are able to think of him. Gilson, cited by Koyré (1922, 11n2), writes of a special method Descartes was forming with Gibieuf where intellectual intuition of the infinite would serve as the basis for our conception of God.[3]

Of course, it is to be expected at this point that some will get anxious about intellectual intuition (intellectual perception). However, to us such an intuition seems to be quite legitimate. It seems evident that we can have what could be called *finite* intellectual intuitions. One just can see that in a triangle, the longest side has to be opposed to the largest angle and that, to use an example of Spinoza, in ½ = 3/x, x has to be six. And for Descartes as well as for Spinoza, and perhaps also for Kant, our idea even of ourselves requires an intellectual intuition of an infinite entity.

That Descartes saw that infinity lies at the core of God gets support from his first demonstration of the existence of God in the Meditations: Before demonstrating the existence of God, the meditator says that by God "aeternum, infinitum, omniscium, omnipotentem, rerumque omnium, quae praeter ipsum sunt, creatorem intelligo," and what he goes on to prove is the existence of an infinite being. This suggests that for Descartes the other perfections flow from his infinity.

Here we will proceed as follows under the assumptions that God's innermost nature is that of the infinite and that this infinite nature is revealed to us through space. We will start from Malebranche and then move to Kant. The rationale for this is that Kant got much from Malebranche, and they form a kind of natural unit.

Malebranche: Space and God

Descartes made a distinction between infinite and indefinite, claiming that only of God can we know that He is infinite, but space is given to us as indefinite. The heart of the distinction lies in the fact that an indefinite entity is one whose limits we cannot discover, but an infinite entity is one which does not have any limits. From the human point of view, God's infinity is knowable, whereas space for us is only indefinite. Malebranche, however, is of the opinion that we know the infinity of space. In the following passage, he nicely distinguishes indefinite from infinite as well as expresses his belief that we know/perceive/see (voyer) that space is infinite:

> Let us imagine a man fallen from the clouds who, on earth, walks continuously in a straight line, I mean, on one of the great circles

> into which geographers divide it; and let us suppose that nothing keeps him from travelling this way. After several days travel could he conclude that the earth is infinite because he has not found an end? If he were wise and cautious in his judgments, he would believe it to be very large, but he would not judge it to be infinite. And when, as a result of walking, he found himself in the same place from which he had set out, he would realize he had in fact gone round it. But, when the mind thinks of intelligible extension, when it seeks to take the measure of the idea of space, it sees clearly that it is infinite. It cannot doubt that the idea is inexhaustible. Let the mind take enough of it to represent the place of a hundred thousand of worlds and again at every instant a hundred thousand more, the idea will never cease to furnish it with all that is required. The mind sees this and cannot doubt it. Yet it is not in this way that it finds out that the idea is infinite. It is on the contrary, because it sees it as actually infinite and knows that it will never exhaust it.
>
> (Malebranche 1992, 156)

Here in the beginning the earth might look infinite to the one fallen from the clouds. However, he would, of course, be wrong in assuming it to be infinite. What is particularly interesting in Malebranche's example is that with respect to the infinity of the intelligible extension ["l'étendue intelligible"] we can be right. We have the capacity to know that it is actually infinite and that we cannot doubt. Moreover, we know the infinity of the intelligible through "seeing," not through infinitizing; i.e., through adding limited spaces and thinking that there is no end to this process. Malebranche's claim that we have such an intuition of the infinite intellectual space is, of course, decisive. We do believe there is such an intuition in us but to argue for it is not necessary and, perhaps, not even possible. The idea of such a space is an ur-intuition of the ur-space. By denying that we, human beings, come into the possession of the infinite space through infinitizing finite quantities, the question about the origin and nature of such an idea becomes acute.

Malebranche undoubtedly sees divine predicates in this infinite intelligible extension:

> You have the idea of space or extension—of a space, I say, that has no limits. This idea is necessary, eternal, immutable, common to all minds—to men, to angels, even to God. The idea, you should note, is ineffaceable from your mind, as is that of being or the infinite, of indeterminate being [*l'être indéterminé*]. It is always present to the mind.
>
> (Malebranche 1992, 155)

It should be noted that Malebranche carefully distinguishes perceptions from ideas. Ideas are the objects of thought and perception, whereas thought and perception are modifications of finite minds. What Malebranche claims is that this space-idea does not have limits; is necessary; eternal; does not change; is common to all minds [*esprits*], as is the infinite-being-idea; and it is always present to the mind.

Malebranche does not identify God with space, even though this passage just cited might suggest just that. What Malebranche wants to say is that even though the idea of space, i.e., what is thought about, is necessary, this does not mean that its object is a necessary existent outside the realm of thought. Malebranche's argument for attributing some kind of necessity to the idea of space seems to be that space is indispensable for human thought and perception. However, space is something that we cannot think away. Somebody may be thinking of a house he plans to build and which now has only contingent being as an object of thought: this thought of the house is in no way constitutive for his thinking and perception. However, his being human requires that his thought and perception is built on space or extension.

Malebranche wants to say that the object of perception and human thought, infinite space or extension, is only an object of thought. His thinking on this matter seems to owe much to Descartes's so-called skeptical doubts.

Earlier in the *Dialogues on Metaphysics*, Malebranche has argued that we are not through our senses in direct contact with the material reality but that the space we are thinking and observing is somehow ideal. Malebranche argues for his view by examples that closely resemble those given by Descartes in the first Meditation. Malebranche's argument is basically that our perceptual situation as such does not tell the source of ideas. God could annihilate all the other beings but leave me, and if he in that situation were to give me the same representations I now have, I would not detect the difference.

> Let us suppose, Aristes, that God annihilates all the beings He created with the exception of you and me, your body and mine. . . Let us suppose, moreover, that God impresses all the same traces on our brains, or rather he presents all the same ideas to our minds which we have in our minds today. On that supposition, Aristes, in which world would we spend the day? Would it not be in an intelligible world?
>
> (Malebranche 1992, 152)

Here Malebranche commits himself to the view that instead of the material non-mental world what is directly perceived are non-material ideas. The argument is what some may infer from Descartes, but it is not at all clear whether Descartes with his dream and evil demon hypotheses was supporting that kind of view. But if it is accepted that our perceptual situation in itself does not tell the foundation of our perceptions, then our perceiving ideas becomes at least defensible. Let us, then, accept Malebranche's view that our perceptions are about non-material ideas. However, it should be warned already here that this does not, in Malebranche, lead to the view that ideas perceived are somehow internal to our minds.

Space is given to us in our perception and so it cannot be a material whole. Moreover, we have already seen that space, according to Malebranche, is seen by us to be infinite. But an idea of an infinite whole cannot be of our own making, Malebranche, echoing Descartes, thinks. A finite mind can comprehend finite things and can mentally construct other finite things with their help, but it can never produce an infinite idea. Space as the object of perception has to be "outside" the mind; i.e., it is not an inner mental object in a finite mind. Thus, this gives an easy and elegant way out of some kind of inner theatre. All perception of spatial objects is about objects in space, and so about objects external to finite subjects. But still these objects are seen in an idea; they are ideas in an idea.

This infinite idea of space needs a thinker and that thinker has to be infinite in respect of its power of thought. Such an infinite thinker it is natural to call God. Thus, through the idea of space we acquire a touch of God's infinity. Not God's essence as such but rather something that expresses that infinity. Thus, space not only reveals to us the external world but also its thinking cause, God. However, via space we do not see God as he is but through a distance. Malebranche writes:

> VI. But, once again, make no mistake about this. You see only confusedly and as from afar what God is. Though you see the infinite or being without restriction, you see it only in a very imperfect way, and so you do not see God as He is in himself [*tel qu'il est*]. You do not see the infinite as a simple being. You see the multiplicity of creatures in the infinity of an uncreated being, but you do not see distinctly its unity.
>
> (Malebranche 1992, 162)

Malebranche follows Descartes in seeing infinity as the core (or, perhaps essence, or even existence) of God; and in his thinking space, in spite of its infinity, is not God himself but an idea God forms. So space is not a separate substance being caused by God, nor is it an accident of God. Were space an (absolutely) infinite substance distinct from God, there would be two absolute infinities, which for Malebranche is impossible. But were space an accident or attribute of God, he would be a spatial and a compositional entity, and thus no God. Space, for Malebranche, is ideal; or to use terminology Descartes made famous: space has only objective reality—i.e., reality as an object of thought. However, the ideality of space does not mean that we are its authors but that it is something that is in a sense given to us. It is given by God's forming an infinite idea of extension in which we all partake.

What lies at the back of Malebranche's mind when he sees the universe as an idea of God is the geometrical intelligibility of the world. In fact, there is something fascinating in Malebranche's train of thought. The world is understandable in geometrical terms: and from this it is in the spirit of the design hypothesis to hold that because of this, the world has to have an intelligible cause. This very soon gives us this picture: there is a superintelligence who has an idea of the universe and goes on to realize it, and we human beings live, perceive, and think the realized world. Malebranche would not deny that we live (and walk) in the world realized by God, but he does not think that we directly perceive it. What we directly perceive is the idea God has of the world which in its infinity is also an expression of God himself. The idea of God, for Malebranche, is like a text we all can read and in such a way we all live in the same reality. The perceived and thought infinite space cannot be something in us or cannot be due to our workmanship but is something that is distinct from us. In thinking of the infinite, we are thinking something that is an intellectual creature of God; something whose existence grows from the most perfect thinking subject; and this is the intellect of God.

Malebranche's argument for the existence of God is extremely interesting. The first premise is that we have a distant idea via space of the infinite. It is distant because we do not conceive the unity and simplicity of God through that idea. This is a bit difficult, but Malebranche helps:

> We see [God's substance] in itself in the sense that we do not see it by means of something finite that represents it. But we do not see it in itself in the sense of attaining its simplicity while distinguishing perfections in it.
>
> (Malebranche 1992, 163)

The first sentence can be seen to exclude God's, i.e., the infinite, being a fiction of human minds. A fiction is made through combining finite representations, but as we have already seen, Malebranche thinks that from finite elements no infinite can be composed. But we understand, perhaps not quite clearly, what is meant by the infinite. It is the second sentence of the quotation that tells in what sense our understanding of the infinite is defective: We do not understand how the perfections are grounded in the simplicity of the infinite.

Thus, infinite is sui generis, or a term Malebranche often uses, an archetype. Malebranche's demonstration of the existence of the infinite (i.e., of God) is, then, simple:

> Since you agree that nothing finite can represent the infinite reality, it is clear that, if you see the infinite, you see it only in itself. But it is certain that you do see it. Otherwise, when you ask me whether God or an infinite being exists, you would be raising a ridiculous

> question involving a proposition the terms of which you would not understand. It is as if you were to ask me whether a "Blictri" exists, that is, a something you know not what.
>
> (Malebranche 1992, 163–164)

This is an excellent argument. The question: "Whether X exists?" when X is a simple concept or idea that is understood in itself is superfluous or is answered in the very posing of the question. In making the question, one has to be acquainted with what X is about. Were it asked whether tigers exist, the question would be whether *space* is so modified that the predicates involved in the concept of tiger are jointly instantiated. However, the only way to understand infinite as a sui generis concept or representation is to be in touch with the infinite.

Kant: Space and God

Kant was a transcendental idealist, as we all know. What this doctrine involves is that space is neither a substance, nor an accident of substance, nor generated by some fundamental things being related to each other. Space, he thought, was ideal. This all comes surprisingly close to Malebranche's teaching, and Kant did not, as we will see, completely hide that. We will here first consider how Kant in his 1763 essay "The Only Possible Argument in support of a demonstration of the existence of God" used space for giving support to the existence of God (or perhaps better, support to an argument for the existence of God). After that we claim that Kant never abandoned the connection between space and time. In addition to the intrinsic importance of the issue, this may also shed some light on what Kant means by the ideality of space in his *Critique of Pure Reason*.

Kant, in his 1763 essay, shares surprisingly much with Malebranche. Kant's treatment of the matter is a bit cleaner than what is offered by Malebranche. However, as for Malebranche, also for Kant space is the source of all possibility, and what is implicit in Malebranche is explicit in Kant, logical possibility (i.e., lack of contradiction) is not enough for possibility. Kant distinguishes in possibility two components—formal and material conditions. The formal condition is simply the lack of contradiction, whereas the material condition is connected to existence: if something is possible, something has to exist, and the material condition is to be found in that which exists. Kant uses as an example that of a fiery body. Fiery bodies are possible and that does not require their existence. A fiery body can be thought irrespective of whether there are fiery bodies. So even though possibility requires existence, it is not the existence of the possible thing itself that is demanded. But then Kant turns to the possibility of body in itself:

> But I then ask is then a body possible in itself? Not being permitted to appeal to experience here, you will enumerate the data of its possibility, namely extension, impenetrability, force, and I know not what else.
> (Kant 1992a, 125; Ak. 2: 80)

There is no contradiction here in the data, and so one might be tempted to claim that this lack of contradiction is just what is needed for the possibility of body. But after this Kant goes on to ask about the possibility of extension, which is a datum of body. The possibility of body requires the possibility of extension, but it seems that extension is not analyzable into a conjunction of elements:

> Suppose that you can now no longer break up the concept of extension into simpler data in order to show that there is nothing self-contradictory in it—and you must eventually arrive at something whose possibility cannot be analyzed—then the question will be whether space and extension are empty words, or whether they signify something. The lack of contradiction does not decide the present issue; an empty word never signifies anything self-contradictory. If space did not exist, or if space was not at least given as a consequence through something existent, the word 'space' would signify nothing at all.
> (Kant 1992a, 125; Ak. 2: 80–81)

Basically, the argument for there being an existent ground for any possibility is that the non-contradiction in the concepts is not enough. While the concepts are analyzed, one of necessity hits upon simple concepts. Whether these are of possible objects cannot be decided through them not being self-contradictory. Moreover, simplicity of a concept cannot be sufficient, because such a concept can be empty and therefore have no meaning at all—it is like an empty word. Remember Malebranche on Blictri cited earlier. Suppose someone says Blictri is possible; can this be defended by claiming that the idea of Blictri is simple and, thus, not contradictory? No, because it should be specified what Blictri signifies. Thus, true or genuine possibilities throw us outside concepts.

The preceding considerations tell that according to Kant possibilities need a fundamentum. Concepts without being anchored to something beyond them cannot be enough for possibilities. From here there is a short route to the necessary existence of such a fundamentum. Possibilities are necessary and eternal; therefore the fundamenta of possibilities have to be necessary and eternal, too. This is not a very convincing argument, however, because why couldn't the destruction of the fundament also be a destruction of the possibilities evolving from it.

However, Kant has another Malebranchian argument to support the necessary existence of the ground(s) of possibilities. What at least since

Anselm of Canterbury has been the target of ontological arguments is that thinking of God is impossible without granting him some kind of being. What this tells is that God cannot be thought away. It seems to make sense that a logical contradiction is an impossibility because it cannot be thought. Let us now consider what it is to think something as non-existent. I may think of a certain body as non-existent because I can think of the space it inhabits as not containing it; and I may think of a body that does not exist but could exist by understanding that space could contain it. In the Kant of 1763, this would mean that when the concept of something is complex we can think of that thing as non-existing by thinking of the marks in the concept as not being instantiated by the world. So there should be a base in thinking something as unintelligible; however, the thinking of the base given by a single concept is impossible. In conceiving, for example, space as non-existent we should think of there being something more fundamental than space and which could somehow have the features implicit in space. But this is impossible: space is not thought through marks but directly.

What has been said previously tells that grounds of possibilities have necessary being, and space for Kant seems to be the ground of the possibility of bodies. Moreover, Kant is of the opinion that God is the only necessary being. So, should we then accept that space is identical with God for the Kant of 1763? No, Kant thinks. For him, will and understanding are the defining features of God, and these he sees incompatible with spatial properties. This places Kant into an awkward position. The examples he gives of the ultimate ground of possibility concern space, but space is not as such the ultimate ground of possibility but God. What, then, is space for Kant? It cannot be a substance outside God, because in being something that cannot be thought away, it should be a necessary but, as just said, there can be only one necessary substance. Moreover, it cannot be a quality or attribute of God, because that would make God spatial.

However, Kant, in attributing an intellect to God and in claiming that God is a mind, certainly holds that God thinks of space. This could lead him to Malebranche's position, where space is God's idea of himself; i.e., in the sense that space *expresses* God's infinity. The Kant of 1763 is quiet about the ontology of space. It is later, in the Inaugural Dissertation (1770), that he a bit shyly declares his debt to Malebranche:

> [Malebranche's] view, the view namely that we intuit all things in God is very close indeed to the one which is expounded here.
>
> (Kant 1992b, 405; Ak. 2: 410)

Kant is making this important comparison in a passage where he tells he is not using the strict method required by metaphysics but is speculating a bit. In any case, the point is that space is a kind of phenomenal omnipresence of the infinite first cause:

> Accordingly space, which is the sensitively cognized universal and necessary condition of the co-presence of all things, can be called PHENOMENAL OMNIPRESENCE.
>
> (Kant 1992b, 404–405; Ak. 2: 410)[4]

It is very interesting that this kind of Malebranchian view is not expounded only in the Inaugural Dissertation. The thought is forcefully presented also in Kant's lectures on rational theology from the 1780s:

> It would be better to say that space is a phenomenon of God's omnipresence, although even this expression is not entirely suitable, though it cannot be avoided on account of the poverty of the language, which lacks words signifying such thoughts, not to mention expressing them clearly. But space is only an appearance of our senses and a relation of things to one another; and the relation between things themselves is possible only insofar as God conserves them through his immediate and inner presence; thus he determines the place of each through his omnipresence; so to this extent God himself is the cause of space, and space is a phenomenon of his omnipresence. The omnipresence of God is consequently not local but virtual; i.e., God's power operates constantly and everywhere on all things; thus he conserves substances themselves as well as governing their state.
>
> (Kant 1996, 435–436; Ak. 28: 1108)

It is understandable that Kant here complains about the poverty of language, for expressing his thoughts correctly. He uses the word phenomenon here perhaps to say that space does not represent God but gives a clue to God's infinity. An alternative way, and the one we favor, would be to say that space is an idea that *expresses* God and his infinity. This seems to come close to what Kant writes here:

> For, indeed, the human mind is only affected by external things, and the world is only exposed to its view, lying open before it to infinity, in so far as the mind itself, together with all other things, is sustained by the same infinite force of one being. Hence, the mind only senses external things in virtue of the presence of some common sustaining cause.
>
> (Kant 1992b, 404; Ak. 2: 409)

It is space that allows us to perceive things as being related to each other, but space in its infinity also expresses the cause of the universe, i.e., God:

> It is rather the case that places exist, that is to say, that relations of substances are possible, because the cause of the universe is intimately [*intime*] present to all things.
>
> (Kant 1992b, 405; Ak. 2: 410)[5]

In the sentence just quoted, there is a kind of refreshing beauty. What this passage, together with the former one, can be seen to say is that through space we arrive as close to God as cognitively possible. In the Malebranchian reading suggested by Kant himself, space would be ideal as an object of God's thought, but it would also be common to all minds; and moreover, as an object of God's thought, it would not be a separate substance nor an accident of God. This would fit quite well with Kant worrying about the ontological nature of space (see B 56).

What both Malebranche explicitly and Kant implicitly, and not true to his own standards of metaphysical rigor, hold is that the closest one can cognitively get to God is through the representation of space. In fact, our cognition of space is cognition of God. Space is one, it is necessary in being necessarily thought by God, and it is also eternal. Space for them is also the universe base, i.e., the ground of all possible worlds, and thus universe-thoughts are also God-thoughts. God is everywhere.

It should be noted here in passing that reading Kant's idealism in a Malebranchian way could give a new guiding thread to the interpretation of transcendental idealism. Kant is easy to interpret in such a way that space as a representation is something that is internal to all human minds, so that each human being would have her own representation of space, which would lead to some kind of psychological idealism. Malebranchian idealism, however, is anchored in space being a common idea to all of us. To use Descartes's language, space is like a mark God has given of himself to all of us.

As has already been said, when Kant presents the connection of space with God he is not arguing with demonstrative rigor. So it is no wonder that this theme is not taken up in Kant's major work *Critique of Pure Reason*, which Kant believes presents a demonstrated doctrine. However, rather surprisingly, the Malebranchian reading might help to find a new way to understand his doctrine of space. In the Transcendental Aesthetic, Kant believes he has shown that space is ideal. Space for him, as has already been said, cannot be a substance nor an accident but must be something else. In the Malebranchian reading, this would mean that space is an idea thought by God: it is a creature of God which has being only as an object of his thinking activity, and it is this object that we finite objects also perceive.

In the *Critique of Pure Reason*, Kant famously claims that space is not an object in itself but is an a priori condition of outer sense. However, in the Metaphysical Exposition, he gives as the last characteristic of our conception of space the following:

> Space is represented as an infinite given magnitude.
>
> (Kant 1998, 175; B 39)

Here, Kant connects infinity and givenness. What this suggests is that the infinity of represented space is not potential—the represented space (i.e., the only space there is according to Kant) is not due to the work of the representing subject. It is something the subject finds: something that from her cognitive beginning is placed opposite to her. It seems to us that given space on a Malebranchian reading makes this tolerably clear. Of course, here we go beyond Kant's strict apodictic reasoning, but it is, as we have seen, Kant's deep conviction throughout his philosophical life that the infinity of space is intimately connected to the only infinite BEING, i.e., to God.

Envoi

Our theme has been the revamping of the quest for the Universe/God, in accounts of both Being and Thinking. This global infinitary ur-unity, The Mother Unity behind all finitary local beings, reigned both ontological and cognitive studies in early modernity (1600–1780). But it seemed to have gone by the board and not been missed. The Universe/God had become an entium non gratum in both science and analytic philosophical thinking, demoted as a mere leftover for ("irrational") religion. The quest for The Universe/God is standardly classed in our time following what many have taken as the Zagat to modernity's "trending" intellectual themes, Kant's *Critique of Pure Reason*, as merely "pseudo-sensible" and "pseudo-rational." We labored to bring The Universe/God back in from the cold of being merely "pseudo"—and in two interrelated ways. We worked through the early modern writers—Descartes, Malebranche, and Kant—to remind ourselves how it was once commonly felt that one can, indeed, one must, engage with the Universe/God, as both the prime being and the prime object of thinking. Inspired by those now forgotten early modern certitudes, we also tried, independently of these old masters, to regain through examples, how each of us non-philosophers, is universe-intoxicated and a day-by-day mundane universe-whisperer. Perhaps our quest is best put by substituting for us mundane-thinkers, what John Berger says so eloquently of Van Gogh, so that his "VG's paintings" and "they" now cover "Our thinkings":

> Once, long ago, paintings were compared with mirrors. Van Gogh's might be compared with lasers. They do not wait to receive, they go out to meet, and what they traverse is not so much empty space as the act of production, the production of the world. Painting after painting is a way of saying, with awe but little comfort: Dare to come this close and see how it works.
>
> (Berger 2015, 269)

Notes

1. Original text: ". . . *et enfin, c'est en usant seulement de la vie et des conversations ordinaires, et en s'abstenant de mediter et d'etudier aux choses qui*

´ exercent l'imagination, qu'on apprend a concevoir l'union de l'ame et du corps."

2. This direct de re thinking relating one being, himself, to another, the infinitary Being/God, is not to be confused with the scholastic, as it were logical-definitional "de dicto" way, of specifying by attributive *propositions* (*dicta*) an entity, whatever it may be, as long as it satisfies the attributes (perfections). Descartes's thinking in his third meditation is de re and *of* (i) himself and that local being's generative relation ("in the image of") to (ii) that ur-infinitary (unlimited) unitary being of whom he is a finite-limitation. As is well known, Descartes pursues in his meditation V a de-attribute path to God. In his Second Replies, Descartes separates this latter as it were "piecemeal," attributive logician path (to Infinity and to God) from the direct object-to-object path to infinity and God, where neither idea is given via attributes.
3. Cf. Gilson (1913, 205):

> Descartes et Gibieuf élaborèrent en commun le programme d'une métaphysique nouvelle et d'une méthode différente de celle de l'école. Méthode moins discursive que celle de l'école, plus préoccupée de s'élever constamment à l'intuition intellectuelle des principes; persuadée que la vérité ne consiste pas dans la chaîne du discours mais dans une vue intuitive et qu'il n'y a de démonstration véritable que celle qui fait apercevoir la conclusion au sein même des principes. La théodicée fondée toute entière sur la considération de l'Infinité et de l'Unité de Dieu . . . il est un point parfaitement assuré. Pendant les trois années . . . il fut en relations avec deux hommes auxquels il doit une conception générale de l'essence divine très différente par l'esprit de celle qu'il tenait de ses maîtres de la Flèche et dont sa doctrine de Dieu devait sortir.

4. The text continues so that there is no doubt that by phenomenal omnipresence, Kant means the phenomenal omnipresence of the general cause = God.
5. Here we deviate from Walford's and Meerbote's translation of "intime" with "inwardly."

References

Berger, John. *Portraits: John Berger on Artists*. London: Verso, 2015.

Descartes, René. The Philosophical Writings of Descartes I-III. Translated and edited by. John Cottingham, Robert Stoothoff, and Dugald Murdoch. (Anthony Kenny is part of the editorial team of volume III) Cambridge: Cambridge University Press, 1991.

Descartes, René. "Meditations." In *Descartes* II (1991).

Descartes, René. "Replies to Objections." In *Descartes* II (1991).

Gilson, Étienne. *La Liberté chez Descartes et la Théologie*. Paris: Vrin, 1913.

Kant, Immanuel. "The Only Possible Argument in Support of a Demonstration of the Existence of God." In D. Walford and Ralf Meerbote, eds. and trans., *Theoretical Philosophy 1755–1770*. Cambridge: Cambridge University Press, 1992a, 107–201.

Kant, Immanuel. "On the Form and Principles of the Sensible and the Intelligible World." In D. Walford and Ralf Meerbote, eds. and trans., *Theoretical Philosophy 1755–1770*. Cambridge: Cambridge University Press, 1992b, 373–416.

Kant, Immanuel. *Religion and Rational Theology*. Translated and edited by Allen W. Wood and George di Giovanni. Cambridge: Cambridge University Press, 1996.

Kant, Immanuel. *The Critique of Pure Reason.* Translated and edited by P. Guyer and A. Wood. Cambridge: Cambridge University Press, 1998.
Koyré. Alexandre. *Essai sur l'Idee de Dieu et les preuves de son existence chez Descartes.* Paris: Editions Ernest Leroux, 1922.
Malebranche, Nicolas. *Philosophical Selections.* Edited by S. Nadler. Indianapolis: Hackett, 1992.

6 Kant's Metaphysics of Nature and Freedom

Michael Friedman

I shall propose and explore an approach to the relationship between nature and freedom in Kant, which has not, as far as I know, been sufficiently explored before. The basic idea is to focus on Kant's conception of the best science of nature—as this is articulated in his *Metaphysical Foundations of Natural Science*—and to explore the relationship of this kind of science to his conception of the demands of morality, that is, the demands of pure *practical* reason. This kind of science represents our ideal of *theoretical* or speculative knowledge, for Kant, and it is science in this sense that must be brought into a satisfactory relation with morally practical knowledge. I propose, then, that juxtaposing passages from the *Metaphysical Foundations* with those from the first, second, and third *Critiques* will shed new light on Kant's system of nature and freedom.

I begin with a well-known passage from the Preface to the second (1787) edition of the *Critique of Pure Reason* concerning knowledge and faith (Bxxx):

> Thus I had to deny *knowledge* [*Wissen*] in order to make room for *faith* [*Glauben*], and the dogmatism of metaphysics—that is, the prejudice that reason can make progress in it without critique—is the true source of all unbelief [*Unglauben*] conflicting with morality, which [unbelief] is always very dogmatic.[1]

The beginning of the preceding sentence makes it clear that the moral or practical faith in question involves "assuming" what Kant will soon, in the *Critique of Practical Reason* (1788), call the three Postulates of Pure Practical Reason (Bxxix-xxx): "Thus, I could not even *assume* [*annehmen*] *God*, *Freedom*, and *Immortality* on behalf of the necessary practical use of my reason if I did not simultaneously *deprive* [*benehmen*] speculative reason of its claim to extravagant insights."

This discussion occurs toward the end of a long paragraph on the positive value of denying (theoretical) knowledge of things in themselves, in accordance with Kant's solution to the Third Antinomy, concerning the apparent incompatibility between the "transcendental" freedom required by morality and the principle of causality. The idea is that, so long as transcendental freedom is logically consistent, as a property of things in themselves, then it is "at least thinkable that [the freedom required by morality] places no obstacle in the way of the *mechanism of nature*, without it being necessary to have any further insight into [this freedom]" (Bxxix). Then, Kant continues (ibid.):

> [T]he doctrine of morality asserts its place, and the doctrine of nature also asserts its own—which, however, would not have occurred if critique had not first taught us our unavoidable ignorance with respect to things in themselves and limited everything that we can theoretically *cognize* to mere appearances. (Italics in original)

So when Kant here speaks of denying knowledge to make room for faith, it seems what he has primarily in mind is our ignorance of things in themselves.

Yet it also seems that Kant did not intend this denial of knowledge to be the whole story. For he had also embarked, during the same period, on a fuller account of the relationship between nature and freedom, beginning with the *Critique of Practical Reason* (1788) and concluding with the *Critique of the Power of Judgment* (1790). This more developed story, I argue, involves a further limitation of our theoretical scientific knowledge within the realm of spatiotemporal appearances, that is, within the realm of phenomena. This same limitation, I argue, is substantially informed by the conception of "*proper* natural science" that Kant articulates in the *Metaphysical Foundations*.

The *Metaphysical Foundations* appeared in 1786 prior to the second (1787) edition of the first *Critique*, and, as I have argued in *Kant's Construction of Nature* (2013), this 1786 treatise is centrally implicated in the changes Kant made in the second edition. What I want to focus on here, however, is the circumstance that, during his work on the *Metaphysical Foundations* and the new edition of the first *Critique*, Kant was already seriously working on the second *Critique*, which was originally intended as an appendix to the second edition of the first. Just a few weeks after completing the manuscript for the second *Critique*, moreover, Kant made it clear, in a letter to Karl Leonhard Reinhold in December of 1787, that the same topics on which he had lately been working so intensively had also led him to the idea of a third *Critique*, in which the three main branches of critical philosophy—"theoretical philosophy, teleology, and practical philosophy" (Ak. 10: 515)—will all be united in one system. It is only in this system that the two apparently incompatible realms of nature and freedom, theoretical science and morality, are to be finally successfully integrated with one another.

The main advance in the second *Critique* is that Kant now asserts that the freedom required by morality is not merely consistently thinkable by theoretical reason; it also has its own kind of *objective reality* as well (real as opposed to merely logical possibility). Such reality is bestowed on the idea of Freedom by pure practical reason, in so far as we are thereby immediately aware of the fact that the moral law is normatively binding on our will. This *fact of reason* provides reality to the idea of Freedom directly and immediately from a purely practical point of view, without any dependence on theoretical reason. And it is here, Kant says (Ak. 5: 5), that "the enigma of the critical philosophy is first explained: how one can *deny* objective *reality* to the supersensible *use of the categories* in speculation and yet *grant* them this *reality* with respect to the objects of pure practical reason."

Once such objective reality has been directly and immediately conferred on the idea of Freedom, moreover, the ideas of God and Immortality acquire objective reality indirectly and mediately, as necessary presuppositions of the normative binding of our will by the moral law that is already in place. In particular, since morality unconditionally commands us to seek the Highest Good—which, as I understand it, amounts to the realization of the Kingdom of Ends here on earth—the ideas of God and Immortality provide us with the conception of a potentially infinite open-ended future in which we have sufficient (practical) grounds for rational hope that it may be progressively approximated indefinitely. Here, however, a new problem of consistency between theoretical and practical reason emerges: how do we know that such a potentially infinite progressive approximation is also really possible—or at least not really impossible—from a theoretical point of view? The required assurance, in the second *Critique*, is provided by what Kant calls "The Primacy of Pure Practical Reason in its Connection with Speculative Reason" (Ak. 5: 119)—whereby, in particular, speculative or theoretical reason cannot deny real possibility to anything that practical reason unconditionally demands.

Yet Kant quickly felt the need for a more detailed account of this supposed impossibility involving three rather than two faculties of the mind. For, as he says in the 1787 letter to Reinhold, Kant had now recognized a separate "faculty of feeling pleasure and displeasure" in addition to the faculties of "cognition" (theoretical knowledge) and "desire" (the will) discussed in the first two *Critiques* (Ak. 10: 514). And it is for this reason that there are now three rather than two branches of the critical philosophy (Ak. 10: 515): "theoretical philosophy, teleology, and practical philosophy." The newly recognized third branch, teleology, mediates between the other two so as fully to explain how the other two branches are necessarily coherent with one another.

The basic idea of the third *Critique* can then be stated briefly. A proper understanding of teleology, for Kant, reveals a bifurcation of the material or phenomenal world into what we might call inorganic and organic physical systems. The former can be understood, in principle, in terms of what Kant calls the "mechanism of nature"—the operation of moving forces subject to fundamental laws of mechanics such as the law of inertia. The latter systems, however, can never be completely understood in these terms, at least by finite intellects like ours, so that only a teleological mode of understanding—appealing to final rather than efficient causes—is then available. The teleological mode of understanding, however, is merely regulative of our scientific knowledge, and not, like the mechanistic mode, properly constitutive of it. Nevertheless, when we pursue the teleological mode of understanding to its logical conclusion, we arrive at the "final purpose [*Endzweck*]" of nature as a whole, which, once again, must be directed towards the realization of the Highest Good by the human species here on earth. Thus, just where genuinely scientific understanding of the phenomenal world (mechanistic understanding) most definitely gives out, the only thing that can then take its place (again from a merely regulative point of view) is the very highest end unconditionally commanded by morality.

Before looking further into the argument of the third *Critique*, however, I shall first clarify the conception of "proper natural science" in the *Metaphysical Foundations*. In the Preface to this work Kant distinguishes between "proper" and "pure" natural science. Pure natural science consists of propositions that are strictly synthetic a priori—such as, paradigmatically, Kant's three "Laws of Mechanics": conservation of the total quantity of matter, inertia, and action equals reaction. Proper natural science, however, is the broader concept, containing pure natural science as a part, but also containing particular *empirical* causal laws—such as, paradigmatically, Newton's law of universal gravitation—which are not strictly synthetic a priori but nonetheless count as necessary in the sense of the category of necessity: it is "determined" from "the actual" given in perception "in accordance with general conditions of experience" (A218/B265–266). In particular, the law of universal gravitation is inferred from Kepler's so far merely inductive "rules" of planetary motion, but, for Kant, it is still (uniquely) determined from these "rules" by geometry and the three Laws of Mechanics (it is not merely a hypothesis for *explaining* these "rules")—and, in this way, it now counts as a necessary and universally valid empirical law of nature.

Pure natural science consists of both a metaphysical and a mathematical part. The three Laws of Mechanics are metaphysical synthetic a priori principles because they result from instantiating the three Analogies of Experience by what Kant calls the empirical concept of matter—the concept of the movable in space. Such metaphysical principles belong to what Kant calls special as opposed to general metaphysics (the transcendental philosophy of the first *Critique*), and, as such, they necessarily involve mathematical synthetic a priori principles as well. Thus, for example, the law of conservation of the total quantity of matter involves a precise quantitative instantiation of the category of substance, the law of inertia involves precise quantitative instantiations of the category of causality and the predicable (derivative category) of force, the law of action equals reaction involves a precise quantitative instantiation of the category of community or interaction [*Wechselwirkung*]. And it is precisely this feature of special metaphysics, which, in the present case, enables a fruitful collaboration between the special metaphysics of corporeal nature and what Kant calls the "mathematical doctrine of motion [*mathematische Bewegungslehre*]"—where the latter is developed, paradigmatically, in Newton's *Mathematical Principles of Natural Philosophy*. Kant, in his treatise, is providing the required metaphysical principles.

An especially significant case of the relationship between the principles of general and special metaphysics is that between the principle of causality in the first *Critique* and the law of inertia in the *Metaphysical Foundations*. Kant states this law in the Mechanics chapter as the proposition (Ak. 4: 543): "All alterations of matter have an external cause." And he begins the following proof with the (parenthetical) statement (ibid.): "From general metaphysics we take as basis the proposition that every alteration has a *cause*, and here it is only to be proved of matter that its alteration must always have an *external cause*." The heart of the proof rests on the idea that all alterations of matter are *motions* (of one body relative to another), and, in the following remark, Kant explains that the inertia of matter therefore consists in its essential *lifelessness*—whereas a living substance, by contrast, can "determine itself to act from an *internal principle*" (Ak. 4: 544).

Kant calls such forces—causal actions exerted by *material* (lifeless) substances—*moving forces* [*bewegende Kräfte*]. Such forces, as explained, satisfy the law of inertia in Kant's sense, and he had already singled out this species of forces even in the first (1781) edition of the *Critique of Pure Reason*. In particular, when discussing the a priori concepts of substance, causality, action, and force in the Second Analogy, Kant appeals to the example of moving forces to provide the particular content, rather than the mere universal form, of an alteration of state (A207/B252): "For this acquaintance with actual forces is required, which can only be given empirically, e.g., the moving forces, or, what is the same, certain successive appearances (as motions), which indicate such forces." And, as Kant explains in a footnote to the main paragraph, the law of inertia is therefore necessarily involved (A207/B252n): "One should well note that I do not speak of the alteration of certain relations in general, but rather of alteration of state. Therefore, if a body moves uniformly it does not alter its state (of motion) at all, but it certainly does if its motion increases or decreases [i.e., accelerates or decelerates]."

Turning now to the argument of the third *Critique*, it is striking, to begin with, that § V of the published Introduction begins by drawing the same distinction between the principle of causality for nature in general in the first *Critique* and the more specific instantiation of this principle for corporeal nature in the *Metaphysical Foundations*.

Here, rather than distinguishing between general and special metaphysics, Kant distinguishes between "transcendental" and "metaphysical" principles (Ak. 5: 181):

> A transcendental principle is that through which is represented a priori the universal condition under which alone things can be objects of our cognition in general. By contrast, a principle is called metaphysical if it represents a priori the condition under which alone objects, whose concept must be empirically given, can be further determined a priori. Thus, the principle of the cognition of bodies as substances and as alterable substances is transcendental, if it is thereby asserted that their alterations must have a cause; it is metaphysical, however, if it is thereby asserted that their alterations must have an *external* cause. This is because in the first case bodies may be thought only by ontological predicates (pure concepts of the understanding), e.g., as substance, in order to cognize the proposition a priori; but in the second case the empirical concept of a body (as a movable thing in space) must be taken as the basis of the proposition—however, as soon as this is done, that the latter predicate (motion only by external causes) belongs to body can be comprehended [*eingesehen*] completely a priori.

Despite the change in terminology, therefore, it is clear that Kant is referring here to precisely the distinction he makes in the *Metaphysical Foundations* between general metaphysics (transcendental philosophy) and the special metaphysics of corporeal nature.

It might appear, however, that the pure natural science articulated in the *Metaphysical Foundations* is at best merely incidental to the problem Kant is addressing in § V of the Introduction to the third *Critique*. In particular, the problem raised in the preceding § IV concerns the relationship between the "universal transcendental laws of the understanding," on the one side, and the potentially infinite manifold of particular empirical laws, on the other (Ak. 5: 179–180):

> [T]here is such a manifold of forms in nature, as it were so many modifications of the universal transcendental concepts of nature that are left undetermined by those laws which the pure understanding gives a priori, since these pertain only to the possibility of a nature (as object of the senses) in general, that there must nevertheless also be laws for the sake of this [manifold] which, as empirical, may seem to be contingent in accordance with the insight [*Einsicht*] of *our* understanding, but which if they are to be called laws (as is also required by the concept of nature), must be regarded as necessary on a principle of the unity of the manifold, even if that principle is unknown to us.

There appears to be no room here for the synthetic a priori principles of the special metaphysics of corporeal nature, so that the quoted passage from the beginning of § V (Ak. 5: 181) may appear merely to illustrate the distinction between "metaphysical" and "transcendental" principles in general.

A closer look at the argument of §V, however, reveals that the reference to the *Metaphysical Foundations* at the beginning is by no means incidental. To be sure, Kant again begins by considering the distinction between universal laws of the understanding and particular empirical laws. But he now considers the latter as more determinate specifications of the universal principle of *causality* (Ak. 5: 183):

> [T]he objects of empirical cognition are still determined or, as far as one can judge a priori, determinable in so many ways besides that formal condition of time determination [succession in time in accordance with the Second Analogy], that specifically distinct natures, besides that which they have in common as belonging to nature in general, can still be causes in infinitely many ways; and each of these ways must (in accordance with the concept of a cause in general) have its rule, which is a law, and hence brings necessity with it, although given the constitution and the limits of our faculties of cognition we in no way comprehend [*einsehen*] this necessity.

So Kant has here narrowed his focus to more determinate specifications of the transcendental principle of causality (the Second Analogy), and what is at issue in the quotation from the beginning of § V is precisely the way in which the Second Law of Mechanics in the *Metaphysical Foundations* provides one such more determinate specification (to causal actions exerted by what Kant calls moving forces). This important example of a more determinate specification of the Second Analogy cannot, therefore, be merely incidental to Kant's argument in § V.

Nevertheless, as I have emphasized, the principles of pure natural science articulated in the *Metaphysical Foundations* are strictly synthetic a priori, but what is at issue in the argument of §V are particular *empirical* causal laws. How are the latter supposed to fit in with the synthetic a priori pure natural science of the *Metaphysical Foundations*? I have, in effect, already answered this question, for it involves precisely the distinction between "pure" and "proper" natural science drawn in the Preface. Proper natural science, as I have explained, is the wider concept, containing pure natural science as a part, but also containing particular empirical causal laws, which are not strictly synthetic a priori but nonetheless still count as necessary in virtue of their relationship to pure natural science. Thus, as I have also explained, Newtonian gravitational force is a particular empirical instantiation of the concept of moving force in general, and it counts as necessary because of the way in which the law of this force (universal gravitation) results from—is determined by—Kepler's merely inductive "rules" by means of Euclidean geometry and the three Laws of Mechanics. In the case of this particular causal law, then, we do comprehend its necessity, and the problem raised in the Introduction to the third *Critique* is that we cannot expect to comprehend the necessity of the whole, potentially infinite totality of such empirical laws in the same way.

In this sense, the later argument of the third *Critique* is quite consistent with what Kant says in the Preface to the *Metaphysical Foundations* regarding "pure" and "proper" natural science (Ak. 4: 468–469):

> Since the word nature already carries with it the concept of laws, and the latter carries with it the concept of the *necessity* of all determinations of a thing belonging to its existence, one easily sees why natural science must derive the legitimacy of this title only from its pure part—namely, that which contains the a priori principles of all other natural explanations—and why only in virtue of this pure part is natural science to be proper science. Likewise, [one sees] that, in accordance with demands of reason, every doctrine of nature must finally lead to [proper] natural science and conclude there, because this necessity of laws is inseparably attached to the concept of nature, and therefore makes claim to be thoroughly comprehended [*eingesehen*].

In both works, therefore, necessity must be predicated of every genuine law of nature, but complete insight into [*Einsicht*] or comprehension of [*Einsehen*] this necessity is only possible for us in cases of proper natural science—and these cases, as it turns out, are relatively rare. Where proper natural science has not yet been obtained, therefore, we must place substantially more reliance on the merely regulative demands of reason and/or reflective judgment.

But what is most important for my argument is that the central role of the *Metaphysical Foundations* is also evident in Kant's later discussion in the third *Critique* of the Antinomy of the Power of Judgment (§ 70). This Antinomy involves an apparent conflict between two (regulative) maxims of reflective judgment (Ak. 5: 387): a thesis, "[a]ll generation of material things and their forms must be judged as possible in accordance with merely mechanical laws," and antithesis, "[s]ome products of material nature cannot be judged as possible according to merely mechanical laws (judging them requires an entirely different law of causality, namely that of final causes)." It is in Kant's discussion of this Antinomy that he finally resolves the apparent conflict between mechanism and teleology, and it also becomes clear, in particular, that the former, for Kant, is essentially framed by his three Laws of Mechanics. To have full insight [*Einsicht*] into causal necessity, for Kant, means to have explained the necessity in question in terms of mechanical moving forces governed by these Laws.

Thus, for example, several pages after Kant states the Antinomy (§ 70), he goes on (§ 72) to consider the "various systems concerning the purposiveness of nature." The main question, he says (Ak. 5: 389–390), is whether the principle of teleology is "merely subjectively valid" (as a regulative maxim), or is also "an objective principle of nature, according to which there would pertain to it, in addition to its mechanism (in accordance with mere laws of motion) yet another kind of causality, namely that of final causes, under which the first kind (that of moving forces) would stand only as intermediate causes." Several pages later Kant considers a "physical" version of the objective conception, and thus a "*realism*" of purposiveness (Ak. 5: 392), which "bases ends in nature on the analogue of a faculty acting in accordance with an intention, on the *life of matter* (in it, or also through an animating inner principle, a world-soul), and is called *hylozoism*." But such a hylozoism, according to the next section (§ 73), is actually quite impossible (Ak. 5: 394): "[T]he possibility of a living matter (the concept of which contains a contradiction, because lifelessness, *inertia*, constitutes its essential characteristic), cannot even be conceived."

Finally, and most importantly, the pivotal section in Kant's resolution of the Antinomy (§ 77) explains "the special character of the human understanding, by means of which the concept of a natural end is possible for us" (Ak. 5: 405). Kant's point, as is well known, is that our (finite) understanding must always comprehend a natural whole in terms of the constitution and interaction of its parts—so that, in cases where this kind of understanding is unavailable, we must conceive of an infinite, divine understanding, which has fashioned such a natural product in accordance with a purpose. What is much less well known, however, is that Kant also explains the limitations of our understanding more pointedly and specifically (Ak. 5: 407; emphasis added): "In accordance with the constitution of our understanding, by contrast [to an infinite, divine understanding], a real whole of nature is to be regarded only as the effect of the concurrent *moving forces* of the parts." This explicit reference to interaction by moving forces, in my view, makes it clear that, when Kant takes mechanism to be our only "objective" (constitutive) mode of understanding natural products, he means to restrict such fully objective understanding to the proper natural science grounded in the *Metaphysical Foundations*.

Such proper natural science, as I have said, involves a mathematically exact realization of the pure categories and principles of the understanding. In particular, the Analogies of Experience, in the *Metaphysical Foundations*, are realized by the quantitative conservation of the total quantity of matter (first Law of Mechanics), the quantitative conservation of momentum in all actions (exertions) of externally acting causality (second Law), and the quantitative equality of action and reaction in all such mutual interactions (third Law). So what Kant requires for full insight into causal necessity, on my reading, is that both the given causally acting (interacting) substances and their given causal actions (interactions) should be thus mathematically representable in accordance with the Analogies. Newton, as Kant understands him, has achieved precisely this in the case of the law of universal gravitation, so that we have thereby achieved full insight into the necessity of at least one especially fundamental empirical causal law.

In the *Metaphysical Foundations*, however, this law and an analogous principle of universal repulsion (modeled on the Boyle-Marriotte law of expansive pressure) are the only empirical causal laws for which we have achieved such complete insight. Even chemistry, according to the Preface, is not yet (and may never be) a science in the proper sense, because we do not have (and may never have) the required insight there. In particular, for distinctively chemical causal actions of matter we do not yet have (and may never have) mathematical laws of attraction or repulsion analogous to the law of gravitation (Ak. 4: 470–471):

> So long, therefore, as there is still for chemical actions of matters on one another no concept to be discovered that can be constructed, that is, no law of the approach or withdrawal of the parts of matter can be specified according to which, perhaps in proportion to their density or the like, their motions and all the consequences thereof can be made intuitive and presented a priori in space (a demand that will only with great difficulty ever be fulfilled), then chemistry can be nothing but a systematic art or experimental doctrine, but never a proper science, because its principles are merely empirical, and allow of no a priori presentation in intuition.

So it is certainly no wonder that, in the third *Critique*, Kant takes the situation in the life sciences to be even worse, since we are incomparably farther there from the kind of mathematically exact laws of causality that Newton has found for the solar system.

Indeed, Kant's model for what is missing in the life sciences is precisely the kind of explanation that he had suggested for the origin of the solar system in his earlier *Universal Natural History and Theory of the Heavens* (1755): a mathematically precise explanation of exactly how the solar system evolved into its present highly ordered state from an initial chaos of matter governed only by the two "Newtonian principles" of attraction and repulsion. We do not have such an explanation, according to the third *Critique*, for the origin of life out of non-living (inorganic) matter. And, Kant thinks, although we are not now and never will be in a position definitively to rule out such an explanation once and for all, it is nonetheless absurd for human beings to pursue a mechanistic explanation of the history and genesis of life as the foundational operative goal of the biological sciences—"to hope that one day, perhaps, a Newton may arise, who will make comprehensible the generation of even a blade of grass in accordance with laws of nature that no intention has ordered" (§75; Ak. 5: 400).

This point is particularly important, for Kant had first rejected the idea of a "Newton of the blade of grass" in the *Theory of the Heavens* of 1755 (see Ak. 1: 230). Moreover, as I said, one of the main lines of argument in this earlier work is that the two Newtonian fundamental forces of attraction and repulsion completely suffice to explain the evolution of the cosmic order—with absolutely no need here to invoke any special appeals to divine teleology (as had been made, for example, by the Newtonians). By contrast, where precisely this kind of mechanical evolutionary story is not available, we do, according to the third *Critique*, need to invoke just such attributions instead—although, as always, from a merely regulative point of view.

Yet what is most decisive for Kant, I think, is his conception of the historical evolution of human culture and civilization—as expressed most clearly in his *Idea for a Universal History with a Cosmopolitan Purpose* (1784). Here Kant self-consciously echoes the title of his earlier *Universal Natural History and Theory of the Heavens*. But now, rather than a mechanical explanation of such evolution based on "Newtonian principles," Kant sketches a fundamentally Rousseauean conception of the natural predispositions [*Anlagen*] inherent in the human species and how these predispositions can be developed to perfection by a "hidden plan of nature." That "the history of the human species in the large can be viewed as the completion" of such a plan, so as to bring about "the only state of affairs in which [this species] can completely develop all of its predispositions [*Anlagen*]"—namely, the creation of a perfected world government ("constitution of states")—is the content of the penultimate eighth proposition of Kant's essay (Ak. 8: 27).

This evolutionary story is thus subordinate to a teleological regulative ideal, and the important question, Kant says, is "whether experience may discover something of such a course [of development] in the purpose of nature" (ibid.). Kant's answer is "a little," and the reason he gives is at first sight surprising (ibid.):

> [F]or this course appears to require such a long time to complete itself that, from the small part that humanity has traversed in this respect, one can determine the shape of its path and the relation of the parts to the whole only as uncertainly as, from all previous observations of the heavens, [one can determine] the course that our sun together with the whole host of its satellites takes in the great system of the fixed stars.

Kant is here invoking his own youthful theory of how the solar system itself orbits around the center of the Milky Way galaxy, also expounded in the treatise of 1755, and he seems to be suggesting that our empirical knowledge of the development of human history is no more uncertain than this kind of astronomical knowledge.

The crucial difference, however, is indicated by what Kant says in the immediately following clause (Ak. 8: 27): "[A]lthough [this latter determination], from the universal ground of the systematic constitution of the cosmic system and from the little that has been observed [so far], is still reliable enough to infer the actuality of such a course." The "systematic constitution" in this case is based on the two fundamental moving forces of universal attraction and repulsion, both of which, according to Kant, have their actuality—and thus their real possibility as well—given in experience. So the real possibility of the orbital trajectory in question is thereby securely established, even if its exact shape, in Kant's time, remained relatively obscure. In the case of the proposed evolutionary trajectory of human history, by contrast, the only ground of its real possibility is an idea of pure practical reason, and so it must ultimately rest on a fact of freedom rather than any empirically given fact of nature.

Kant makes just this point in his concluding ninth proposition (Ak. 8: 29): "*A philosophical attempt to work out a universal world history in accordance with a plan of nature, which aims at the complete civic unification of the human species, must be viewed as possible and even as furthering this purpose of nature*" (italics in original). Such an attempt may appear fanciful, Kant says, more like a "novel" than a genuine "history." However, he continues (ibid.):

> If, nevertheless, one may assume that nature does not proceed without a plan and final aim [*Endabsicht*], even in the play of human freedom, then this idea [of a cosmopolitan history] could still very well become useful; and, even though we are too short-sighted to see through to the secret mechanism of its organization, this idea may still serve as a guiding thread for presenting an otherwise planless *aggregate* of human actions, at least in the large, as a *system* (italics in original).

Kant's point, therefore, is that universal human history—unlike the history of the heavens—must be guided by a moral and political ideal (in this case that of a world government) even to make sense as a coherent intellectual project. So it is precisely here, in the end, that Kant's attempt to strike a balance between a very strict—and therefore limited—conception of genuine or proper natural science, on the one side, and a very expansive—and thus overriding—conception of pure practical reason, on the other, is supposed to bear its intended fruit.

My understanding of Kant's ultimate solution to the coherence of nature and freedom, theoretical science and morality, is further clarified, finally, by his essay *Concerning the Use of Teleological Principles in Philosophy* of 1788. Much of this essay is a defense of Kant's theory of the different human races—such as the "Negroes," the "Indians," the "Americans," and the "Whites"—as all contained potentially in the form of "original predispositions" or "seeds" implanted in every member of the much more comprehensive human species. And which such predispositions are realized then depends on climate and geography. Kant's more important methodological point, however, is that natural *history* is quite different from mere natural *description*. The former is not only empirical but also *systematic*, where the systematic organization, in this case, depends on teleological principles rather than what Kant calls "physical-mechanical" principles. And the latter principles are only applicable where there are fundamental forces—such as, paradigmatically, the moving forces—whose actuality (and therefore real possibility) is given to us in experience and is not, by contrast, merely hypothetical. Thus, Kant here takes the same line on the difference between "physical-mechanical" principles and teleological principles that he will soon develop at great length in the third *Critique*—where his conception of "physical-mechanical" principles, in both works, is clearly indebted to the *Metaphysical Foundations*.

The 1788 essay is clearly connected to the *Universal History* of 1784 through the idea of "original predispositions [*ursprüngliche Anlagen*]," which can express themselves differentially in different external circumstances. In the 1784 essay, however, the predispositions in question belong to cultural and political history rather than natural history—that is, they belong to human history properly so-called. More generally, comparing the *Universal Natural History* of 1755, the *Universal History* of 1784, and the *Teleological Principles* of 1788, we see that there are three levels of historical evolution distinguished in Kant's thought. The evolution of the cosmos can be completely explained by "physical-mechanical" principles, whereas the evolution of biological species and sub-species requires teleological original predispositions governing biological characteristics—such as, in the case of the human species, skin color, bone structure, and the like. The evolution of human culture and civilization, finally, requires what we might call "moral" original predispositions, whose expression, in this case, essentially involves the socio-political organization of human communities and institutions.

In the end, therefore, Kant's metaphysics of nature and freedom involves no mysterious division of reality into two different "worlds"—phenomenal and noumenal. It instead involves two different standpoints—theoretical and practical—on the same natural (phenomenal) world. These two standpoints, moreover, are inextricably connected by a teleological evolutionary conception wherein the approximation of the Kingdom of Ends by humanity necessarily appears as a really possible continuation of the history of life on earth.

Note

1. All translations are by the author.

References

Friedman, Michael. *Kant's Construction of Nature: A Reading of the Metaphysical Foundations of Natural Science*. Cambridge: Cambridge University Press, 2013.

Kant, Immanuel. *Allgemeine Naturgeschichte und Theorie des Himmels* [*Universal Natural History and Theory of the Heavens*]. Berlin: Academy Edition. Vol. 1, 217–368. Ed. by J. Rahts. Berlin: Georg Reimer, 1910. (Cited as Ak. 1.)

Kant, Immanuel. *Kritik der reinen Vernunft: Erste Auflage 1781* [*Critique of Pure Reason, A Edition*]. Berlin: Academy Edition. Vol. 4, 5–252. Ed. by B. Erdmann. Berlin: Georg Reimer, 1911a. (Cited as "A" with page number following.)

Kant, Immanuel. *Kritik der reinen Vernunft: Zweite Auflage 1787* [*Critique of Pure Reason, B Edition*]. Berlin: Academy Edition. Vol. 3, 2–552. Ed. by B. Erdmann. Berlin: Georg Reimer, 1911b. (Cited as "B" with page number following.)

Kant, Immanuel. *Metaphysische Anfangsgründe der Naturwissenschaft* [*Metaphysical Foundations of Natural Science*]. Berlin: Academy Edition. Vol. 4, 467–565. Ed. by A. Höfler. Berlin: Georg Reimer, 1911c. (Cited as Ak. 4.)

Kant, Immanuel. *Kritik der praktischen Vernunft* [*Critique of Practical Reason*]. Berlin: Academy Edition. Vol. 5, 3–163. Ed. By P. Natorp. Berlin: Georg Reimer, 1913a. (Cited as Ak. 5.)

Kant, Immanuel. *Kritik der Urteilskraft* [*Critique of the Power of Judgment*]. Berlin: Academy Edition. Vol. 5, 167–485. Ed. by W. Windelband. Berlin: Georg Reimer, 1913b. (Cited as Ak. 5.)

Kant, Immanuel. *Briefwechsel. Band I, 1747–1788* [*Correspondence*]. Berlin: Academy Edition. Vol. 10. Ed. by R. Reicke. Berlin and Leipzig: Walter de Gruyter, 1922. (Cited as Ak. 10.)

Kant, Immanuel. *Idee zu einer allgemeinen Geschichte in weltbürgerlicher Ansicht* [*Idea for a Universal History with a Cosmopolitan Purpose*]. Berlin: Academy Edition. Vol. 8, 17–31. Ed. by H. Maier. Berlin and Leipzig: Walter de Gruyter, 1923a. (Cited as Ak. 8.)

Kant, Immanuel. *Über den Gebrauch teleologischer Principien in der Philosophie* [*On the Use of Teleological Principles in Philosophy*]. Berlin: Academy Edition. Vol. 8, 159–184. Ed. by H. Maier. Berlin and Leipzig: Walter de Gruyter, 1923b. (Cited as Ak. 8.)

7 From Nothing to Something—Why Metaphysics Cannot Be Reduced to Logic

Camilla Serck-Hanssen

I. Introduction

That the *Critique of Pure Reason* (CPR) is concerned with the possibility of metaphysics is obvious to any reader. Already in the Preface to the A-edition Kant states:

> by [the critique of pure reason] I do not understand a critique of books and systems, but a critique of the faculty of reason in general, in respect of all the cognitions after which reason might strive independently of all experience, and hence the decision about the possibility or impossibility of a metaphysics in general.
>
> (Axii)

In the longer and more elaborated Preface to the B-edition, we find that the A-edition's locution "metaphysics in general" is replaced by a distinction between a first and a second part of metaphysics. The first part "concerns itself with concepts a priori to which the corresponding objects appropriate to them can be given in experience" (Bxix). Kant argues that this part of metaphysics is not only possible, but that his altered way of thinking, i.e., the assumption that objects must conform to our cognition rather than the other way around (Bxvi), can

> very well explain the possibility of a cognition a priori, and what is still more, we can provide satisfactory proofs of the laws that are the a priori ground of nature, as the sum total of objects of experience—which were both impossible according to the earlier way of proceeding.
>
> (Bxix)

The second part of metaphysics concerns objects that transcend possible experience. This part of metaphysics, which strives for a priori cognition of objects beyond the boundaries of possible experience and which in doing so is "precisely the most essential occupation of this science" (Bxix), is according to Kant impossible in the theoretical domain.[1]

In raising the question "How is metaphysics possible?" (B22), the aim of the CPR is not only to do metaphysics, but to inquire into metametaphysical questions about the goals, limits, and methodology of metaphysics. In Kant's own words "Philosophy needs a science that determines the possibility, the principles, and the domain of all cognitions a priori" (A2/B6). This inquiry is normative (Ak. 9:14, A54/B78). It is a critique, which aims at establishing the goals, limits, and methods metaphysics ought to abide by.[2] In doing so, it also brings out why metaphysics should be conceived as a special science, which is irreducible to the empirical sciences as well as to mathematics and logic.

The aim of this chapter is to provide a novel reading of Kant's critique of "the second part of metaphysics," a critique that takes place in the Transcendental Dialectic of the CPR. According to Kant, this second part of metaphysics is nothing but speculative and failed attempts to arrive at knowledge of objects beyond possible experience. Kant offers a number of complex and difficult arguments in support of his verdict, most of which depend on what he takes himself to have established in the first part of the Critique. These include claims such as: the alleged proponents of speculative metaphysics are transcendental realists; they conflate things in themselves and appearances; they apply the categories beyond their proper limits; they conflate the pure and the schematized category; they fail to engage in transcendental reflection.

My reading focuses on a different aspect of the shortcomings, namely, on why the attempts to reduce speculative metaphysics to logic fail and in what sense "general logic, which is merely a canon for judging, has been used as if it were an organon for the actual production of at least the semblance of objective assertions, and thus in fact . . . has thereby been misused" (A 61/B 85). In doing so, I draw primarily on resources found in Kant's logic. This strategy enables a line of argument that is less circular than those which depend on the success of the previously mentioned premises taken from the first part of the CPR.

In a nutshell my reading goes like this: From Kant's logic we can extract an argument against speculative metaphysics that is as beautiful as it is simple, namely, that their reasoning is invalid because it is based on inferences in which they conflate two different concepts of negation. The senses are those expressed by respectively negative judgment (S is not P) and infinite judgment (S is non-P). While the first concept of negation can be used to express truths, e.g., "the world is not finite," even if the subject term fails to refer and in this sense is empty, empty subject terms will yield false judgments if the negation is of the second kind (e.g., "the world is non-finite").

Now, if these two forms are conflated in a syllogism, which I will show is the case with one of the proofs about the thinking I or soul, it can result in formally invalid inferences with ambiguous middle terms. But the two forms must be kept strictly apart for other logical reasons as well. Due to the difference in the condition of existential import, the move from a negative judgment (e.g., the soul is not a composite) to an infinite judgment (the soul is non-composite) is formally invalid. As we shall see, this affects proof strategies found in the paralogisms. Finally, the distinction between negative and infinite judgement is also crucially involved in Kant's resolution of the antinomies and his concomitant dismissal of the apagogic method in metaphysics, i.e., the use of indirect proofs by assuming the opposite or *reductio ad absurdum*. For while an affirmative judgment (S is P) and its negation (S is not P) are contradictories and cannot both be false, an affirmative judgment (S is P) and the corresponding infinite judgment (S is non-P) are contraries and both judgments will be false in the case of empty subject terms. Hence, since the theses and the antitheses of the antinomial conflicts instantiate the latter forms (affirmative and infinite judgment), both of them will be false if the subject term is empty. This implies that the proponents' use of apagogic method is illicit unless they have safeguarded themselves against the use of empty subject terms, a condition Kant believes they cannot meet.

To see in more detail how the distinction between negative and infinite judgment can be used to reconstruct Kant's critique of speculative metaphysics, let us now turn to an examination of the Transcendental Dialectic itself.

II. Kant's Critique of Speculative Metaphysics

The Common Formal Fallacy

In the Preface of the CPR, Kant argues that the failure of speculative metaphysics serves as "an experiment providing the checkup on the truth of the first assessment of our rational cognition a priori" (Bxx), to wit, that a priori cognition is only possible with respect to appearances. For the attempt to extend a priori cognition to objects of mere thought (of reason) fails as it results in a contradiction, and this contradiction can be solved if one accepts the Copernican revolution and the concomitant distinction between appearances and things in themselves.

When we turn to the Transcendental Dialectic itself however, which contains Kant's elaborate critique of reason's attempts to derive a priori cognitions of objects that transcend possible experience, we realize that the Preface's description of the flaws of speculative metaphysics is at best imprecise. First, it is only the arguments about the world, i.e., the antinomies, that lead to contradictions. Moreover, strictly speaking it is only the two last pairs of these arguments, the so-called dynamical antinomies, that function as the announced experiment. For only here does Kant's solution require the distinction between things in themselves and appearances, as both the theses and antitheses are supposed to be true albeit about different domains (things in themselves and appearances).

A fundamental error in all four antinomies is, however, that their arguments are grounded in a *sophisma figurae dictionis*, i.e., the fallacy of an ambiguous middle term (A500/B528). In such a dispute

> no means is left for ending [. . .] in a well-grounded way and to the satisfaction of both sides, unless through the fact that they can do such a fine job of refuting each other they are finally won over to the view that they are disputing about nothing, and that a certain transcendental illusion has portrayed a reality to them where none is present.
>
> This is the path on which we will now set forth in settling a dispute that cannot be decided by a final judgment.
>
> (A501-2/B529-30)

In the Transcendental Dialectic itself, it seems that this diagnosis is the most general one: the attempted proofs of speculative metaphysics fail because they are formally invalid inferences that involve the employment of ambiguous middle terms.[3]

The idea that all attempted proofs of speculative metaphysics suffer from the same kind of underlying fallacy might seem far-fetched. I will neither attempt to defend it as an accurate reconstruction of the metaphysical tradition that Kant criticizes, nor will I be able to discuss each of the inferences to show in what sense they can be argued to be formally invalid—a claim that has been met with criticism.[4] Nevertheless, given the logic he ascribed to, I believe that Kant was indeed right in characterizing the proofs of speculative metaphysics as being formally invalid, albeit not only in the sense of falling prey to the fallacy of an ambiguous middle term. Within the scope of this chapter, I will limit myself to discussing some examples from the alleged proofs. I chose these examples because I find them especially enlightening, as they bring out in a clear and elegant manner what goes wrong with the attempt to reduce metaphysics to a logical exercise.

Let me start with the so-called Paralogisms. These are dialectical syllogistic inferences which give rise to four ideas of the soul (the idea of the substantial soul, the simple soul, the numerically identical soul, and the materially independent soul). This part of the Critique was totally rewritten in the second edition, and it is the version in B-edition that will function as my first example.[5] Kant claims that the whole science of rational psychology is based on the following paralogism:[6]

> What cannot be thought otherwise than as subject does not exist otherwise than as subject, and is therefore substance.
>
> Now a thinking being, considered merely as such, cannot be thought otherwise than as subject.
>
> Therefore it also exists only as such a thing, i.e., as substance.
>
> (B410–411)

According to Kant this inference is formally invalid. It allegedly commits the mistake of the kind *sophisma figurae dictionis*, in which one operates with ambiguous middle terms. Thus, despite the appearance of being a valid three term categorical syllogism of the form:

All S are P
a is S
a is P

Kant holds that the form of the paralogism is the invalid four-term kind:

All S are P
a is R
a is P

In contemporary logic, a fallacy of equivocation would normally not be thought of as a *formal* fallacy, since it depends on the meaning of terms. However, as we shall now see, the peculiar character of this inference is that the ambiguity of the middle term can be construed as pertaining to a logical term, namely the term "not."

In the categorical syllogism rendered previously, i.e., the paralogism of pure reason, the middle term is "What cannot be thought otherwise than as subject." This term appears as the subject term in the major premise and as the predicate in the minor. It thereby functions as the medium or ground that allows one to draw the conclusion in which the minor term ("a thinking being considered only as such") and the major term ("substance") become connected.

At first glance, there appears to be no ambiguity here. Notice however that this middle term involves the use of negations. In addition to the explicit negation of possibility (of thinking something): "what *cannot* be thought," there is also another negation, namely, the one pertaining to the way the object (referred to by "what") is thought. For "otherwise than as subject" is just another way of saying "not as subject." Now in this context, this is equivalent to "as predicate."[7] By combining the two negations in the original middle term, we get middle term*: "what must be thought not as predicate." This middle term is however ambiguous. For according to Kant the word "not" can refer to two different concepts of negation—those expressed by respectively negative and infinite judgment. To understand the formal invalidity of the paralogism, we therefore first need to turn to his understanding of this distinction.

Infinite Judgment and Its History

In the CPR, Kant's comments on the table of judgment centre on the distinction between affirmative and infinite judgments. From the point of view of general logic, both are affirmatives. Nevertheless, an infinite judgment has a weaker cognitive status than the affirmative. For the former only predicates negative properties of the subject whereby no new positive determinations are achieved (A71/B97). This characterization of infinite judgment will, among other things, serve as a crucial tool for Kant's solution to the third antinomy, since it allows him to claim that a version of the thesis can be true (roughly put: "the causal series of appearances has a non-empirical cause") without overstepping the bounds of his own theory. This feature is most likely partly what he has in mind when he cryptically adds to his comments in the early parts of the Analytic that: "the function of understanding [function of infinite judgment] that is hereby exercised may perhaps be important in the field of its pure a priori cognition" (A73/B98).

However, for my purpose in this chapter, it is Kant's distinction between negative and infinite judgments, a distinction which is spelled out in Kant's logic, that is the most important. Here we find two differences. The first is the scope of the negation. In a negative judgment, the negation affects the copula, whereas in the infinite judgment it affects the predicate (Ak. 9:104n.; Ak. 24:764; Ak. 24: 930). The second concerns existential import: "[I]n the negative [judgment] the subject is posited outside the sphere of the [predicate][8] and in the infinite it is posited in the sphere of a concept that lies outside the sphere of another" (Ak. 9: 104). To put the point in a less Kantian way: While infinite judgment is a form that asserts what something x is (although it only ascribes a negative property to it: non-F), a negative judgment negates the copula and says only that it is not the case that this something x is F. The truth of the latter but not the former is compatible with empty subject terms.[9]

The idea that there are two kinds of negations, one that negates propositions and another that negates predicates, is not new with Kant. We find infinite negation already in Aristotle, and in medieval philosophy the logic of infinite negation became a topic of great interest as they discussed the proper interpretation of the O form in the traditional square of opposition.[10]

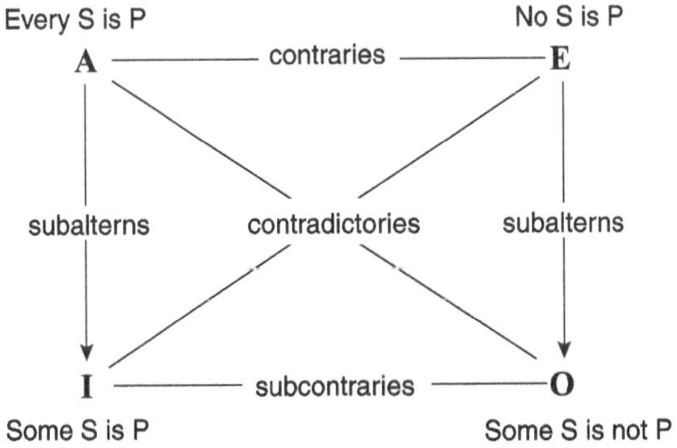

For our purposes, the most interesting point is that medieval logicians tried to develop translation rules for infinite negation. One example is contraposition: 'Every S is P' is equivalent to 'Every non-P is non-S.' Other examples are rules of obversion such as: 'Some S is P' is equivalent to 'Some S is not non-P' and 'Some S is non-P' is equivalent to 'Some S is not P.' In the thirteenth century, the great logician Buridan showed, however, that many of these rules were invalid, as they would lead from a truth to a falsehood in the case of empty terms. These invalid rules include obversion when the direction is from the negative to an infinite or affirmative judgment. One case that will be relevant for us later is the move from 'Some S is not P' to 'Some S is non-P.' One of Buridan's funny counterexamples to this form is the obviously invalid move from 'A chimera is not a man' to 'A chimera is a non-man.'[11] Another relevant case is the invalid form where one attempts to move from 'Some S is not non-P' to 'Some S is P.'

The subtle discussions and insights of medieval logic were however lost. From the middle of the sixteenth century and onwards, spurious principles of contraposition and obversion were back on the scene, without any sign of worry about empty terms. As Parsons notes, the highly influential Port Royal Logic's

> discussion of the O form is so vague that nobody could pin down its exact truth-conditions, and there is certainly no awareness indicated of problems of existential import, in spite of the fact that the authors state that the E form entails the O form (4th corollary of chapter 3 of part 3).
>
> (Parsons 2017)

The same deficiencies are apparent in the eighteenth and nineteenth century. Meier's logic, an excerpt of which made out the textbook in logic (Meier 1752/2016) that Kant had to teach for almost 40 years, shows that Meier is blatantly unaware of the problems of empty terms. As is clear from the two following quotes, he embraces the invalid direction of obversion and takes the move from 'Some A is not B' to 'Some A is non-B' to be valid:

> If, in particular negative judgments one applies the negation to the predicate, then they become particular affirmative judgments [§294], and then they can be converted absolutely.
>
> (Meier 1752/2016, §351, 79)

> Thus, one can transform all negative judgments into affirmative ones, if one moves the negation from the concept of combination to the predicate. E.g. the soul is not mortal, the soul is immortal.
>
> (Meier 1752/2016, §294, 68)

I have not found any place in which Kant explicitly criticizes Meier for having proposed invalid rules. Nevertheless, his own lectures on logic show that he avoids these mistakes. As we have already seen, he takes the infinite form of negation to have existential import while the ordinary negation does not. Hence, pace Meier, for Kant the truth of "the soul is not mortal" does not imply the truth of "the soul is non-mortal (i.e., immortal)."

Moreover, Kant also appears to avoid Meier's endorsement of the invalid rule of contraposition[12]: 'All S are P' implies 'All non-P are non-S.'[13] While the Jäsche logic is unclear on whether the move is from an affirmative to an infinite judgment or from an affirmative to a negative, the Philippi logic, and even more explicitly the Pölitz logic, show that the principle of contraposition that Kant presents is a different and indeed valid one, which moves from an affirmative to a negative judgment: 'All S are P' implies 'All non-P are not S.'[14] In doing so, Kant appears to endorse the non-standard form of contraposition that we find in Aristotle.[15] If so, it is likely that it is precisely the problem of empty terms and existential import which makes him eschew the version of contraposition that was widely accepted at his time.[16]

The Ambiguity of Negation in the Paralogism

We have now extracted the resources from Kant's logic, which enable us to return to an interpretation of what goes wrong in the paralogism. Recall that the argument runs like this:

> What cannot be thought otherwise than as subject does not exist otherwise than as subject, and is therefore substance.
>
> Now a thinking being, considered merely as such, cannot be thought otherwise than as subject.
>
> Therefore it also exists only as such a thing, i.e., as substance.
>
> (B410–411)

What we are trying to understand is the claim that this inference is formally invalid due to an ambiguous middle term. In a footnote Kant suggests that it is the word "thought" that is ambiguous (B411n), but he adds a number of other alleged conflations as well, which makes his argument overloaded and confusing. Not surprisingly, his explanation of the purported fallacy has been subject to extensive discussions and criticism in the commentary literature, which I cannot go into here. I therefore suggest that we first take a step from Kant's arguments, and concentrate instead on the middle term which in fact is the whole phrase "What cannot be thought otherwise than as subject."[17]

As I have already argued, this middle term involves two negations and can be rendered "what must be thought not as predicate." Equipped with

Kant's distinction between negative and infinite judgment terms, we are however now in a position to see that this middle term can be interpreted in two ways. It can either be read with an instantiation of the infinite kind of negation, i.e., the negation of a predicate (viz. "predicate"): "what must be thought as a non-predicate." Or, it can be read as instantiating mere negation, i.e., the negation of the copula: "what must be thought as (it is) not a predicate."

Now, in the major premise the middle term should clearly be read as containing the infinite kind of negation. As we have seen, if a judgment of this form is to be true, which the major premise is, according to Kant, it presupposes non-empty subject terms. And indeed, Kant makes this condition explicit by adding to the major premise "then it *exists* also as . . . ." In the minor premise however, although the rational psychologist might believe that he has the right to use the middle term in the same sense as in the major, this is not correct, according to Kant. For the thinking being about which it is true to say that it must be thought as not a predicate, is merely the logical I of the principle of apperception (B407). But from this I, no existence claim can be inferred. Thus, on the basis of this I alone, which according to Kant is the "sole text" of rational psychology (A343/B401), the middle term in the minor premise must be read as instantiating mere negation: "what must be thought as (it is) not a predicate." This reasoning yields the following inference:

> What must be thought as a non-predicate does not exist otherwise than as subject, and is therefore substance.
>
> Now a thinking being, considered merely as such, is what must be thought as (it is) not a predicate.
>
> Therefore, it also exists only as such a thing, i.e., as substance.

Unless the minor premise of this inference can be "inflated" into an infinite form of negation, the inference is invalid. And this "inflation" is a tall order for the logical approach to metaphysics. For as we saw earlier, as opposed to Meier, Kant sees that due to the condition of existential import, negative judgments cannot simply be converted into infinite judgments by means of *obversion*. This means that unless the rational psychologist has earned the right to assume that the I of apperception is more than a grammatical I and is indeed an existing object (the soul), there is no logical route to the metaphysical insight he desires, to wit, that the thinking being (soul) is a substance.

Although my reading deviates from Kant's own focus on the word "thought" which supposedly is used about objects (in general) in the major premise and about a subject, namely the logical subject I of the I think, in the minor premise, I believe that my reading nevertheless captures the logical gist of Kant's argument. For when the ambiguity is cashed out as an ambiguity between a negative and an infinite reading of the middle term, these two forms instantiate, according to Kant's logic, two different rules of thought. In order to be potentially true, infinite judgment requires a cognitive relation to an existing object of some kind, i.e., an object in general. A negative judgment has, however, no such restriction built into it and can thus also be used about the I of the I think. Finally, on my reading the locus of the ambiguity is the term "not." This fits well with Kant's characterization of the paralogism as a formal fallacy.

Another Formal Fallacy in the Paralogisms

We have now seen one example of how the failures of speculative metaphysics can be understood as logical errors that involve taking existence claims for granted. In doing so, "general logic, which is merely a canon for judging, has been used as if it were an organon for the actual production of at least the semblance of objective assertions, and thus in fact . . . has thereby been misused" (A61/B85).

I will now turn to another example, taken from the reasoning in the second paralogism of the A-edition. This time my point is not that the syllogism is formally invalid due to an ambiguous middle term (which it is).[18] Rather my point is that at least the way Kant argues on his behalf, the proponent applies the illicit form of obversion, as he attempts to move from 'Some S is not non-P' to 'Some S is P.'

The second paralogism runs like this:

> That thing whose action can never be regarded as the concurrence of many acting things, is simple.
>
> Now the soul, or the thinking I, is such a thing. Thus etc.
>
> (A351)

In the following prose, however, Kant presents the argument in another form.

> This is the Achilles of all the dialectical inferences of the pure doctrine of the soul, nothing like a mere sophistical play that a dogmatist devised in order to give his assertions a fleeting plausibility, but an inference that seems to withstand even the sharpest testing and the greatest scruples of inquiry. Here it is.
>
> Every composite substance is an aggregate of many, and the action of a composite, or of that which inheres in it as such a composite, is an aggregate of many actions or accidents, which is distributed among the multitude of substances. . . . Yet with thoughts, as accidents belonging inwardly to a thinking being, it is otherwise. For suppose that the composite were thinking; then every part of it would be a part of the thought, but the parts would first contain the whole thought only when taken together. Now this would be

> contradictory. . . . Thus it is possible only in one substance, which is not an aggregate of many, and hence it is absolutely simple.
> (A352)

Now let us look only at the logical structure of the argument and disregard the content of the arguments that Kant presents on behalf of the rational psychologist. The first part of the argument says in effect that every composite thing is such that its actions must be regarded as aggregates. As Van Cleve has noticed, this suggest that the syllogistic inference Kant actually criticizes is not the syllogism as first presented, but rather a version in which the original major is replaced by Major* by means of contraposition:[19]

> Major*: If something is non-simple (i.e., a composite), its action can be regarded as the concurrence of many acting things.

The next part of the argument says that it is contradictory to assume that the actions of the soul, viz. its thinking, is an action of a composite. This line of reasoning gives us the minor premise of the inference:

> Minor: The action of the soul cannot be regarded as the concurrence of many acting things.

From this we can draw the conclusion by modus tollens:

> Conclusion: The soul is not non-simple (i.e., composite).

So far so good. The problem is however that this conclusion is not what the rational psychologist wants. He wants to show something more, to wit that the soul is simple. But that requires the move from the negative conclusion obtained through this chain of reasoning: "it is not the case that the soul is non-simple," to an affirmative judgment: "the soul is simple." However, as we already know, such moves of obversion would bring you from a truth to a falsehood in the case of empty terms. They are therefore formally illicit unless you add the premise that the soul exists. The vulnerable heel of the rational psychologist is therefore once again disclosed: He either begs the question or commits logical mistakes.

And in fact, it is not only in the second paralogism that we find Kant suggesting that the rational psychologist makes the mistake of moving in the illicit direction of obversion. When Kant reconstructs the line of thought of the first paralogism in the *Prolegomena* (Ak. 4: 335), he also appears to present it as a *modus tollens* argument:

> Major: What is non-substance (i.e., a property) can be thought as non-subject i.e., as a predicate of something else.[20]

Minor: The I cannot be thought as non-subject, i.e., as predicate.[21]
Conclusion: Therefore, the I is not a non-substance (a property).

Again, so far the argument is logically speaking unproblematic. The problem arises when the rational psychologist attempts to move from here to the affirmative conclusion he desires, to wit that:

The I is a substance.

Again, this move is formally illicit unless you add the premise that the I exists. And according to Kant this premise cannot be provided. For although it "appears that . . . the object, namely the *absolute subject* itself, is given in experience . . . this expectation is disappointed . . . the I is not a concept at all, but only a designation of the object of inner sense insofar as we do not further cognize it through any predicate" (Ak. 4:334). Now, even if one were to find Kant's argument wanting, the challenge to the rational psychologist would still stand: His argument relies on the "I" not being empty, so either he must provide an argument to that effect, or he is caught in illicit reasoning.

The rational psychologist is however not alone in making such mistakes. In their application of the apagogic method, proponents in the cosmological debate also attempt to move from the falsity of S is non-P to the affirmation of S is P. The lesson that a negation of a proposition that itself is an infinite judgment (as opposed to a negative one) does not add up to an affirmative if the subject term is empty can therefore also be directed against the proof strategy of rational cosmology, the topic to which I now turn.

Kant's Dismissal of the Apagogic Method

According to Kant the apagogic method, i.e., the application of *reductio ad absurdum* is illicit in metaphysics. The following quotes from the Methodology chapter are instructive:

> The third special rule of pure reason, if it is subjected to a discipline in regard to transcendental proofs, is that its proofs must never be apagogic.
>
> (A789/B817)

> Apagogic proof, however, can be allowed only in those sciences where it is impossible to substitute that which is subjective in our representations for that which is objective, namely the cognition of what is in the object. Where the latter is the dominant concern, however, then it must frequently transpire that the opposite of a certain proposition either simply contradicts the subjective conditions of

> thought but not the object, or else that both propositions contradict each other only under a subjective condition that is falsely held to be objective, and that since the condition is false, both of them can be false, without it being possible to infer the truth of one from the falsehood of the other . . . In mathematics this subreption is impossible; hence apagogic proof has its proper place there. . . . The transcendental attempts of pure reason, however, are all conducted within the real medium of dialectical illusion, i.e., the subjective which offers itself to or even forces itself upon reason as objective in its premises.
>
> (A791–792/B819–820)

Putting the Kantian details aside for the moment, the problem with apagogic proofs is that one has to be sure that one deals with contradictory judgments, not contraries that can both be false. It is common to notice this restriction in relation to Kant's treatment of the first two antinomies, since here both the thesis and the antithesis are claimed to be false. But the close connection between Kant's dismissal of apagogic proofs in metaphysics, the distinction between negative and infinite judgments, and the concomitant point about existential import, has not been sufficiently attended to.[22]

The latter omission is most likely a result of the example that Kant discusses before he tackles the antimonies, an example about smelling bodies:

> If someone said that every body either smells good or smells not good, then there is a third possibility, namely that a body has no smell (aroma) at all, and thus both conflicting propositions can be false. If I say the body is either good-smelling or not good-smelling (*vel suaveolens vel non suaveolens*), then both judgments are contradictorily opposed, and only the first is false, but its contradictory opposite, namely that some bodies are not good-smelling, includes also those bodies that have no smell at all.
>
> (A503/B531)

The point in this example is apparently not about empty subject terms. Rather the point seems to be that if a body does not meet the condition for being divided into good-smelling or non-good smelling as it falls beyond the domain of objects which carry any olfactory properties (e.g., a number), then good-smelling and non-good-smelling are contrary and not contradictory predicates. And contraries can of course not be used in this way in apagogic proofs. You cannot prove that S is P by assuming that S is non-P and show that it leads to an impossibility, or prove that S is non-P, by assuming that S is P and derive an impossibility.

One reason why Kant thinks it is proper to use this example might be that it presents a category mistake. For in his subsequent critique of the antinomies, he will proceed to argue that the attempts to determine properties of the world, when conceived as an unconditioned object in four specific ways, fail in analogous ways. Just like someone who erroneously believes that all objects carry olfactory properties would fail to see that good-smelling and non-good-smelling are contrary predicates and thus would be prone to logical mistakes, the proponents of rational cosmology are also victims to logical errors because they believe that all objects are spatial or temporal.

The details of this line of argument are, however, extremely complex. It has been subject to a vast number of readings and criticized among other things for being circular. I cannot go into a discussion of these readings and objections here. For my purposes, it suffices to point out that there is another way of capturing what goes wrong in the antinomies which construes the problem along the lines of empty subject terms. One of the virtues of this reading, to which I now turn, is that it goes to the heart of the logical failures in which the rational cosmologists get trapped without presupposing the truth of Kant's transcendental idealism. And although it is not the most obvious reading of Kant's line of criticism, we shall see that it surely fits Kant's example, given the logic Kant subscribes to.

As we already know, the crucial difference between the two negations is that while negative judgments can be true even if their subject term is empty, infinite judgments cannot be so. This also has a bearing on the applicability of the apagogic method. For, just like an affirmative judgment, the falsity of an infinite judgment 'S is non-P' opens two paths: it can be false if the subject term is empty or because S is not non-P. Hence, in the case when S is an empty term it is neither true that S is P nor is it true that S is non-P. In short, in the case of empty terms 'S is P' and 'S is non-P' are contraries, not contradictories.

Now, in such cases the apagogic proof method cannot be applied. For the desired move from the falsity of 'S is non-P' to the truth of 'S is P' is blocked. This point we saw previously, as it is the illicit form of obversion found in the Achilles argument. Moreover, also the move from the falsity of 'S is P' to the truth of 'S is non-P' is blocked in the case of empty terms. For this is another illicit form of obversion that moves from 'S is not P' to 'S is non-P', the form that Meier explicitly embraces but Kant avoids. The logical problem for the debates of rational cosmology, then, is that it is the affirmative and the infinite forms of judgment that stand opposed in all of the antinomies, not the affirmative and the negative form (which would yield contradictories). This much is clear from the beginning of Kant's resolution to the antinomies:

> if I say that as regards space either the world is infinite or it is not infinite (non est infinitus), then if the first proposition is false, its contradictory opposite, "the world is not infinite," must be true. Through

> it I would rule out only an infinite world, without positing another one, namely a finite one. But if it is said that the world is either infinite or finite (not-infinite [*nichtunendlich*]), then both propositions could be false.
>
> (A503/B531)

And these forms do not allow the use of the apagogic method unless the parties have earned the right to assume that their subject term ("the world") is not empty. Moreover, since a negative judgment cannot count as a thesis in the metaphysical sense, it is no option for the proponent to revert to this form. Just like the rational psychologist then, the reasoning of the rational cosmologists is either logically fallacious or they beg the question concerning the existence of the objects of their proofs.

As is well known, Kant goes further than this in his criticism of the cosmological proofs, as he thinks he can show that the four subject concepts in the dialectical debate, i.e., the four cosmological ideas of reason, are empty[23] and but the result of invalid syllogistic inferences.[24] He also argues that this emptiness is concealed to us because the ideas have an illusory nature. Just like optical illusions which represent e.g., the oar as being itself bent, although I know that it only appears so to us, the ideas of reason force themselves upon us, as if they represent an independent reality (A339/B397).[25] Hence, "both parties in the conflict are disputing about nothing, and . . . a certain transcendental illusion has portrayed a reality to them where none is present" (A502/B530).

This is why, unlike in mathematics, the proofs of speculative metaphysics "must never be apagogic . . ." (A789/B817). For even if you can make your opponents aware of the logical mistake you commit if you use the apagogic method in instances of empty subject terms, in speculative metaphysics one would still be prone to use the method illicitly. For in this field, the merely "subjective" ideas appear as being objective, and, thus, the problem of empty terms is concealed: "The transcendental attempts of pure reason, however, are all conducted within the real medium of dialectical illusion, i.e., the subjective which offers itself to or even forces itself upon reason as objective in its premises" (A791–792/B819–820).

Kant's arguments about the origin of the ideas of reason and their illusory status are as intriguing as they are complicated, and they depend on a host of premises drawn from the earlier part of the CPR. For the purpose of my argument in this chapter, we need not, however, take a stance as to whether these arguments succeed or not. It suffices to make the weaker point: Both sides in the antinomial disputes either beg the question about the existence of their object or fall prey to logical mistakes. As such, they are no better off than the rational psychologist.

III. Concluding Remarks

By means of examples taken from the arguments Kant presents in the Transcendental Dialectic, I have shown that at least some of the purported proofs of speculative metaphysics are formally invalid, and that their invalidity can be traced back to shortcomings in their logic of negation. By focusing on this shortcoming, my reading brings out that Kant provides the resources for a less circular challenge to speculative metaphysics than is commonly held to be the case: As their proofs depend on non-empty subject terms, they either commit logical errors, or they take for granted that the objects to which they purport to refer exist and thus beg the question.

Kant clearly holds that the proponents of speculative metaphysics cannot establish principles on which all true existence claims are grounded without following the tenets of his transcendental idealism. Until they take this turn, they will therefore also lack the required resources for distinguishing between mere figments of thought on the one hand, and real and existing objects on the other. As Kant says frequently about the alleged objects of speculative metaphysics, they are no more than objects of thought (*Gedankendinge*) or nothing (*ens rationis*) in the sense of empty concepts without objects (A490/B517; A543/B571; A566/B594; A292/B348; A681/B709).[26]

However, I believe this latter part of Kant's critique can and should be distinguished from his logical challenge to speculative metaphysics, and this is the line I have pursued in this chapter. By focusing on the logical shortcomings in the arguments of speculative metaphysics, one sees more clearly why metaphysics cannot be reduced to a mere logical exercise and in what sense "general logic, which is merely a canon for judging, has been used as if it were an organon for the actual production of at least the semblance of objective assertions, and thus in fact . . . has thereby been misused" (A61/B85). Again, the crucial point is that the logical route to metaphysical insight only works if the objects to which the subject terms purport to refer, exist and the subject terms of speculative metaphysics succeed in referring to them.

To lay down the principles for all true existence claims is part of the task of a positive metametaphysics, a task that Kant undertakes in the Postulates of the first Critique. A discussion of the Postulates falls beyond the scope of the present chapter. Nevertheless, I believe that my reading brings out why that part of the Critique is pivotal to understanding Kant's full answer to the question "How is metaphysics possible?"[27] Finally, although I cannot substantiate this claim here, I also believe that my reading reveals that Kant is not a thinker of an empiricist bent who simply took for granted that the only way we can access objects is through our senses.[28] He was rather brought to his position of metaphysical humility by a careful reflection on negation, the problem of empty terms and the need for proper constraints on existence claims, reflections that started already at the time of the *Only Possible Argument* (1763).[29]

Notes

1. Kant does, however, believe that we can have a kind of cognition of such objects through practical reason.
2. Kant refers to this science as transcendental logic, cf. A57/B81.
3. I here follow Grier's reading. Cf. Grier (2001, 156–157).
4. Both Bennett (1974, 72–73) and Van Cleve (1999, 173), for instance, object that the proofs concerning the soul, i.e. the Paralogisms, are obviously formally valid.
5. For my reading of the second paralogism in the A-edition, see Serck-Hanssen (2017).
6. As opposed to in the A-edition, Kant now refers to *a* paralogism in the singular. This single paralogism is a slightly different version of the *first* paralogism in the A-edition. Presumably, Kant's idea is that it suffices to show that this inference is invalid, as the further attempts to determine the soul's properties (simplicity, unity, etc.) all depend on the soul being a substance.
7. This move is more explicit in the major premise in the A-edition version: "That the representation of which is the absolute subject of our judgments, *and hence cannot be used as the determination of another thing*" (A348, my italics).
8. Kant also says that in the negative judgment the "subject is removed from the sphere of the predicate" (Ak. 9:118).
9. Kant's understanding of empty terms and his concomitant theory of existence is a very complex topic that I cannot go into here. For the purpose of this chapter, it suffices to show that the use of certain logical forms and inferences take non-empty terms for granted.
10. Aristotle, *De Interpretatione*, 2 and 3. The reference is taken from Parsons (2017). What I say in the following paragraph is extracted from the same entry.
11. Cited in Parsons (2017).
12. Cf. Meier (1752/2016, §352).
13. Buridan showed that this rule is invalid, as it can be used to infer a falsehood from a truth: All humans are beings, therefore, all non-beings are non-humans. Cf. Parsons (2017).
14. Ak. 24.2:585: "es wird nämlich ein negativer Satz. . . . Alle Menschen sind sterblich. Was nicht sterblich ist, ist kein Mensch."
15. Parsons (2017n10).
16. In correspondence with Martin Hammer, I have been made aware that it is highly likely that Kant's improvement of Meier's understanding of the logic of negation was influenced by Lambert, whose works Kant knew well. Cf. Hammer (2019). However, since my point in this chapter concerns not so much the *origin* of Kant's logic of negation, but rather the systematic role this logic plays in his critique of speculative metaphysics, we need not here dwell on this otherwise interesting question about the genesis of Kant's ideas.
17. In the corresponding footnote, Kant singles out the term "thought" as the problem and argues that it is used about objects (in general) in the major premise and about a subject, namely, the logical subject of the I think, in the minor premise, and, thus, the conclusion does not follow (B411n). This explanation of the purported fallacy has been subject to extensive discussion

and criticism in the commentary literature, which I cannot go into here. Nevertheless, I believe that my reading captures the gist of Kant's argument and brings it out in a clearer way. On my reading, the ambiguity is between a negative and an infinite reading of the middle term. This fits well with Kant's focus on the term "thinking," as thinking in the sense of applying the form of infinite judgment requires a cognitive relation to an existing object of some kind. Thinking in the sense of applying the form of negative judgment has, however, no such restriction built into it.

18. For a reading that brings out the formal paralogistic fallacy of the second paralogism, see Serck-Hanssen (2017).
19. Van Cleve (1999, 176).
20. This principle implies that there is—in the strictest sense of "substance"—no substance at all in nature: "we should take nothing that we can attain for a final subject, and that the substantial itself could never be thought by our ever-so-deeply penetrating understanding, even if the whole of nature were laid bare before it; for the specific nature of our understanding consists in thinking everything discursively, i.e., through concepts, hence through mere predicates, among which the absolute subject must therefore always be absent. Consequently, all real properties by which we cognize bodies are mere accidents for which we lack a subject" (Ak. 4: 333).
21. The reason is that "all the predicates of inner sense are referred to the *I* as subject, and this *I* cannot again be thought as the predicate of some other subject" (Ak. 4: 334).
22. In the German literature, there has been more interest. Cf. e.g. Ishikawa (1990).
23. Kant also argues that his theory allows the last two concepts of the world to be ameliorated and thus rendered possibly not empty, by splitting them into four: two that refer to the series of appearances in space and time and two that refer to things in themselves. Thus, the theses and antitheses can both be true and are now conceived neither as contradictories, nor as contraries, but as subcontraries. Notice however that these four concepts of the world are not the same concepts as the ones that occur in the original theses and antitheses.
24. This time the inferences have the form of the hypothetical inference and are supposedly all based on the following schema: "If the conditioned is given, then the whole series of all conditions for it is also given; now objects of the senses are given as conditioned; consequently, etc." (A497/B525). For the purpose of this chapter, we need not enter into a discussion of how this form is supposed to yield the four cosmological ideas which make out the subjects of the respective antinomies.
25. For a thorough treatment of this theory, see Grier (2001).
26. In putting it this way, Kant is however not claiming that he has proven that the purported referents of these concepts do not or could not somehow exist. His position is not dogmatic, nor skeptical, but that of a critique by which our "ignorance in regard to all possible questions of a certain sort—are not merely suspected but are proved from principles" (A762/B790). Speculative metaphysics is however uncritical and dogmatic and victim to an illusion in which mere ideas of reason are taken to give access to (or even themselves be) mind-independent objects (A339/B397; A792/B820). Kant also holds that the concepts of reason, i.e., the transcendental ideas, have a regulative role in directing and organizing the cognitions of the understanding. Finally, according to Kant's practical philosophy, at least some of the ideas of reason, namely those of the immortal soul, freedom, and God have practical reality. However, on my view, these ideas are not identical with those which are subject to critique in the transcendental dialectic.

27. See, e.g., Kannisto (2016), (2018); Vanzo (2014).
28. The most explicit proponent of the opposed reading is perhaps Rae Langton. Cf. her Receptivity Thesis in Langton (1998).
29. Notice that in *The Only Possible Argument*, just before Kant starts his famous discussion of existence, he refers to the problem of empty terms (Ak. 2:81). For an interesting reading of these passages, see Almog's and Koistinen's chapter in this volume. Moreover, at the beginning of the crucial section: THIRD REFLECTION: OF ABSOLUTELY NECESSARY EXISTENCE, Kant distinguishes between the negation of a predicate and the negation of existence (Ak. 2:82).

References

Bennett, Jonathan. *Kant's Dialectic*. Cambridge: Cambridge University Press, 1974.

Grier, Michelle. *Kant's Doctrine of Transcendental Illusion*. Cambridge: Cambridge University Press, 2001.

Hammer, Martin. "Lamberts Theorie der Termini Infiniti." In H. Nowitzki and G. Stiening, eds., *Johann Heinrich Lambert (1728–1777) und die Mathematisierung der Aufklärung*. Berlin/Boston: De Gruyter, 2019.

Ishikawa, Fumiyasu. *Kants Denken von einem Dritten: Das Gerichtshof-Modell und das unendliche Urteil in der Antinomienlehre*. Frankfurt am Main: Peter Lang, 1990.

Kannisto, Toni. "Positio Contra Complementum Possibilitatis: Kant and Baumgarten on Existence." *Kant-studien* **107** (2016), 291–313.

Kannisto, Toni. "Kant and Frege on Existence." *Synthese* **195** (2018), 3407–3432.

Kant, Immanuel. *Critique of Pure Reason*. Translated and edited by Paul Guyer and Allen Wood. Cambridge: Cambridge University Press, 1998. (Cited, in standard A/B edition pagination with 'A' and/or 'B' and page numbers following.)

Kant, Immanuel. *Gesammelte Schriften*. Königliche Preussische (later Deutsche) Akademie der Wissenschaften (Ed.). Berlin: Georg Reimer, later Walter de Gruyter, 1900–. (Cited as Ak., followed by volume number. All translations are from the Cambridge edition.)

Langton, Rae. *Kantian Humility: Our Ignorance of Things in Themselves*. Oxford: Clarendon Press, 1998.

Meier, Georg Friedrich. *Excerpt from the Doctrine of Reason*. Translated by A. Bunch. London: Bloomsbury, 1752/2016.

Parsons, Terence. "The Traditional Square of Opposition." In E.N. Zalta, ed., *Stanford Encyclopedia of Philosophy*, 2017. <https://plato.stanford.edu/archives/sum2017/entries/square/>.

Serck-Hanssen, Camilla. "Fighting Achilles." In G. Motta and U. Thiel, eds., *Immanuel Kant: Die Einheit des Bewusstseins*. Berlin: De Gruyter, 2017, 130–147.

Van Cleve, James. *Problems from Kant*. New York: Oxford University Press, 1999.

Vanzo, Alberto. "Kant on Existential Import." *Kantian Review* **19** (2014), 207–232.

8 Transcendentally Idealistic Metaphysics and Counterfactual Transcendental Arguments

Toni Kannisto

Although transcendental arguments have been heavily debated in analytic philosophy, there exists no formalization that could distinguish them from other arguments either employing or seeking to ground necessity—e.g., analytic and mathematical. In this chapter I seek to remedy the lack by using counterfactuals to develop a novel formalization of them that captures their most prominent formal characteristic: the inverted order of inferring to a necessary condition rather than to a consequence of a given premise. Furthermore, unlike the standard way of defining transcendental arguments through certain presupposed non-formal features, the formalization allows for the more desirable route of deriving rather than presupposing these features. In this chapter I will demonstrate this for the most ubiquitous feature of transcendental arguments: that they involve and rely on a reference to mental or cognitive capacities. I will further use the formalization and the reference to cognitive capacities to support Barry Stroud's (1968) contention that *metaphysical* or *objective* use of transcendental arguments must resort to idealism. Contrary to Stroud, I do not take this as a sign of their weakness: I believe it was Kant's intention to originally develop transcendental arguments as the proper method for transcendentally idealistic metaphysics that has no validity in non-idealistic metaphysics.

I. Transcendental Arguments and Reinforced Modus Ponens/Tollens

While it is notoriously difficult to define transcendental arguments precisely, they are easy to characterize loosely: they infer from a universally accepted feature of one's subjective mental life to its (initially disputed) necessary conditions or presuppositions. Typically, then, a transcendental argument infers from a premise shared by everyone to something not accepted by the opponent as its necessary condition. Various types of conditions can be pursued. In the strongest and most metaphysical (and arguably traditional) sense of transcendental arguments, the condition is an objective property of the mind-external world. So-called *modest* transcendental arguments only seek to establish further subjective conditions for representing the world (see section V).[1]

The defining formal characteristic of transcendental arguments is that they do not infer from a given fact to its consequences but rather inversely to its enabling conditions or presuppositions. For example, Wittgenstein's Private Language Argument (1953, §§243 ff.) states that a mind-external linguistic community (argued-for objective feature) is an enabling condition for individuals to have language capacities (accepted subjective feature), refuting solipsism. (See note 5 for more examples.) This inverted direction has not been successfully captured in existing formalizations of transcendental arguments, which construe them either as modus ponens or modus tollens reinforced by a necessity-operator.[2] That is, their first (major) premise is a *strict conditional*:

(NMP) "modus ponens"	*(NMT) "modus tollens"*
necessarily, if p, then q,	necessarily, if not-q, then not-p,
p,	p,
∴ q.	∴ q.

Variations exist,[3] yet they have the same shortcoming: they do not explain *why* q (necessarily) follows from p. Since transcendental arguments are employed to exactly *establish* the necessary connection between p and q, this is a major flaw. If on the one hand the connection is presupposed, it hardly requires any special 'transcendental' argument to infer q from p. If on the other hand the necessary connection is relinquished, the *NMP/T* says nothing about how one is supposed to establish it. By assuming rather than justifying the first premise these formalizations disregard precisely what makes transcendental arguments *transcendental.* I am, of course, not claiming that the *NMP/T* formulations are *wrong*—an inference indeed cannot justify its own premises—but that they are *insufficient* for capturing transcendental arguments, as they should involve the justification of the major premise of *NMP/T*. Thus, they presuppose an underlying transcendental argument rather than explicate it.

Not only does the *NMP/T* fail to penetrate the process of establishing the necessary connection, it also (exactly for this reason) cannot distinguish transcendental arguments from other inferences that employ necessity. Consider e.g., the following analytic and mathematical inferences that—albeit atypical—are formally *NMTs*:

Necessarily, if *a* is not colored, then *a* is not red,
a is red,
∴ *a* is colored.

Necessarily, if $x \neq 1$, then $x + 2 \neq 3$,
$x + 2 = 3$,
∴ $x = 1$.

A formalization that cannot tell transcendental arguments apart from analytic and mathematical inferences is severely lacking—or else there is no formally distinct class of transcendental arguments. Given the impotence of the standard formalization, it is hardly surprising that transcendental arguments are regularly defined using extra-logical features—e.g., that they involve mental capacities or have an anti-skeptical aim (Stern 2011). Yet it would be hasty to abandon all attempts at formalization. Indeed, I will show that although transcendental arguments do constitute a sub-class of *NMP/T*, they also exhibit a deeper, unique logical structure.

Note that although one reason for introducing the *NMT* version is to avoid merely presupposing the necessary connection and to more faithfully capture the argumentative direction of transcendental arguments, it fails on both accounts. Arguing from not-q to not-p might appear to presuppose less, as it seems that the connection between p and q is not given, but since the contrapositives '$p \rightarrow q$' and '$\neg q \rightarrow \neg p$' are logically equivalent, there simply is no formal difference between the *NMP* and *NMT*. Hence the shortcomings of one translate to shortcomings of the other.

II. Transcendental Arguments and Counterfactuals

I begin by observing that transcendental argumentation is essentially counterfactual.[4] It sets off with a universally accepted fact and asserts that *were* something not the case, this fact *would* be impossible. Since this something—the desired conclusion—is a necessary condition of a universally accepted fact, it too is established as (universally) necessary.[5] Thus I will formalize the main premise of the transcendental argument as 'necessarily, if q did not hold, then p would not hold', i.e., '$\Box (\neg q \;\Box\!\!\rightarrow \neg p)$', where the standard theory of counterfactuals by David Lewis (1973) will be used.[6]

Lewis uses the notion of similarity between possible worlds to develop semantics for counterfactuals. Based on their similarities and differences, possible worlds are so ordered that those most similar to the actual world are closest to it. '$p \;\Box\!\!\rightarrow q$' is true if and only if in every closest world in which p is true (p-world), q is also true. For example, the counterfactual 'if tapirs had longer snouts, they would be elephants' is true if and only if in all the closest possible worlds in which tapirs have longer snouts, they are elephants. By denoting the set[7] of p-worlds closest to the actual world by W_P, we get: '$p \;\Box\!\!\rightarrow q$' $=_{df}$ '$\forall w \in W_P$: q.' I will denote counterfactuals about other worlds w than ours by '$p \;\Box_w\!\!\rightarrow q$': 'for the possible world w, if p held, then q would hold.'

Lewis shows that counterfactuals cannot be formalized via strict conditionals (1973, 4–13). This is crucial, as attempts to define transcendental arguments via strict conditionals were criticized earlier. Thus, their counterfactual formalization differs from previous attempts and may (and will) avoid their problems. Indeed, as will be demonstrated shortly, counterfactuals have one particularly decisive feature, namely that the rule of contraposition does not hold for them—unlike for strict conditionals (Lewis 1973, 31–36). That is, '$p \Box\!\rightarrow q$' and '$\neg q \Box\!\rightarrow \neg p$' are not logically equivalent.

III. The Logical Form of Transcendental Arguments

Since in transcendental arguments one seeks not to show that with minimal changes, the universally accepted premise would not hold without the proposed condition, but that it is impossible for it to do so, no matter how 'wildly counterfactual thought experiments' (Westphal 2004, 3) we conjure up, its major premise must involve necessity, and the inference itself is as follows:

> (**NCMT**) Necessarily, if not-*q* held, then not-*p* would hold,
> *p*,
> ∴ *q*.

Although this *necessary counterfactual modus tollens* captures the counterfactuality of transcendental arguments more faithfully, it nonetheless shares one shortcoming with the *NMP/T*: it does not tell us how the major premise is grounded. Transcendental arguments were supposed to establish rather than presuppose the necessary connection. But unlike for the *NMP/T*, a simple analysis of the major premise of the *NCMT* solves the problem.

That the counterfactual '$p \Box\!\rightarrow q$' is necessary means that it is true in every possible world. Let $W = \{w_1, w_2, \ldots, w_n\}$ so that $w_1, \ldots, w_n$ are all possible worlds.[8] Establishing the truth of '$p \Box_w\!\rightarrow q$' for all the worlds $w \in W$ demonstrates the necessary counterfactual '$\Box(p \Box\!\rightarrow q)$.' Similarly, through a series of counterfactuals '$\neg q \Box_w\!\rightarrow \neg p$' we can establish that '$\Box(\neg q \Box\!\rightarrow \neg p)$.' We can express this by two series of counterfactuals (for *W*):

(CSMP) Counterfactual series "modus ponens"		*(CSMT) "modus tollens"*
C_1:	$p \Box_{w1}\!\rightarrow q$	$\neg q \Box_{w1}\!\rightarrow \neg p$
C_2:	$p \Box_{w2}\!\rightarrow q$	$\neg q \Box_{w2}\!\rightarrow \neg p$
⋮		
C_n:	$p \Box_{wn}\!\rightarrow q$	$\neg q \Box_{wn}\!\rightarrow \neg p$
	$\therefore \Box(p \Box\!\rightarrow q)$	$\therefore \Box(\neg q \Box\!\rightarrow \neg p)$

It is through such a series that the major premise of the *NCMT* is established. This is the formal equivalent of going through 'wildly counterfactual thought experiments' (Westphal 2004, 3) in order to establish the impossibility of *q*, no matter how alien a *p*-world we imagine—and so to determine *q* as a necessary condition of *p*.[9]

One might question the importance of this result, since similarly a series of material implications establishes the strict conditional. Thus, the

NCMT would not appear to differ from the *NMP/T* after all. Yet there is indeed a crucial difference that motivates the whole counterfactual formalization. Since contraposition is not valid for counterfactuals, the individual counterfactuals of the series are sensitive to direction: 'if p held, then q would hold' does not equal 'if q did not hold, then p would not hold.' This difference between the *CSMP* and *CSMT* is tantamount to a formal exposition of the argumentative direction that is the hallmark of transcendental arguments. Hence counterfactuals provide a formalization of them, superior to material implications.[10]

Importantly, although, as mentioned previously, strict conditionals and counterfactuals are not equivalent, strict conditionals and *necessary* counterfactuals are: '$\Box(p \rightarrow q)$' $\Leftrightarrow$ '$\Box(p \mathrel{\Box\!\!\rightarrow} q)$.' (For proof, see appendix.) Thus, since contraposition therefore is valid for the necessary counterfactual in the conclusion of the *CSMT*, the resulting *NCMT* is logically equivalent to the *NMP/T*. This shows that transcendental arguments conform to the *NMP/T*, yet have a deeper logical structure: the counterfactual analysis remedies the lack of justification for the major premise of the *NMP/T* by showing that their unique form lies in the inference *to*, not *from*, the strict conditional.

This explains why the argumentative direction disappears when transcendental arguments are formalized using strict conditionals, for it makes no difference whether we express the major premise as '$\Box(p \rightarrow q)$' in the *NMP* or '$\Box(\neg q \rightarrow \neg p)$' in the *NMT*. By using strict conditionals (or material implications) one effectively abstracts from the level on which their argumentative direction is manifest. The counterfactual formalization not only avoids this failure but also explains its source.

One might object that if this logical form were all there is to transcendental arguments, a myriad of arbitrary arguments would count as transcendental. Take any counterfactual with a necessarily false antecedent, say: 'if $2 + 3 = 5$ were false, then we would not have perceptions.' Granting that counterfactuals with impossible antecedents are vacuously true (Lewis 1973, 24), and given that we do perceive, this 'transcendental argument' proves that $2 + 3 = 5$. But it would be peculiar to demonstrate mathematical—let alone analytic or purely logical—truths this way or to call the demonstration transcendental. Do analytical and mathematical arguments count as transcendental arguments in my analysis after all?

I grant the objection but contest its severity. It just goes to show that mathematical and analytic arguments do not *need* counterfactuals to justify the necessity in the major premise since it is already established—as our willingness to grant the impossibility of the antecedent testifies. Transcendental arguments, on the contrary, *rely* on counterfactuals for justification. Thus the *CSMT*-grounded *NCMT* represents them more faithfully than the *NMP/T*, and it can be used to distinguish them from both analytic and mathematical inferences. Nor is it problematic that counterfactuals nevertheless *can* be used for supporting logical, analytic, and mathematical truths. That an analytic proposition such as 'Triangles have three sides' is arguably a priori (*pace* Quine) does not mean that one could not *try* to demonstrate it a posteriori. Attempting to prove it by drawing triangles may be silly, but it is not *wrong*: using (empirical) induction to argue for truths that could be proven a priori is weak but not faulty argumentation. (Indeed, it would be strange if induction did not *support* necessary truths—which is not to say that they could *establish* them.) Furthermore, the argumentation is weak only if the truths *really* can be proven a priori: if not, it is only reasonable to settle for the 'weaker' and more uncertain a posteriori proof.

Similarly, one *can* use transcendental arguments to prove logical, analytic, or mathematical truths, but it is superfluous if a stronger proof already exists. This is not their shortcoming. Quite the contrary, were one to doubt the *prima facie* necessity of e.g., mathematical truths (or the immutability of mathematical principles, see note 13), one could very well seek to ground their necessity via a transcendental argument.[11] For such uses and for such reasons transcendental arguments can have considerable philosophical significance, depending on what tools one is and is not willing to presuppose without additional proof.

IV. Transcendental Arguments and Cognitive Capacities

The preceding objection underscores the fact that there is more to an argument than its inferential form: the justification of its premises. On the one hand transcendental arguments are unique in that they *must* rely on the *CSMT* to justify the major premise of the *NCMT*. On the other hand this reliance raises a worry about validity: the *CSMT* seems to have an inductive rather than deductive form.[12] Granted, if we can determine the truth of all members of the series of counterfactuals, the *CSMT* will be deductively valid.[13] Yet I noted that transcendental arguments have little utility in mathematical, analytic, or purely logical contexts where one can already presuppose the necessity of their truths, and so—if one does not want to settle for the uncertainty of empirical induction for transcendental arguments—there ought to be yet another way to go through the series of the *CSMT*. This brings us to cognitive capacities, with which transcendental arguments are almost universally associated.

The logical structure of transcendental arguments helps us see that they are not merely arbitrarily bound to such mental features as our capacity to experience, cognize, perceive, and to use language. Rather, without this bond they would either fall short of deductive certainty (having the justificatory force of mere empirical induction) or be rendered superfluous by the strength of mathematical, analytic, and purely logical proof. Either way, there would be little room for meaningful use of transcendental arguments, and it is such use that the cognitive capacities can offer.

In essence, due to our intimate familiarity with them, our mental capacities allow us to run through the *CSMT*. If we find it impossible to conceive of (represent/imagine) e.g., language without a linguistic community, it seems reasonable (if not incontestable) to conclude that it *is* impossible for beings of similar cognitive make-up.[14] This is best put in terms of *accessibility*, a relation that in modal logic limits the set of possible worlds. (Necessity as truth in all possible worlds is as flexible a concept as the set of all possible worlds—denoting e.g., logical, metaphysical, physical, or epistemic necessity.) Because we have such direct access to our mental capacities, we can arguably use them to determine the sphere of cognitively accessible worlds—what is cognitively possible (and necessary) *for us*.

To determine the cognitively accessible worlds, a transcendental argument starts off with a universal premise p about our cognitive capacities. It then uses the series of counterfactuals in the *CSMT* to ground the impossibility of representing p without q, and so establishes q as equally necessary within the accessible worlds determined by p. If, by using our cognitive capacities, we could not represent p without q, it seems that for us p (something we can represent) would be unrepresentable without q—which indicates that q is necessary for our capacity to represent p, i.e., cognitively necessary. If q and p are not connected to our cognitive capacities, the argument is suspect. It seems irrelevant e.g., to the possibility of a particle simultaneously having both a positive and negative charge whether or not we can imagine it (Lewis 1986, 114). Thus, we could with fairly strong (albeit not absolute) certainty infer (in an inductive-like manner) from arbitrarily many failures to think something to its impossibility in all cognitively accessible worlds—to cognitive necessity.

Importantly, although such inference would be inductive, it differs from empirical induction exactly by being essentially counterfactual: it is an experiment based not on how the world empirically *is* but on how the world counterfactually *could have been*.[15] Arguably (see Williamson 2007, 179 ff., 207), as counterfactual possibility cannot be tested empirically, the only kind of experiment that can demonstrate it is a *thought* experiment—and when does a thought experiment have more credibility than when it is about our cognitive capacity to make them in the first place?[16] Perhaps even more importantly, the inference differs from mathematical (analytic) induction as well: unlike the latter, the former is not based on stipulated or presupposed axiomatic and immutable principles. It is not a matter of definition whether we are capable of cognizing or representing something: determining this does constitute a kind of experiment where the result may surprise us. We can dub the induction in transcendental arguments *transcendental*—following closely Kant's definition of transcendental as concerning the necessary conditions of the possibility of cognition (A11/B25; see also Hintikka 1972, 274). The case can thus be made that transcendental arguments not only form a formally distinct class of arguments but also involve a unique kind of inductive certainty, falling between empirical and mathematical induction (deduction).[17]

Finally, as *p* is a universal cognitive capacity, it is impossible to represent not-*p* to begin with—so the question arises: how then could one represent '$\neg q \Box\!\!\rightarrow \neg p$'? But what is at stake is whether *p* can be represented without *q*: we think up a case in which *q* is false and then, with these restrictions, try to represent *p*. If we fail, the counterfactual is true. We are hence not trying to represent not-*p*, but *p*. And there are three possible outcomes:

1. We think up a world in which *q* is false and we successfully represent *p*, proving that *q* is not a necessary condition of *p*. E.g., *q* could be 'we perceive light' and *p* 'we represent space.' Since I can use my hearing for spatial representation, I can imagine a world in which I am blind (not-*q*) and still represent *p*. Thus the transcendental argument is unsound.
2. We are unable to think *q* as being false, in which case it is a necessary truth (e.g., of logic). Here the transcendental argument is vacuously sound yet superfluous, as argued previously.
3. We are incapable of representing *p* in worlds in which *q* is false. Thus *q* is a (cognitively) necessary condition of *p* and the transcendental argument is sound.

This brief sketch suggests that transcendental arguments and their *CSMT* *can* be valid and non-superfluous for cognitively accessible worlds. By this I do not seek to prove that they *are*, but rather to motivate the frequent and well-founded attempts of philosophers to use transcendental arguments in connection with cognitive capacities. The very cognitive capacities, the necessary conditions of which are to be established, at the same time restrict cognitive accessibility and ensure that their conditions are cognitively necessary. Naturally, nothing in how transcendental arguments have here been construed establishes whether the conditions hold *objectively*: the fate of strong transcendental arguments remains undecided. This echoes the admission of many scholars (e.g., Strawson 1985; Stroud 2000; Stern 2004) that transcendental arguments are valid only modestly of representations, not metaphysically of reality. But, arguably, there are ways to strengthen them, if one is willing to pay the price: I will close by using the preceding results to support Stroud's (1968, 2000, p. 159) claim that objective or metaphysical application of transcendental arguments requires accepting (some form of) idealism. Specifically, I wish to offer transcendental argumentation as a valid method of transcendentally idealistic metaphysics.

V. Strong Transcendental Arguments and Transcendental Idealism

On the one hand, the need to fall back on cognitive necessity to make transcendental arguments valid suggests that their validity may be only modest, yet on the other hand it points to a way to overcome the limitation. Namely, to make a modest transcendental argument strong, a move from cognitive necessity to *metaphysical* necessity is required—a move that is nigh-on inconceivable without some form of idealism. If one can rely on idealism, however, then (at least some) cognitively (and subjectively) necessary mental features of representations of objects will also be metaphysically (and objectively) necessary features of the objects themselves. Without such background theory—given that the restriction of transcendental arguments to cognitive accessibility holds—it is difficult to see how one could derive metaphysical necessity from cognitive necessity (see Stroud 2000, 158–159).

The leading question of strong transcendental arguments is: How can subjective representability establish objective truth? That we cannot conceive of, e.g., a temporal order of events without causality does not mean that causality is an objective feature. Following Hume, perhaps causality is merely a subjective feature—a habit—of our cognitive capacities rather than a fact of objective reality. This possibility can be undermined by supplementing transcendental arguments with idealism: that our subjective cognitive capacities do not only restrict representations but also objects.[18] More technically, idealism would bridge the gap between cognitive representability and metaphysical possibility by allowing us to assume that all metaphysically accessible worlds belong to *cognitively* accessible worlds as their subset—and hence that cognitive necessity implies metaphysical necessity: what is true of all cognitively accessible worlds pertains also to all metaphysically accessible worlds. By accepting that some metaphysically possible worlds can be cognitively impossible, a realist cannot sanction such an implication.

I have shown elsewhere (Kannisto 2012, 252–262) that Kant's transcendental proofs conform to the counterfactual formalization presented here. First, Kant explicitly uses the *NCMT* form with the correct direction when characterizing them:

> In the Transcendental Analytic we drew, e.g., the principle "Everything that happens has a cause" from the unique condition of the objective possibility of a concept of that which happens in general, namely that the determination of an occurrence in time, and consequently this (occurrence) as belonging to experience, *would be impossible if it did not stand under such a dynamical rule.*[19]
>
> (A788/B816, my emphasis)

Second, to justify the necessary counterfactual of the *NCMT*, transcendental proofs carry out the *CSMT* in what Kant calls *transcendental deductions*: the metaphysician must produce "a legitimate [transcendental] proof through the transcendental deduction of its premises" (A794/B822, translation altered). Such deductions consist in binding representations to their enabling cognitive capacities, i.e., in showing their necessity within the relevant cognitively accessible worlds.[20]

Finally, Kant explicitly connects cognitive and metaphysical accessibility in his "supreme principle of all synthetic judgments," which draws on transcendental idealism to render transcendental proofs metaphysical:

> The [necessary] conditions of the *possibility of experience* in general are at the same time [necessary] conditions of the *possibility of the objects of experience*, and on this account have objective validity in a synthetic judgment *a priori.*
>
> (A158/B197; emphasis in original)

While this may show that Kant does employ strong transcendental arguments with a counterfactual form to establish transcendentally idealistic metaphysics, it is worth emphasizing that nothing here constitutes a proof for transcendental idealism or for strong transcendental arguments. Transcendental idealism is just a way to extend transcendental arguments to objects (as appearances, not as things in themselves). Otherwise, strong transcendental arguments fall prey to the plethora of criticism levelled at their objective validity. Thus, transcendental argumentation also cannot be employed to prove transcendental idealism (*contra* e.g., Hookway 1999, 173). Without a theory that cognitively necessary worlds are also metaphysically necessary—i.e., that the necessary features of our cognitive capacities do not pertain only to thought of objects but also to the objects themselves—strong transcendental arguments would not get off the ground. They presuppose rather than prove idealism: there are no sound transcendental arguments *for* idealism, only *from* it. Whether idealism is acceptable in any form, and hence whether this result constitutes an endorsement or a *reductio ad absurdum* of strong transcendental arguments, is a question for a future article.

VI. Conclusion

I have shown that transcendental arguments have a unique logical form for establishing a necessary connection between the presupposition and the conclusion. The formalization via counterfactuals captures the argumentative direction of transcendental arguments and distinguishes them formally from other necessity-wielding arguments. This confirms and clarifies the connection between transcendental arguments and cognitive capacities: (only) within the confines of cognitive accessibility can they be simultaneously sound, non-superfluous, and have more justificatory force than mere empirical induction. This furthermore highlights their reliance on idealism to support their strong, metaphysical use—or, if idealism is rejected, forces us to remain satisfied with moderate, subjective transcendental arguments. Since Kantian transcendental idealism makes metaphysically accessible worlds a subset of cognitively accessible worlds, strong transcendental arguments constitute a valid method in transcendentally idealistic metaphysics.

Acknowledgement: I am grateful to Bianca Ancillotti, Robert Hanna, Frode Kjosavik, Paul Mucichescu, Sami Pihlström, Henrik Rydenfelt, Camilla Serck-Hanssen, and the audiences in Helsinki and Luxembourg for helpful comments and acute critique on previous versions of this article. I also wish to express my gratitude to the Centre for Advanced Study in Oslo for facilitating my work on it in 2015–2016.

Notes

1. Robert Stern recognizes three types of modest transcendental arguments: belief-, experience-, and concept-directed (Stern 2004, 10–11). One can, e.g., use an accepted belief to establish another belief as its necessary condition.
2. These formalizations are often implicit. For explicit instances, see e.g. Bhaskar (1978, 257), Bird (1999, 43–44), Forster (2008, 41), and Walker (1978, 10).
3. One might make it explicit that *q* is a necessary condition of the *possibility* of *p*: 'necessarily, if p is possible, then *q*.' (E.g. Bird 1999, 43–44.) As will be made clear, these tweaks do not deviate sufficiently from the standard formalization to warrant separate analysis, however. The same goes for formulations that derive the necessity rather than the truth of *q*.
4. Despite being clearly visible in numerous examples, the counterfactual nature of transcendental arguments has received relatively little attention. Timothy Williamson comes close by logically equating "metaphysical modal thinking" with "a special case of counterfactual thinking" (2007, 158), yet he does not connect it to transcendental arguments. Similarly, Kenneth Westphal, who emphasizes Kant's transcendental method of identifying "some of our cognitive capacities by using wildly counterfactual thought experiments" (2004, 3), leaves the relationship between counterfactuals and transcendental arguments undeveloped.
5. For example, the Private Language Argument states that were there no other language users, I would not have language (which I, however, do). Similarly, Kant argues in his Refutation of Idealism (B274–279) that were there no outer experience (of mind-external objects), inner experience (of mental states) would not be possible—refuting (skeptical) idealism. Kant's claim that if objects did not follow causal laws, then we could not order our experiences temporally (A189/B232ff.), has also been construed as a transcendental argument. Strawson (1963, 95ff.), in turn, argues that if one did not distinguish other minds, one could not conceive of one's own.
6. I will not here concern myself with recent debates about the correctness of Lewis's view, as it would take us too far afield without having a material impact on my view.
7. The set may consist of one world, of several equally similar worlds, or be empty.

8. For simplicity, the number of possible worlds is assumed to be finite. The infinite version is easy to develop, although the problems of induction (see ahead) will have to be considered.
9. Obviously, one does not slavishly go through the series but seeks to cover vast spheres of worlds by focusing on relevant differences—like the scientific method employs hypotheses and well-designed experiments to avoid blind inductive repetition. (See induction, following.)
10. Although Williamson (2007, 157 ff.) uses counterfactuals rather than material implications to ground metaphysical necessity, he recognises neither an essential difference between counterfactual contrapositives nor between counterfactuals and necessary counterfactuals.
11. Arguably, this is exactly what Aristotle does when he grounds the necessity of the law of non-contradiction on an argument often construed as transcendental (*Metaph.*, 1005b35–1006a28). Similarly, Kant uses transcendental arguments to (synthetically) prove rather than (analytically) presuppose the necessity of metaphysical truths (Kannisto 2012, 239–262).
12. However, see Williamson (2007) for a view according to which counterfactual thought experiments have a deductive rather than inductive form. (Cf. induction following.)
13. This is why mathematical induction is actually a species of deduction: the immutability of mathematical principles allows us to infer from n cases to all cases.
14. See Williamson (2007, 163–165) for a discussion and defence of imaginability as a valid (or at least reliable) test for possibility.
15. Counterfactuals can be tested empirically in the sense of taking something (e.g. a metal) to another context (e.g. from heat to cold) to confirm e.g. the counterfactual 'had this metal not been heated, it would not have expanded.' The 'essential counterfactuality' of transcendental arguments excludes this: the consequent of '$\neg q \;\Box\!\!\rightarrow \neg p$' cannot by definition actually occur, since p is accepted as necessary, so no experiment can confirm the counterfactual (see Williamson 2007, 155–169).
16. One could object that it is still a kind of empirical experiment whether one can represent certain ways the world could have been—just a psychological one about one's own capabilities. In other words, one would not 'experiment' on possibility but on our actual psychological or cognitive capacity to conceive of possibility. Such an objection—which is here acknowledged but set aside for brevity—can be found in Williamson (2007, 165–169, 179 ff.).
17. The analytic/synthetic and a priori/a posteriori distinctions provide more rigour. Empirical induction is a posteriori synthetic and mathematical induction (as deduction) is a priori analytic (based on axiomatic definitions of terms and operations). How to classify transcendental arguments? Although the result of the (thought) experiment is synthetic, since it adds to our knowledge and is not based on pre-determined definitions, it is arguably a priori in the (Kantian) sense of not being grounded on actual but merely possible states of affairs (that are impossible to experience). (Or, given the Williamsonian objection, it constitutes at least a special kind of a posteriori experiment that is neither standard a posteriori nor a priori.) This harmonizes well with Kant's method of grounding synthetic a priori propositions not on experience but on the (conceived) possibility of experience itself (A157–158/B196–197, A111; see section V.) However, to clarify and analyze—not to mention defend—this suggestion would be a task for another chapter. (See also Kannisto 2012, 232–262.)
18. It is interesting, then, to see that in his response to Hume's causal skepticism, Kant exactly grants that if "appearances were things in themselves," i.e. if transcendental idealism were not presupposed, "then no human being would

be able to assess from the succession of representations how the manifold is combined in the object" (A189/B235). As Kant's subsequent argument is a prime example of a transcendental proof, he would hardly need to make such a concession if he thought that transcendental proofs were valid within a transcendentally realistic framework.

19. Kant recognizes the importance of the direction of the counterfactual: inverting it would lead to a failure of the transcendental proof (see Kannisto 2012, 256–257). That Kant actually argues for the principle of causality using appropriate counterfactuals is shown in Kannisto (2017, 503–507).
20. On the notion of deduction, see Baum (1986), Henrich (1989), and Leppäkoski (1993).
21. As $\Box(p \rightarrow q)$ and $\Box(p \mathrel{\Box\!\!\rightarrow} q)$ are both vacuously true, hence trivially equivalent, if p is false in W, we assume a non-empty S_w for the remainder of the proof.
22. Since $S_w \subseteq W$, it suffices to consider W only for the universal quantifier.

References

Aristotle. *Metaphysics*. In *The Complete Works of Aristotle*. 2 vols. Edited by J. Barnes. Princeton: Princeton University Press, 1984. (Cited as *Metaph.*)

Baum, Manfred. *Deduktion und Beweis in Kants Transzendentalphilosophie*. Königstein: Hain Verlag, 1986.

Bhaskar, Roy. *A Realist Theory of Science*. Sussex: The Harvester Press, 1978 [1975].

Bird, Graham. "Kant and the Problem of Induction." In R. Stern, ed., *Transcendental Arguments—Problems and Prospects*. Oxford: Clarendon Press, 1999, 31–46.

Forster, Michael N. *Kant and Skepticism*. Princeton: Princeton University Press, 2008.

Kannisto, Toni. *From Thinking to Being—Kant's Modal Critique of Metaphysics*. PhD-thesis, University of Oslo, 2012.

Kannisto, Toni. "Kant on the Necessity of Causal Relations." *Kant-Studien* **108** (2017), 495–516.

Kant, Immanuel. *Critique of Pure Reason*. Translated and edited by Paul Guyer and Allen Wood. Cambridge: Cambridge University Press, 1998. (Cited in standard A/B edition pagination, as 'A' and/or 'B,' with page number following.)

Henrich, Dieter. "Kant's Notion of a Deduction and the Methodological Background of the First *Critique*." In E. Förster, ed., *Kant's Transcendental Deductions*. Stanford: Stanford University Press, 1989, 29–46.

Hintikka, Jaakko. "Transcendental Arguments: Spurious and Genuine." *Noûs* **6** (1972), 274–281.

Hookway, Christopher. "Modest Transcendental Arguments and Sceptical Doubts: A Reply to Stroud." In R. Stern, ed., *Transcendental Arguments—Problems and Prospects*. Oxford: Clarendon Press, 1999, 173–188.

Lewis, David. *Counterfactuals*. Oxford: Blackwell, 1973.

Lewis, David. *On the Plurality of Worlds*. Oxford: Blackwell, 1986.

Leppäkoski, Markku. *The Transcendental How—Kant's Transcendental Deduction of Objective Cognition*. Stockholm: Almqvist & Wiksell International, 1993.

Stern, Robert. *Transcendental Arguments and Scepticism—Answering the Question of Justification*. Oxford: Clarendon Press, 2004.

Stern, Robert. "Transcendental Arguments." *The Stanford Encyclopedia of Philosophy* (Spring 2011 Edition). Edward N. Zalta. <http://plato.stanford.edu/archives/sum2013/entries/transcendental-arguments/>

Strawson, Peter F. *Individuals: An Essay in Descriptive Metaphysics*. Garden City, NY: Doubleday, 1963.

Strawson, Peter F. *Skepticism and Naturalism: Some Varieties*. London: Methuen, 1985.

Stroud, Barry. "Transcendental Arguments." *The Journal of Philosophy* 65 (1968), 241–256.

Stroud, Barry. "Kantian Arguments, Conceptual Capacities, and Invulnerability." In B. Stroud, ed., *Understanding Human Knowledge: Philosophical Essays*. Oxford: Oxford University Press, 2000 [1994].

Walker, Ralph C. S. *Kant: The Arguments of Philosophers*. London: Methuen, 1978.

Westphal, Kenneth R. *Kant's Transcendental Proof of Realism*. Cambridge: Cambridge University Press, 2004.

Williamson, Timothy. *The Philosophy of Philosophy*. Oxford: Blackwell, 2007.

Wittgenstein, Ludwig. *Philosophical Investigations*. Translated by G.E.M. Anscombe. Oxford: Blackwell, 1953.

Appendix

Proof that $\Box(p \rightarrow q) \Leftrightarrow \Box(p \;\Box\!\!\rightarrow q)$.

Definitions:

(D1) Let W be the set of all possible worlds, and

(D2) $S_w \subseteq W$ a sphere of worlds containing at least one p-world, ordered according to similarity to a given world $w^* \in W$ and centred on it.

(D3) $\Box p =_{\mathrm{df}} \forall w \in W: p$

(D4) $p \;\Box\!\!\rightarrow q =_{\mathrm{df}} \exists S_w: \forall w \in S_w: \neg(p \wedge \neg q)$[21]

Proof that '$\Rightarrow$':

(1)	$\Box(p \rightarrow q) \rightarrow \Box(p \;\Box\!\!\rightarrow q)$	*Premise*
(2)	$\neg\Box(p \rightarrow q) \vee \Box(p \;\Box\!\!\rightarrow q)$	(1), *conditional*
(3)	$\neg(\forall w \in W: p \rightarrow q) \vee$ $(\forall w \in W: p \;\Box\!\!\rightarrow q)$	(2), (D3)
(4)	$\neg\forall w \in W: \neg(p \wedge \neg q) \vee$ $\forall w \in W: \exists S_w: \forall w \in S_w: \neg(p \wedge \neg q)$	(3), (D4), *conditional*
(5)	$\neg\forall w \in W: \neg(p \wedge \neg q) \vee$ $\forall w \in W: \neg(p \wedge \neg q)$	(4), (D2), *resolution*[22]

Since (5) is a tautology (Law of Excluded Middle), '$\Rightarrow$' is proven.

Proof that '$\Leftarrow$':

(6)	$\Box(p \;\Box\!\!\rightarrow q) \rightarrow \Box(p \rightarrow q)$	*Premise*
(7)	$\neg(\neg\exists w \in W: \forall S_w: \exists w \in S_w: (p \wedge \neg q)) \vee$ $\forall w \in W: \neg(p \wedge \neg q)$	Repeating steps (2)—(4)
(8)	$\neg\forall w \in W: \neg(p \wedge \neg q) \vee$ $\forall w \in W: \neg(p \wedge \neg q)$	(7), *distribution*, *resolution*

Since (8) is a tautology (LEM), '$\Leftarrow$' is proven.
QED.

9 Phenomenology as Constitutive Realism

David Woodruff Smith

I. Introduction: Reality in Relation to Our Consciousness

Having launched his conception of phenomenology in the *Logical Investigations* (1900–01), in *Ideas* I (1913) Edmund Husserl laid out his "transcendental" formulation of phenomenology, his "new science," the "science of consciousness" (§33). In this science, we would study consciousness by the method of *epoché*, or "transcendental reduction," where we "bracket" the question of the actual existence of the world around us, turning our focus to "pure" or "transcendental" consciousness (§§27ff.). Thereby we study the domain of our experience as a "new region of being": consciousness, wherein the essence of an experience features properties distinctive of "acts" of perception, judgment, will, etc. (§§33ff). We find that such experience is characteristically a "consciousness *of* something": that is, each act of consciousness is "intentional," a consciousness of some object presented in some way, say, where I see "this blooming tree."

Husserl's doctrine about the relation between pure consciousness and our surrounding world is customarily called "transcendental idealism": a neo-Kantian term Husserl deployed prominently in his later *Cartesian Meditations* (1929), though he abandoned the term "idealism" altogether in his late writings gathered in the *Crisis* (1935–38). Transcendental idealism is the trickiest part of Husserl's conception of phenomenology. The heart of Husserl's phenomenology is his theory of intentionality, the trickiest part of which is his theory of *noema*. Here we shall explore in close detail how Husserl characterizes the noema of an act of consciousness, in key passages in *Ideas* I. In Husserl's phenomenological theory of intentionality, featuring the "noematic" content of experience, we can see a particular metametaphysics (in today's current idiom).

In metaphysics we, being conscious subjects, analyze fundamental structures of the *world* around us: tautologically, the world in which we come to experience and know things around us. In phenomenology we, reflecting subjects, analyze the fundamental structure of our *consciousness* of things in the world around us. Our deepening analysis of the *relation* between our phenomenology and our metaphysics defines a *phenomenological metametaphysics.*

Briefly, it is only in our intentional experiences that we posit the existence of things and their natures, including our consciousness in and of the world. In short, we are both the medium and in part the message of our account of the world. This *metatheoretical* point is the point of intersection between phenomenology and ontology or metaphysics. And therein lies a Husserlian approach to metametaphysics. Though Husserl called his phenomenological doctrine "transcendental idealism," a better name is "constitutive realism," as Husserl holds that in consciousness "real" things are "constituted" as real and existing. What "constitution" amounts to is central to Husserl's closely formulated notion of noema, and so to the doctrine of "transcendental idealism" in Husserl's philosophy.

As we unfold the details of Husserl's doctrine, we shall find: An *act of consciousness*, a phenomenal intentional experience (such as perception), includes a component called *noesis*, a dependent part of the experience (occurring in "inner" time). Thanks to its noesis, the experience *entertains a noema*, an ideal form of meaning (*Sinn*). The noema is not the object of consciousness, but that *by virtue of which* the act is *intentional*, i.e., a consciousness *as of* some object.

We appraise this intentional structure in *phenomenological reflection.* At that level of abstraction, we hold that our consciousness-of-things is consciousness of things in an *existent real world*, and that very consciousness is *grounded* in—and so both facilitated and constrained by—ideal noematic meaning. "Real" things are "constituted" insofar as they are or may be experienced through a complex of such noematic meanings. Moreover, many if not all of the forms of meaning we live through in everyday life, on which we reflect in phenomenological analysis, have a *genesis* in prior collective or intersubjective activities of consciousness, including our forerunners' judgments in philosophical and scientific theorizing.

Here, then, is a "transcendental" phenomenological ontology that defines *conditions of the possibility* of *meaningful* metaphysics: a distinctively Husserlian metametaphysics. As our study unfolds, we begin with an overview of transcendental idealism reconfigured as *constitutive realism* (section II). Then we present the fundamentals of Husserl's theory of intentionality, epoché, noema, and constitution (section III). We then dig into the details of Husserl's nuanced doctrine of noema (section IV). The constitution of everyday objects we then frame in terms of a structure of noemata, which are themselves constituted only through phenomenological reflection (section V). Finally, we draw upon Husserl's metaphor of a "zigzag" movement between world and phenomenology, between the "real" world and its "constitution" in consciousness (section VI). As we shall see, Husserl's *metatheory*—a philosophical metalogic—governs this phenomenological metametaphysics.

(Note: Analysis of Husserl's concept of noema divides two prominent interpretations of Husserl's phenomenology: the so-called West Coast and East Coast interpretations. Here we detail a West Coast "semantic" formulation of Husserl's theory of intentionality via noema (qua *Sinn*) and its implications for transcendental idealism. Zahavi 2017 develops an interpretation of Husserl's transcendental idealism based in an East Coast reading of Husserl's notion of noema (qua "object-as-transcendentally-given," with a Kantian flavor); cf. Smith 2019b, a review of Zahavi's interpretation. Cf. Smith 2007/2013 for a recent study of Husserl's systematic philosophy, emphasizing Husserl's metatheory, drawing on the West Coast view. That view began with lectures by Dagfinn Føllesdal at Harvard and Stanford, relating Husserl to both Bolzano and Frege. Cf. Føllesdal 1969/1982; Dreyfus and Hall 1982. Cf. Smith and McIntyre 1982 on connections of Husserl's theory of intentionality with semantic theories in Frege, Carnap, Hintikka. Cf. Føllesdal (2019) for a recent account of the West Coast view of noema. Cf. Yoshimi et al. (2019) on the history of California Phenomenology. The East Coast view of noema, developed by Robert Sokolowski and John Drummond, takes the noema to be the object *as given in transcendental reflection* i.e. where bracketing its actual existence. Cf. Drummond (2015) for a recent discussion of his interpretation.)

II. Metametaphysics in Husserl's "transcendental idealism": A Form of "constitutive realism"

Dagfinn Føllesdal has argued that Husserl's doctrine of "transcendental idealism" is properly seen as a form of realism, albeit a realism grounded in phenomenological critique of "reality." As Føllesdal has posed the problem:

> A central issue in metaphysics is the opposition between realism and idealism. The great metaphysicians, such as Kant, often professed to be idealists and regarded realism as an untenable position. In 1906, Husserl declared himself an idealist, but in 1934 he wrote, "Nobody could be more of a realist than I." [We should see] that Husserl's basic position remained the same all these years, but that he came to appreciate that what he had called "idealism" was just a carefully thought through realism, where one has reflected on what is meant by "existence" and "reality". This reflection brings out the core features of phenomenology, as "idealist realism" or, to use Richard Tieszen's apt phrase, "constitutive" realism. Notably, there are ideas in this direction in James' *Principles of Psychology*, which Husserl studied in 1894.

(The passage is from a brief note leading to the present chapter.)

Everyday realism holds that things in our surrounding world really exist, and exist for the most part independently of our experience. By contrast, subjective idealism à la Berkeley holds that everyday things—bodies including tables, trees, and human bodies such as mine and yours—are in fact just bundled ideas in minds. A variant form of idealism holds instead that such everyday things are dependent for their existence as tables, trees, etc., on our ideas of them. A "transcendental" form of that dependence lies in the way our very concepts of things around us in space and time frame their existence, concepts without which we could not even consider whether such things exist. Accordingly, transcendental idealism à la Kant holds that our cognition of things in space and time cannot, as it were, get off the ground except as our own (human) forms of "intuition" are "transcendentally" shaped by a system of conceptual categories of the understanding. On that tack, we cannot fully conceive of spatiotemporal things without this conceptual apparatus. And so, arguably, the question of the external existence of such *things in themselves* does not fully *make sense* for us—once we take the Kantian transcendental turn, so it seems, along Husserl's lines.

Such, in a gloss, is the philosophical backstory into which Husserl enters. However, Husserl's doctrine of "transcendental idealism" is its own thing. Husserl consistently rejected all "subjective" forms of "idealism," whereby things around us (trees, rocks, rivers, animals, etc.) are held to be either bundles of ideas in minds, or things created by minds, or things dependent for their existence on mental activities. Husserl regularly rejected the Kantian notion of *Ding an sich*, yet he sometimes adopted the Kantian label "transcendental idealism." What then shall we make of Husserl's doctrine? What did Husserl mean by "transcendental," and what did he mean by "idealism," when he spoke of phenomenology as developing a form of "transcendental idealism?"

A better name for Husserl's own doctrine is arguably "constitutive realism," as Føllesdal has suggested, drawing on Richard Tieszen's "constituted Platonism" for ideal mathematical entities. According to Husserl's phenomenological theory of intentionality: certain *forms of meaning* called "noemata" are as it were the conduit of our consciousness, the "transcendental" *conditions of the possibility* for our conscious intentional experience of things in the world around us—and ultimately for our understanding of the world in contemporary mathematical sciences of nature and even perhaps of phenomenal consciousness correlated with neural activities (according to "neurophenomenology," in Francisco Varela's term). Of course, we find things like trees and horses and automobiles as "real" things in our surrounding world: "realism" so far. Yet we apprehend such things only in our conscious intentional experiences, and only through the *sense* things have for us in experience: here is the "transcendental" crux in Husserl's doctrine.

In Husserl's transcendental phenomenology, detailed in *Ideas* I (1913), things in the world around us are "constituted" in our consciousness insofar as they are "intended" by virtue of appropriate forms of *meaning*, called "noematic content" or "noema," in a duly logical or semantic "manifold" of meaning or sense (*Sinn*). These structures of meaning form the *phenomenal intentional content* of our everyday experience of things in our surrounding world. To say a tree is "constituted" in our experience is not to say the tree itself is constructed in our consciousness. Rather, for Husserl, noematic meaning facilitates and is "constitutive" of our experience in, say, seeing the tree: without meaning, we do not and cannot *experience* the tree, this "real" thing before us. Within that structure of meaning, the sense <this tree> is linked with a "horizon" of further "motivated" forms of sense which present further aspects of the given tree. The "constitution" of the tree, in Husserl's phenomenological analysis, lies in this complex structure or *field* of meaning regarding the "intended" tree. (See Smith and McIntyre 1982 on Sinn and horizon. And see Walsh 2017 on ramifications of Husserl's theory of motivation, horizon, and phenomenality in intentional experience.)

Husserl characterizes noemata as a type of meaning, or *sense*, i.e., *Sinn*. We may say noemata are *ideal* in that they are shareable by different concrete experiences, much as universals are sharable by different particulars. As the ideal property Tree is shared by different individual trees, the ideal sense <tree> is similarly shareable by different experiences presenting trees. For Husserl, such ideal entities contrast with "real" entities in that ideal entities are not themselves located in space-time like "real" trees or even elements in the temporal flow of "real" experiences. Accordingly, Husserl speaks of the "ideal existence" of *sense* and, apropos of logic, "judgment-content," i.e., *propositions* in today's parlance. (Cf. Husserl 1929/1969, §89.)

Husserl speaks of "ideas in the Kantian sense," meaning regulative ideals. These "ideas" are not mental entities or episodes, as in Berkeleyan "idealism." Still, Husserl is not primarily following Kant, and the "ideals" Husserl posits are forms of *Sinn* adapted from a broadly Platonistic logical tradition, notably including Bernard Bolzano and Gottlob Frege. However, for Husserl, these ideal forms of meaning inhabit our *lived experiences*; they do not reside in a Platonic heaven. As noted, Husserl calls the relevant forms of meaning "noemata," and Husserl contrasts his phenomenological system with both Berkeleyan and Kantian "idealisms." (The title of Husserl's *Ideen* uses "*Idee*" in his own way for regulative ideals of phenomenology itself.)

When we turn to Husserl's texts, we need to bear in mind Husserl's specialized use of certain terms, not always apparent in English translations. A "thing" (*Ding*) is an object in the "real" (*reale*), spatiotemporal, world of nature. An experience (*Erlebnis*) is a "*reelle*," temporal, "act" (= activity) in the stream of consciousness. A noema is "correlated" with an intentional experience, borne "in" the experience. However, a noema is "not *reelle*," not temporal, not occurring in "inner time" in consciousness. Also, a noema is "*irreale*," not "*reale*," not occurring in nature, in space-time. Rather, for Husserl, a noema is a type of ideal meaning, or *Sinn*, with its own structure.

Today such meanings would be called *abstract* entities. We shall call them *ideal* entities, not occurring in time or in space-time, yet guiding the intentionality of an experience that flows off in "inner" time. Husserl calls dependent parts "abstract parts," parts (= "moments") abstracted in theory from the wholes of which they are parts. But that is not what philosophers today mean by "abstract." As noted, Husserl spoke of the "ideal existence" of sense, *Sinn*, and we shall speak accordingly of *ideal* meaning, or sense.

At a certain level of abstraction, certain forms of meaning—propositions (*Sätze*)—form the content of our intersubjectively developed "theories" of things in the world, including our increasingly mathematized theories in physics and beyond. What Husserl called "theory of theories"—"pure logic," i.e., metalogic, or metatheory—structures a metametaphysics of the world *in relation to* our consciousness of the world. The principles of this meta-theory define a phenomenological critique of "theory" itself, and so of ultimate metaphysics, i.e., the theory of reality. Husserl's conception of metatheory—the "theory of theories," ideally a "*mathesis universalis*"—shapes the *Logical Investigations* (1900–01) and takes on a "transcendental" slant in *Formal and Transcendental Logic* (1929).

The foundation of such a metametaphysics—in a metatheoretical critique of the metaphysics or ontology of mind-and-world—is Husserl's phenomenological *theory of intentionality*. A linchpin of that theory is Husserl's theory of the "content" of consciousness, which he called "noematic content," or simply "noema"—from the ancient Greek perhaps akin to what we have "in mind," what is "minded" or, in Husserl's idiom, what is "intended" in consciousness.

There is a telling conception of noema that we can use here to open the door to the metametaphysical issues before us. As Føllesdal has long argued in lectures and writings, Husserl's development of phenomenology and the phenomenological theory of intentionality grew in part out of logical theory in the nineteenth century, notably featuring work of Bernard Bolzano and later Gottlob Frege, with the Platonistic logic of Hermann Lotze also in the background. In particular, Husserl explicitly emphasized the *noema* of an act of consciousness as a type of *meaning*, or *Sinn*. And Husserl's invocation of Sinn pointed to logical semantics in the tradition of Bolzano and Frege—as opposed to the more purely epistemic tradition of Descartes and Hume concerning appearance-and-reality. (Note: The West Coast reading of Husserl has been called "Fregean," but this label is misleading. It was not only Frege, but importantly Bolzano, and other Platonistic logical theorists, who were influential for

Husserl's "anti-psychologistic" conception of phenomenology, starting in the *Logical Investigations* of 1900–01. See Føllesdal (2019) on Bolzano's and Brentano's influence on Husserl. Cf. Beyer (1996) on Bolzano's influence on Husserl's phenomenological theory of meaning.)

This logical or *semantic* aspect of Husserl's phenomenology moves beyond Kant's "transcendental idealism," in so far as Husserl's own conception of a "transcendental" phenomenology is based in logical notions akin to those of Bolzano and Frege and moreover to ideas in mathematical logic developing among Husserl's contemporaries and colleagues—including David Hilbert, Hermann Weyl, Georg Cantor, and Rudolf Carnap, with Alfred Tarski and Kurt Gödel soon to develop now-famous results.

As we explore the links between Husserl's transcendental phenomenology and certain logical or semantical notions, then, we can draw out a unique form of metametaphysics. Occasionally Husserl spoke of a "zigzag" (*Zickzack*) method in understanding as we move back and forth between different constellations of meaning, e.g., in earlier and later formulations of a theory such as geometry—which characterizes the spatiotemporal form of things in our natural surroundings. Indeed, in Husserl's method of "bracketing," or epoché, we move back and forth (or up and down in abstraction) between *things* in the world and our *sense* (*Sinn*) of them. Thus, in the practice of phenomenological reflection, we move between two forms or levels of consciousness: (1) as we experience everyday *things* in perception, and (2) as we turn in "transcendental" reflection toward our *perceptual experience* as of things in our surrounding world. Reality and consciousness meet as we crisscross them in the structure of intentionality.

(Note: In the *Crisis*, §9, Husserl speaks of a "zigzag" movement between our everyday geometry and our "mathematized" geometry, noting also Riemannian geometry as used in Einstein's relativity theory: see §9 on Galileo's mathematization of nature, and the subsection lettered "l" on method. Similarly, in *Ideas* I, §65, Husserl speaks of phenomenology's "relating back" (*Rückbeziehung*) to itself. Thus, the "new science" of phenomenology relates back to everyday experience and to empirical "psychology": this backward movement is practiced *within* the methodology of epoché that also leads forward to "pure" phenomenology. Similarly, the image of feedback loops guides the methodology of "reflective equilibrium" in axiomatic theory and also in ethical and political theory, as Dagfinn Føllesdal has observed in lecture. Cf. Føllesdal 2005.)

Thus, at work in Husserl's *phenomenological* meta-theory we may see a particular style of metametaphysics, linking *reality* and our experienced *sense of reality*. Here is a novel view discernible in the details of Husserl's theory of intentionality, as mapped out in the following. We may well call Husserl's theory the "semantic" conception of intentionality, drawing on the logical conception of semantics that developed largely after Husserl's lifetime. In particular, Alfred Tarski's "semantic" conception of truth, developed in the 1930s and beyond, can be seen as a "mathematization" of a Husserlian "semantic" conception of intentionality.

(Note: The "semantic conception of truth," as presented in Tarski 1944, can be seen as enjoining a similar back-and-forth between an asserted sentence 'p' and its articulated truth-conditions: "'p' is true if and only if it is the case that p." Cf. Smith 2016a, 2016b on Husserl's method of epoché vis-à-vis Tarski's semantic conception of truth.)

By way of terminology, Husserl developed a complex system of "formal ontology" shaping "material ontologies" of nature, consciousness, and social reality. The term "metaphysics" Husserl reserved, idiosyncratically, for the particular factual existence of concrete things (according to forms of "naturalism," "subjective idealism," and "historicism" that Husserl resisted). Today the terms "metaphysics" and "ontology" are often used interchangeably, and this should be born in mind as we proceed in the present concerns of "metametaphysics." (Husserl's system of ontology, featuring "formal" and "material" categories, is reconstructed in Smith 2007/2013.)

III. Intentionality, Epoché, Noema, Constitution

The heart of Husserl's phenomenology is his detailed theory of intentionality. Following the preceding lead, we should begin to see a metametaphysical focus in Husserl's *logical* conception of meaning as realized in *consciousness*. Briefly, our experience and understanding of the world around us is *dependent* on ideal forms of meaning realized in our lived experiences of the things around us. The *formal structure* of intentionality, for Husserl, is schematized as a four-element structure:

I—act—content —> object.

As Føllesdal has long argued, Husserl drew inspiration first from Bolzano's conception of logic as *Wissenshaftslehre*, yielding Husserl's notion of "theory of theories," and further from Frege's conception of logical semantics as featuring the role of Sinn in reference (*bedeuten*) and truth. Husserl carried this semantic conception of "content" into the theory of consciousness, in a way that was less salient in nineteenth-century logic, and that perspective was key to Husserl's "anti-psychologism" as he sought to reform Brentano's initial conception of "descriptive psychology" as "phenomenology."

Husserl's long train of theory in the *Logical Investigations* (1900–01) digs into the *acts of consciousness* that underlie phenomena of meaning, reference, and truth in logical semantic theory. Husserl appraises the fundamental structure of intentionality in consciousness as a relationship among act (by a subject "I"), content ("*reelle*" and "intentional"), and object (if the act has an existing object). Notably, Husserl's conception of "content" is shaped, from the beginning, by an *ontology* of meaning already taking shape in logical theory—albeit in the wake of Kantian, Humean, and Cartesian epistemological theory, and indeed Brentano's conception of the "phenomena" of consciousness. Thus, Husserl's concern was the role of *ideal meaning* in *lived experience*. By Husserl's lights, we need a distinctive "science" of consciousness: phenomenology! It is not enough to focus on sensory "phenomenology" as what it is like to experience varieties of sensory consciousness (a current theme bearing Humean motivations). For Husserl, we need to focus also on the lived character of experience bearing meaning (*Sinn*). We move then beyond predicative "concepts" expressible in language (such as "tree") to include as well basic "demonstrative" forms of *perceptual meaning*, which are expressible (with limits) by "occasional" demonstrative expressions (such as "this"), and which capture the subject's spatiotemporal connection with things in the surrounding world.

As frequently remarked, around 1906 Husserl began to fold into his conception of phenomenology a neo-Kantian conception of "transcendental" philosophy. The transcendental idiom appeared in his 1907 lectures called *The Idea of Phenomenology*, introducing his method of "transcendental" or "phenomenological" "reduction": not an ontological reduction of world to consciousness, but a "reduction" of focus to our *consciousness* as of things in the world around us. This technique of *epoché*—"bracketing" the question of the actual existence of our *Umwelt*—was to refocus our concern upon the phenomenological structure of our experience. We are meant to experience our acts of consciousness in a new way, reflectively, with a guided focus on their *meaning*.

And so, by 1913 Husserl ramified his evolving fundamental theory of intentionality, in *Ideas* I, through a focus on "noematic content," or "noema," revealed in reflection by the method of epoché. Moreover, in *Ideas* II (written in 1912 along with *Ideas* I) Husserl analyzed the role of our sense of *embodiment* and of *intersubjectivity* in our experience of things around us, the things of our *Umwelt*, our surrounding world, or *Lebenswelt* (as Husserl would later say). The noematic content of our everyday experiences, on Husserl's closer analyses in *Ideas* II (and *Ideas* III), will bear familiar forms of meaning that implicate our embedment in our living bodies and in our social relationships—with rich implications for the metametaphysics we are seeking in Husserl's phenomenology.

On Husserl's analysis, it is by virtue of the noema entertained or realized in an act of consciousness that an object is presented or "intended" in a *meaningful* ("*sinnvolle*") way. And thereby is the object "constituted" in that experience, together with a "horizon" of further potential acts of consciousness presenting the object in other "motivated" ways. The "constitution" of the object in consciousness is thus defined by the relevant manifold of noematic meanings ideally realizable in consciousness. However, as Husserl insisted, this is not to say that the object itself is *caused* to exist, or to *be* what it is, by the ways it is "intended." Rather, to say an object is "constituted" in a certain way in consciousness is simply to say it is correlated with a meaningful pattern of noematic meaning, or *sense*, actually or potentially realized in our varied acts of consciousness. The relevant pattern of *noematic content* is structured as a "horizon" of meaning correlated with "motivated" forms of further possible experience: as where, in Husserl's guiding example, I see a tree from one side and my very *sense* of the tree indicates its back side, to which I might turn in further possible perception. (See Walsh 2017 on the crucial role of motivation in the "horizon" structure of phenomenal intentional experience.)

By invoking the ancient skeptics' idiom of "epoché," Husserl echoed the themes of radical skepticism that launched Descartes' dualism, subsequently followed by Hume's skeptical epistemology and then by Kant's "transcendental idealism." However, as Husserl emphasized, we do not *doubt* the external world when we bracket the question of its factual existence. Hence, Husserl's theory of intentionality is misleadingly allied to the problem of "mere appearance" in the lineage of Descartes, Hume, Kant, and even Brentano. For the "pure phenomena" that Husserl studies are precisely the *phenomenal-cum-logical* structures that Husserl called "noemata," which enter into our lived intentional experiences.

To be clear, in the *Logical Investigations* and again in *Ideas* I, Husserl developed a detailed phenomenological foundation for epistemology. Yet Husserl's primary concern there is not simply a response to radical skepticism or to subjective idealism, but rather the fundamental role of meaning, or *Sinn*, in all intentional activities, including those of perception and judgment bearing intuitive evidence supporting knowledge.

Accordingly, Husserl's "transcendental idealism" must be understood in the special context of Husserl's integration of logical theory with the theory of consciousness. Husserl's new "science" of phenomenology then takes on a metatheoretical character of its own. And the crux of that meta-theory is precisely what Husserl termed "noematic content," or "noema," as elaborated in *Ideas* I.

We should note that in *Ideas* I (1913) Husserl speaks of "transcendental consciousness" and "transcendental reduction" practiced by epoché, but he does not there speak of "transcendental idealism" by name. The term "transcendental idealism" Husserl employed in his later works, *Formal and Transcendental Logic* (1929/1969) and *Cartesian Meditations* (1929/1960), but soon thereafter he abandoned the term "idealism." Our concern is the fundamental theory running through Husserl's writing, even as he struggled with the term "idealism" in the milieu of ideas surrounding him.

We shall map out crucial details from Husserl's own prose in *Ideas* I in what follows, as Husserl lays out the role of noema in the intentionality of an experience of perception. Before turning to the texts themselves, however, let us look to the phenomena themselves—*Zu den Sachen selbst!*, as Husserl liked to say.

It is easy to lose the forest for the trees in Husserl's ontology of noema, so let us focus on a simple case of perception and its experiential content.

Consider a case similar to that in Husserl's discussion (in *Ideas* I, §§ 88–89). In my backyard is a huge Podocarpus tree. As the spring rains have encouraged new growth in the tree, I look upward at the tree. And I remark to the gardener:

> "Wow, I see the Podocarpus has a lot of new greenery."

Now, in phenomenological reflection, I observe just how the tree is visually given to me on this occasion. Thus, I say, more formally, in reflection:

> "I see this tall Podocarpus tree with fresh green needle-leaves."

Assuming Husserl's account of the structure of intentionality, we should recognize several distinct formations in the intentional relation between my visual experience and its object.

In living through the experience, I *experience* the noematic content of the experience, which we may articulate as the meaning structure:

> < I now here see this tall Podocarpus tree with fresh green needle-leaves >.

This is not to say that I put up my transcendental periscope and look around in my Cartesian theatre of mind. I do not *observe*, and describe or analyze, that ideal noematic meaning itself when, in the experience, I am so visually presented this tree. Only in reflection do I separately and subsequently turn to the meaning itself, *abstracting* the ideal meaning from the lived experience on which I reflect.

Accordingly, in my new experience of reflection, I judge that my visual experience entertains the noematic Sinn < . . . this tall Podocarpus tree . . . >. In my ontologically-informed judgment, I further hold that this noematic Sinn is instantiated or realized in a "*moment*," or dependent part, of my lived visual experience, a part Husserl calls "noesis," the "sense-giving" component in the experience on which I am now reflecting.

And there is more. In the practice of phenomenological reflection, guided by Husserl's technique of "bracketing," or "epoché," I can move back and forth between *the tree itself*, which I see or just saw, and *my visual experience* of the tree, that is, *as the tree is given* in my original visual experience. This back-and-forth movement, between world and consciousness, is practiced precisely as I distinguish *the tree itself* from *the tree-as-perceived*, and so as I track the *intentional relation* between my experience and the tree.

In epoché, however, the only item of content I bracket is my positing of the actual *existence* of the tree as given in my experience. That item will be a particular element in the noema: part of what Husserl calls the "thetic" or "positing" character in the experience's noema, a character implicit in the act-character <see>.

With this overview in mind, let us trace out Husserl's theory in his own exacting formulation: To the texts themselves.

IV. Husserl's Phenomenological Ontology of Noema: the Fine Details

A distinctively Husserlian approach to the integration of phenomenology and ontology—whence a metametaphysics—takes shape with Husserl's formulation of the notion of noema in the key sections of *Ideas* I.

To appreciate the distinctive form of "transcendental" philosophy in Husserl's hands, we need to dig into the role of ontology in Husserl's account of epoché and intentionality.

Consider a basic intentional experience wherein, say, I see the Podocarpus tree before me in the garden. Here, we may assume, the tree and I and my visual experience all exist as part of the *Umwelt* of my experience. In phenomenological reflection, we then bracket the question of the actual existence of the presumably "real" tree. Thereby we refocus our reflection on the intentional structure of my initial act of visual consciousness. That structure we may depict so:

I—my visual experience—the noema—> [the tree].

Here, in phenomenological reflection, we focus on the noema "in" my experience, whose structure we may articulate as follows:

< I now here see this large Podocarpus tree >.

This noema is an ideal meaning structure, realized in my concrete ("*reelle*") experience as of a "real" tree before me, i.e., regardless of whether the "intended" tree is actually "real" ("*reale*," actually existing in space-time). (The brackets indicate the "suspension" of the assumption of the veridicality of the experience.)

Now, in metametaphysical reflection, we begin our "critique" of my relation to the tree. What are we entitled to say about that relation in light of the structure of intentionality revealed by phenomenological reflection? Such is the story we now undertake, following Husserl's long train of argument or analysis in *Ideas* I. Only in this metatheoretical analysis can we track the "critique" of intentionality that Husserl takes as a "transcendental" analysis: an "*erkenntnistheoretische Kritik*" (critique of theory of cognition), in Husserl's own and Kantianesque formulation (§ 33, following §§ 27–32 elaborating on epoché).

In Husserl's system—surprisingly perhaps—epoché does not avoid ontology as we bracket the question of the *actual* existence of things in our surrounding world. Indeed, the metametaphysics we are pursuing in Husserl's transcendental phenomenology turns on the way ontology and phenomenology are interdependent. (That interdependence is reconstructed in detail in Smith 2007/2013.)

Ideas I opens with a compressed chapter on "fact and essence" (titled "*Tatsache und Wesen*," §§ 1–17). The realm of "fact" includes actual concrete things in spatiotemporal nature as well as actual historical "communities" of "persons" interacting. These concrete entities have "essences" (*Wesen*), or *eidos*. In preparation for his presentation of phenomenology, Husserl outlines a concise ontology of categories, distinguishing "formal" and "material" categories. For Husserl, "formal ontology" features formal essences, or "formal categories," specifically: the formal types of Individual, Property, State of Affairs, Number, etc. By contrast, for Husserl, "material" ontology features "material" essences or types, including the "regions" of Nature, Consciousness, and *Geist* (social formations).

Husserl launches his conception of phenomenology with the method of epoché (in §§ 27ff), where we "bracket" the question of the existence of the surrounding world, or *Umwelt*. Thereby we develop the new "science of consciousness": phenomenology (§ 33).

Yet, as he proceeds, he continues to make use of his detailed ontology of formal categories applied to the material region of Consciousness. In Husserl's system of *categories*, the formal types structure the material types. And so, in due course, he introduces the notion of noema in §§ 88–89, drawing on formal ontology regarding the structure of consciousness. (Husserl restricts the term "category" to formal high-level essences, but our purposes are served if we use "category" to designate both formal and material essences at the highest level, distinguishing formal categories from material categories. Cf. Smith 2013, chapter 4.)

In § 59 Husserl explicitly addresses the use of formal ontology, joined with "pure logic as *mathesis universalis*," in transcendental phenomenology. We are to bracket material ontologies positing "transcendent" material essences such as the essence Tree; in particular, "things" in nature, such as trees, are by essence "*reale*," i.e., spatial or spatiotemporal. Further, Husserl says, we are to bracket the disciplines of formal ontology and formal logic per se, yet we continue to make use of formal essences such as Object, Property, State of Affairs, etc. The constraint of epoché, Husserl holds, is that we follow the "norm" of phenomenology: as we turn to "pure" consciousness, we may describe an experience formally as an "object," with its key "properties" including intentionality, but these structures must be explored in the "evidence" of "intuition" about our lived experiences. What form of *intuition* would that be, for Husserl? My perception of the tree carries the support of hyletic or sensory data: constitutive of my *perceptual intuition* of the tree. By contrast, as I turn to my visual experience in *phenomenological reflection*, my act of reflection carries the support of *eidetic intuition* of the essence of the visual experience on which I reflect. Thus, in phenomenological reflection I grasp the *essence* of my *experience* of the tree (which involves having a noema with sensory support). My act of reflection itself gains intuitive support (as needed) from "eidetic variation" on similar visual experiences (each with its own noema with sensory base). Now, whereas evidence in perception is ultimately founded on sensory hyle, evidence in phenomenological reflection is ultimately founded instead on eidetic insight. Reflecting on the experience as an *object* and its lived character as an *essence*, for Husserl, draws precisely on formal ontology: with duly noted intuitive force.

In § 65 (as noted previously) Husserl allows that phenomenology "relates back" to itself, considering the "psychology" of everyday experience as leading by epoché into "transcendental" reflection on everyday experience. It's not as though we forget our everyday experience that posits a factual "real" world around us. And it's not as though we forget our various "sciences" from mathematics to mathematical physics to formal logic and formal ontology. This methodological back-and-forth is part of Husserl's systematic philosophy. (Here we see the "zigzag" image that Husserl sometimes invokes.)

In § 85 Husserl turns to the formal structure of intentionality in consciousness. Thus, Husserl distinguishes two fundamental *parts* of an act of perception: a *sensory* "moment," and a *noetic* "moment." In Husserl's ontology, a *moment* (*Moment*) is a dependent part of a whole. So a perceptual experience is a fusion of sensory and noetic parts, where each could not exist apart from the other and apart from the whole act. Here Husserl adapts Aristotle's hylomorphic model of a concrete object (matter-with-form), applied now to the concrete sensory-intentional experience (sensory hyle with noetic form). The experience is "*reelle*," occurring in time. And its "noetic" moment, or "noesis," is the distinctively "*reelle* component" (*Komponent*) in the experience.

In § 86 Husserl explicitly articulates two key aspects of noesis. First, the noeses in an act of consciousness perform the "function" that makes the act a consciousness "of" something, and this they do as they "harbor" or entertain sense (*Sinn*) "in themselves" (*in sich zu bergen*). That is, *ideal sense* is realized in, or drawn into, the noesis or noeses that are parts of the "*reelle*" temporal experience. In Husserl's evocative idiom, the noesis "harbors," or secures, the ideal sense *within* the real noetic component of the experience, the "sense-giving" component of the act. Second, this function of noesis effects the "constitution" of the object of consciousness: as the noesis draws in the sense, the experience becomes "sense-ful" and thereby a consciousness "of" the object.

In Husserl's phrasing, we find a fundamental model of "constitution" via sense, or Sinn:

> Yet the greatest problems [for phenomenology] are the *functional* [*funktionellen*] *problems*, i.e., those of the "*constitution of objectivities of consciousness*" ["*Konstitution der Bewusstseinsgegenständlichkeiten*"].

> They concern the way, as for example with respect to nature, noeses . . . bring about consciousness of something.
>
> "*Function*" in this sense (a sense totally distinct from that of mathematics) is something totally unique, in that which is grounded [*Gründendes*] in the pure *essence* [*Wesen*] of the noeses. Consciousness is precisely consciousness "of" something, it is its essence to harbor "sense" ["*Sinn*"] in itself [*in sich zu bergen*], so to speak the quintessence of "soul", "spirit", "reason" ["*Seele*", "*Geist*", "*Vernunft*"].
>
> The viewpoint of function is the central viewpoint of phenomenology, that of . . . its investigations spanning . . . the complete phenomenological sphere.
>
> (§ 86, my translation, Husserl's emphases)

In the full passage Husserl rejects empiricist terms of "bundles" of "sensations," even as he adapts neo-Kantian terms of "soul," "Geist," "reason" (following Kant's "critical" philosophy). Still, Husserl lays out his own account of phenomenological structure where ideal Sinn enters into real-time consciousness. Already in § 55 Husserl had emphasized the "sense-giving" ("*Sinngebung*") aspect of consciousness, and insisted that this sense-giving does not throw us into "the arms of a 'Berkeleyan idealism'," that is, with the function of "intentional 'constitution'" (cited in § 54).

So the function of noesis in an experience is to entertain Sinn. In § 88 Husserl specifically defines the notion of "noematic content," or "noema," as an "intentional correlate" of the noetic moment, or noesis, in an intentional experience. The "*reelle*" noesis and the "intentional" noema show up, for Husserl, within the practice of epoché, or "transcendental reduction" (see § 88). Thus, in reflection, the ontological structure of consciousness appears within the phenomenological structure of the act of consciousness.

Husserl's exact phrasing, in § 88, is as follows (in my translation, with Husserl's own terms, emphases, and quotation marks):

> Every intentional experience [*Erlebnis*] is, thanks to its noetic moments [*Momente*], precisely noetic; that is to say, it is [of] its essence [*Wesen*] so to harbor [or entertain] in itself [*in sich zu bergen*] something as a "sense" ["*Sinn*"] and perhaps manifold sense [*mehrfältigen Sinn*], upon the ground of this sense-giving . . . to become therewith precisely "sense-ful" [meaningful: "*sinnvolle*"]. . . . As just now this series of exemplary moments [*Momente*] point to real components [*reelle Komponenten*] of the experience, so they also point, namely through the title sense [*Sinn*], to the *not real* [*nicht reelle*].
>
> According [*entspricht*] overall with the manifold data [*Daten*] of the real, noetic content [*Gehaltes*] is a manifold [*Mannigfaltigkeit*] of data [*Daten*], demonstrable in actually pure intuition [*Intuition*], in a correlative "noematic content" ["*noematishen Gehalt*"], or in short

"noema" ["*Noema*"]—terms which we shall constantly use from now on.

Perception, for example, has its noema, most basically its perceptual sense [*Wahrnehmungssinn*], i.e., the *perceived as such* [*Wahrgenommene als solches*].

We need to appreciate the *ontology* Husserl assumes in this *phenomenological* characterization (under epoché) of noema as the *ideal intentional content* of an experience such as a simple perception of a tree before one. The experience itself includes a "*reelle*" *noetic component*, or *noesis*: a part of the experience, a part flowing in time (along with the experience) and in that way "real." This noetic part is a "moment" (*Moment*) of the experience, a *dependent part* of the experience, a part that cannot exist apart from the experience, and so a part that itself flows along in time with the temporally flowing experience. But the act's noema is not "*reelle*": it is, rather, an ideal sense drawn *into* the "*reelle*" act thanks to the noetic component of the experience. (Husserl's notion of "*Moment*" is detailed in his theory of parts and wholes in the *Logical Investigations*, Investigation III.)

Thus, "according with" the noetic component in the experience there is a "correlative" (*korrelativen*) form of *sense*, or Sinn, that Husserl christens *noema*. Husserl chooses an evocative idiom for how the experience is related to its noema: the experience "harbors," or "secures," "in itself" the correlated noema. The German idiom, "*in sich zu bergen*," has connotations of securing or keeping in a safe place. So we may think of the experience as "harboring" or "entertaining" its noema securely "in itself." Husserl specifically emphasizes that the noema of the experience is "not real," i.e., not a "*reelle*" temporal part of the experience (along with the component noesis in the experience). Rather, the noema is the "correlative" (*korrelativen*) noematic *content* "in" the experience: itself a form of ideal *sense*, or Sinn. We should hear echoes of the logical theory of *Sinn* and *Satz* on which Husserl is drawing, even as phenomenal experience is his focus.

(Note: Husserl's use of the German verb "*bergen*" has been translated as "harbor," which is evocative, or as "contain" or "include," which are more prosaic. Here we use the English "entertain," meaning "to hold in mind": connoting a secure relation between the experience and its noema, without suggesting containing as an ontological part—whereas the noesis is properly a dependent part, or "moment." For Husserl, aside from the idiom "*bergen*," the relation between an experience and its noema is strictly one of "*Korrelation*" as a "*korrelativen*" ideal entity. To give a name to the specific type of *formal relation* between an experience and its noema, we may say the experience *entertains* its correlative noema. By etymology, "entertain" literally means to "hold" something "in" mind. But note that it is the experience, not the subject "I," that stands in the special relation to the noema: I see the tree, not the noema that is entertained in my visual experience. "Entertain" is the term used in Smith and McIntyre 1982.)

Husserl's conception of a noema as a kind of Sinn is resonant with Aristotle's theory of *in re* universals: a noema is realized in a particular experience insofar as a noesis in the experience instantiates, or draws into the experience, that ideal meaning, the noema. Indeed, this neo-Aristotelian point mirrors Husserl's ontology of Sinn in the *Logical Investigations*, where ideal meaning is treated as an "intentional essence" that may be realized in an experience precisely as an ideal intentional type. By contrast, however, in *Ideas* I Husserl gives pride of place to the category of *Sinn per se* (duly emphasized in the passage we quoted earlier). Fundamentally, in the refined ontology of *Ideas* I, Husserl distinguishes noema from essence. Accordingly, the *noema* correlated with an experience is not identical with the *essence* of the experience: rather, "it is of the essence" of the experience to "entertain in itself" a noema. Crudely, an *essence* gathers similar types of objects, whereas a noematic *sense* has a specifically logical force of "meaning" or "intending" something. For Husserl, the noematic sense "in" an experience is thus distinguished from the essence realized "in" the experience. But also, for Husserl, the ideal noematic sense is distinguished from the ideal *essence* of the "real" *object* of the experience (if it has such an object). (Husserl explicitly distinguishes noema from essence in *Ideas* III, §16, "*Noema und Wesen*." Also, Føllesdal has noted that Husserl links the notions of noema and sense in an unpublished manuscript titled "Noema und Sinn.")

For Husserl, what distinguishes forms of meaning, under the category Sinn, is precisely their "function" in *intentional consciousness*. More than an ideal type or universal *per se*, a meaning has logical or *semantic* force: here lies the logical side of Husserl's conception of noema. But there is more.

The ideal character of noema Husserl explicitly lays out in a striking passage in § 89 (in my translation, the emphases and quotations marks in Husserl's original):

> "In" the reduced perception (in the phenomenologically pure experience [*Erlebnis*]) we find, as belonging inextricably to its essence, the perceived as such, to be expressed as "material thing", "plant", "tree", "blossoming", and so forth. The *quotation marks* [*Anführungszeichen*] are obviously significant; they express that change in sign, the corresponding radical modification of the meaning [*Bedeutungsmodifikation*] of the words. The *tree simpliciter* [*schlecthin*], the thing [*Ding*] in nature, is anything but [*ist nichts weniger als*] this *perceived tree as such*, which as perceptual sense [*Wahrnehmungssinn*], belongs inseparably to the perception. The tree simpliciter can burn up, be resolved into its chemical elements, etc. But the sense [*Sinn*]—the sense *of this* perception, an [entity] belonging necessarily

> to its essence—cannot burn up; it has no chemical elements, no forces, no real [*realen*] properties.

The phrase "*ist nichts weniger als*" is notoriously flexible. Some translations of this passage use the literal "is nothing less than," while other translations use the contrary "is anything but." On the one translation, Husserl is declaring that the tree itself and the Sinn of the perception are *identical*; on the other translation, he is declaring that these are precisely *not* the same entity. However, the extended passage makes clear which of the two uses Husserl intends. The tree itself can burn up, but perceptual Sinn, the perceived-tree-as-perceived, cannot. So the tree and the Sinn cannot be *identical* because they have radically different properties. (See Føllesdal (2019) on the key phrase. Cf. Drummond (2015) agreeing on the translation yet seeking an accommodation with his alternative understanding of noema.)

Indeed, the flow of argument in §§ 88–89 specifies that:

* the tree itself is a "*reale*," spatiotemporal "thing" in nature;
* the noesis in the perception is a "*reelle*," temporal part of the experience;
* the noema "harbored" as *Sinn* "in" the experience is neither a "*reelle*," temporal part of the experience, nor a "*reale*," spatiotemporal thing in nature (say, a part or aspect of the tree itself).

Thus, the noema is the ideal "intentional correlate" of the noesis that is a temporal "moment" of the experience.

Accordingly, given Husserl's categorial ontology in *Ideas* I, we should see that: the tree falls under the material region Nature, while the experience and its noetic moments fall under the material region Consciousness, but the noema of the experience falls under the *logical* or "semantic" category Sinn. Arguably, Sinn plays a different fundamental role in Husserl's ontology than either the "formal" essences or the "material" essences. In Husserl's developed system, Sinn forms in effect a third fundamental category distinct from both "Fact" (governed by the material regions of Nature, Consciousness, Geist) and "Essence" (including formal categories of Individual, Property, State of Affairs, Number, etc.). In the long course of Husserl's argument in *Ideas* I, the ontological force of Sinn—the "function" of "constitution"—emerges in detailed formulation only well after Husserl's condensed ontology in the opening chapter of the book. The "transcendental" force of Sinn was elaborated only with the full doctrine of noema. But remember that conceptions of ideal meaning were in the air Husserl breathed. (See the reconstruction of Husserl's categorial ontology in Smith (2013), chapter 4 following chapter 3 on logical structure, and note ramifications in chapter 9.)

The doctrine of noema and constitution emerged in connection with the newly discovered "region of being" that is the essence Consciousness.

This does not entail that meaning is an entity falling under the essence Consciousness. As the formal essence Individual applies to an act of consciousness, so the logical category Sinn applies to a "correlated" noesis that is a dependent part of an act of consciousness. In other words, in Husserl's systematic ontology, formal essences (e.g., Individual) cross material essences (e.g., Consciousness), and so too logical forms (e.g., Noema) cross formal and material essences (as a *noema* is "in" an *individual* entity that is an *experience*).

Now, while Husserl says the noema falls under the general title "Sinn," he also reserves the term "Sinn," strictly speaking, for a proper part of the noema of an act. The structure of an act's noema includes the "noematic Sinn" and a modifying component of "thetic character." The Sinn presents the object intended, as an object X with various properties; the thetic character modifies the Sinn as perceptual, judgmental, volitional, etc., i.e., with the character of the relevant type of intentional experience. For example, in the case where I have a visual experience of seeing a Podocarpus tree in my back yard, the noema of the experience may be depicted as follows:

< I now here see this tall Podocarpus tree >.

The noema is thus a specific *form of meaning*, structured as so indicated.

In this case, the Sinn of the experience is the sense <this tall Podocarpus tree>, the basic thetic character is the modifying element of noematic content <see>, which qualifies the experience as one of visual perception. Normally, this thetic character includes a character of "positing" (*Setzen*) with a commitment to the actual existence of the object so intended.

Within the structure of the noema, we note as well the sense of *subject* <I.> For Husserl has characterized the *form* of an intentional experience as that of the *cogito*, or "I think" (§81).

Also, we note the sense of spatiotemporality <now here>. Well before §§ 88–89, Husserl noted that the experience has a temporal structure indicating a place in the subject's "stream of experience" (§ 83).

And in *Ideas* II, Husserl noted the subject's sense of embodiment in a perceptual experience, where the object is presented in a spatiotemporal environment wherein "I" am the "origin" of the spatiotemporal domain or field within which "I" experience the visual presentation of "this . . . tree." This form of the experience was well amplified as the "phenomenal field" of the perception, in Merleau-Ponty (1945, 2012).

In works prior to *Ideas* I, Husserl had written extensively on both the sense of "time-consciousness" (*Zeitbewusstsein*) and the sense of "space" (*Raum*). Both are well studied; here we simply note the sense <I> and the senses <now> and <here>, and locate them in the overall structure of the noema as indicated previously. Importantly, for Husserl, this structure in the noema defines the form of experience flowing off in the stream of consciousness. What lies beyond is a further matter, "transcendent" of the experience per se—in particular, what there is in the actual surrounding world that answers to the content <I> or to the content <this . . . tree> or indeed to the content <now here see> (e.g., the neural substrate of the subjective visual experience).

What is new and revolutionary in Husserl's theory of intentionality, as we look toward a metametaphysics in phenomenology, is the confluence of two essential aspects of noematic content: the logical and the phenomenal. First, the noema is a form of Sinn: an *ideal meaning* of a certain ontological type, whose function is to effect "intention" and thus "constitution." Second, the noema or Sinn in the act is *phenomenally experienced* by the subject in living through the act: not as an object of consciousness, given in reflection, but as the medium of "intention" of an object (if the act has one). Only in phenomenal intentional experience of things in our surrounding world do we make meaningful, or "sense-giving," connection to the world of "real" existing things surrounding us. Thus, only in our own "sense-ful" consciousness do we reach the *real world* in which we find ourselves, a world given or "constituted" thanks to a complex of noematic meaning.

According to Husserl's program in *Ideas* I, we become *aware of* this lived phenomenal *intentional relation* in a distinctive form of *phenomenological* reflection, attained by the methodological device of epoché. However, when we go to lunch or when we trim the tree in the garden, we drop the brackets of epoché, as Husserl clearly taught when he distinguished the tree simpliciter from the tree-as-perceived. In this way we move back and forth between our world—our *Umwelt* or *Lebenswelt*— and our consciousness of the world. Here we see a particular meta-theory in Husserl's program.

Even in everyday life, however, we may move back and forth between world and experience. Husserl offers a memorable example. In the first-person: At a wax museum, I see a woman on the stair, waving at me, and suddenly I realize it is a wax figure I see. In such a case, Husserl says, the content of my perception "explodes," and the noema in my initial perception (<this woman>) is replaced by a different noema in the second perception (<this wax figure>). Well, prior to any practice of epoché, I am aware of a shift in "what I see." And this contrast indicates a shift in meaning, a shift we all understand in our life-world, a shift that paves the way for Husserl's fully developed theory of noema.

Husserl's meta-theory for his conception of phenomenology turns crucially on his theory of noema and its function in "constitution." Let us draw out some further implications of Husserl's theory of "constitution" at this higher meta-level of reflection.

V. Constitution of and by Virtue of Ideal Noemata

The noema realized "in" an act of consciousness is, for Husserl, an ideal entity. Husserl began his career as a mathematician, and in his mature ontology, in *Ideas* I, we find some three types or categories of ideal entities: mathematical entities (numbers, sets, manifolds, geometric spaces); universals (kinds, properties, relations); and meanings (forms of *Sinn*, *Satz*, *Bedeutung*, noema). Starting with a broadly Platonic notion of "essence" (*Wesen*), or *eidos*, Husserl developed a structured system of types of ideal entities. All are *ideal*, but each has its own categorial niche in the world, neither "*reelle*" (temporal) nor "*reale*" (spatiotemporal). Only noemata, however, have the "function" of facilitating "constitution," as Husserl elaborated his full theory of noema in *Ideas* I.

In the *Logical Investigations* Husserl called the ideal content of an experience its "intentional essence," which he took to afford meaning to linguistic expression: *Sinn* as realized in experience, *Bedeutung* as expressed in language grounded in experience. In *Ideas* I Husserl took meaning, Sinn, in a further direction as "noema," which he took as a form of ideal meaning with its own categorial niche. The simpler ontology of the Investigations assumed a type/token, or universal/particular, distinction: a concrete act's intentional content just *is* its intentional type or "species." However, in *Ideas* I, as Husserl wrote, the *essence* of an experience is not merely its type, or "species": rather, the experience is such that "it is [of] its essence so to harbor [or entertain] in itself [*in sich zu bergen*] something as a 'sense' ["*Sinn*"]," namely, its noema, the ideal correlate of the act's noesis. So, for Husserl, the essence of the experience entails its unique character of entertaining a *meaning*, viz., a noema. The noema is ideal, but not identical with the act's essence, or "species." The function of the act's intentional essence is thus to draw "into" the act a still different type of ideal entity, the act's noema, whose own function is to help to "constitute" the intended object.

In *Ideas* I, as in the Investigations, Husserl has been taken to embrace a "Platonism" regarding ideal entities, whether essences or meanings or numbers, sets, etc. However, Husserl's Platonism is minimal. For Husserl, ideal entities do not reside in a Platonic heaven, or even in a Fregean third realm. Rather, they are *categorially distinct* from "real" entities (whether "*reale*" or "*reelle*"): that is all. Thus, the tree in the garden is spatiotemporal, and my visual presentation of the tree is temporal (flowing in "inner" time); yet the noema "in" my experience is neither temporal nor spatiotemporal, but *ideal*. Again, whereas the kind or essence Tree is realized "in" my particular Podocarpus tree, the sense <tree> is realized "in" my visual experience but not "in" the tree itself. Thus, we distinguish these two distinct types of "in"-relations, different types of realization or inherence of the ideal in the real. And what of numbers? The number 2 is presumably realized "in" a pair of birds on a wire (to follow an example

Husserl used), but where numbers fit in Husserl's formal ontology is an interesting question. Given the complexity Husserl seems to allow in his ontology: arguably, the ideal entity 2 need not be identified either with an essence (an "ideal species") realized in the pair of birds, or with the set of all pairs (2-tuples), including the pair of birds. Husserl often speaks of "manifolds" as mathematically complex entities, viz., structured wholes. Here we must leave for another day Husserl's exact ontology of different types of mathematical entities. But whatever their exact categorial niche, these entities are ideal.

Looking to the status of ideal entities in mathematics, Richard Tieszen has drawn on Husserl's phenomenology as deployed by Kurt Gödel. On Tieszen's interpretation of Gödel and Husserl, mathematical entities—numbers, sets, manifolds, etc.—are both *ideal* (nonspatiotemporal) and *real* (existent) in that they are part of the structure of the world, yet they are "*constituted*" in the forms of mathematical intuition and judgment that produce our mathematical theories. Tieszen calls this Gödelian-Husserlian view "constituted Platonism" regarding mathematical entities.

(Note: Tieszen's nuanced view is elaborated in Tieszen 2011, *After Gödel: Platonism and Rationalism in Mathematics and Logic*. Cf. his detailed studies of phenomenology and philosophy of mathematics in Tieszen 2005, considering several mathematicians who interacted with Husserl, including Cantor, Weyl, Hilbert.)

Husserl had studied with Weierstrass and Kronecker before he developed phenomenology, and he interacted with Cantor, Hilbert, and Weyl during his career. Gödel's famous results in the 1930s arrived as Husserl was in his last years, even as his views on phenomenology and mathematics continued into his writings of 1935–38 gathered posthumously as the *Crisis*. The "constitution" of mathematical entities was on Husserl's horizon already in his pre-phenomenological *Philosophy of Arithmetic* (1891), and it was a driving theme in Husserl's posthumous *Crisis*, so it is important that Husserl interpretation recognize the continuing role of mathematics in Husserl's phenomenology even after his "transcendental" turn.

However, as we have emphasized, Husserl's ontology recognizes different types of *ideal* entities. And this complexity invites a ramified form of "constitutive realism"—following in the spirit of Tieszen's take on Gödel's take on Husserl's phenomenology of the "constitution" of ideal entities. The more general doctrine we are here calling "constitutive realism" would cash out Husserl's "transcendental idealism" as a *realism* about ideal entities such as numbers or sets, as well as "real" entities such as trees. And that *realism*, about both "real" and "ideal," is enabled by intentional "*constitution*." Let us draw together several threads in this doctrine.

In Kantian transcendental philosophy *concepts* of things in nature are not easily distinguished from the *properties* of those things as they appear in our cognition of such things. Even in semantic theory in Husserl's wake, e.g., in Rudolf Carnap's work of "logical empiricism," we may find properties identified with concepts, where concepts are of our own making. Husserl's doctrine is more complex. (Cf. Carnap 1928/2003, known as the "*Aufbau*," where Carnap developed a "logistic" formulation of "*Konstitutionstheorie*," footnoting Husserl's *Ideas* I along the way.)

Husserl's ontology is remarkable for its role in his transcendental phenomenology as detailed in *Ideas* I. Thus, for Husserl, the tree "simpliciter" is a plant that is blooming, while the noematic Sinn in my seeing the tree is the ideal meaning <this blooming tree>, an entity that is not itself blooming. For Husserl, the noematic sense <tree> is not identical with the botanic property Tree: the essence Tree and the sense <tree> are both ideal entities, but fundamentally *different types* of ideal entity. (Cf. *Ideas* III, §16, "Noema and Essence," amplifying *Ideas* I.)

Now, Husserl long held that ideal entities like numbers are "constituted" in mathematical intuition and judgment: in effect, that was what Husserl was trying to say in his *Philosophy of Arithmetic* (1891), though he did not then have a working theory of intentionality (as Dallas Willard has observed in his translator's introduction to Husserl 1891). If noemata are ideal meanings, as Husserl held in *Ideas* I, then we should expect that, for Husserl, noemata are "constituted" in appropriate acts of consciousness. Indeed, the point of practicing epoché, we should see, is precisely to shift our focus from *objects* in the world around us to our *consciousness* of such objects (their factual existence in brackets). And so, in phenomenological reflection, we thereby focus on the *noemata* thanks to which we are conscious of objects. And that form of reflection entails that, for Husserl, noemata themselves are "constituted" in our acts of *reflection* on the ideal correlates of consciousness, as opposed to the objects intended by virtue of these noemata.

Here we find in Husserl a "constitutive realism" about ideal *meanings* in relation to "real" things in nature.

We find in Husserl's transcendental phenomenology, then, a doctrine of constitution on different levels. The tree I see is "constituted" in my visual experience by virtue of ("thanks to") the noematic sense entertained in the experience. Further, that noematic sense I contemplate in phenomenological reflection on my visual experience. Accordingly, the noematic sense entertained in my visual experience is itself "constituted" in my acts of *reflection* on that meaning entity itself. Reflection is a higher-order level of experience, in which we *abstract* the noema of my visual experience.

The noesis in my visual experience, Husserl holds, is a "moment" in the visual act: a dependent, or "abstract," part, which cannot exist apart from the visual act but can be abstracted from the act in reflection. Correlated with the noesis is the act's noema, which we abstract in reflection by judging how that noema could be shared by other visual experiences including their own noeses with which the relevant noema is correlated. That noema is then "constituted" in the structure of our reflections—reflective judgments—on possible experiences that may share the same noema. Instead of a structure of visual experiences directed toward the same *object* by virtue of a manifold of noemata, we find in reflection a structure of reflective acts directed toward the same *noema* by virtue of the noemata correlated with such reflective acts. This structure of higher-order noemata of reflection defines the "constitution" of the first-order noema: in our example, the noematic content <I now here see this tall Podocarpus tree>. Following out the Husserlian theory of constitution for *ideal meanings*, then, we find a structure of higher-order noemata entertained in reflection on the first-order noema of my seeing the tree.

Thus, noemata are both "*entertained*" in experiences intending everyday objects (e.g., trees) and "*constituted*" in acts of reflection intending such experiences bearing noemata (i.e., in reflection via the practice of epoché).

Return to the example case of my seeing the Podocarpus tree. On one level of consciousness, in *everyday perception*: the *tree* I see is "constituted" thanks in part to the noematic sense correlated with my visual experience, viz., the ideal sense <this tall Podocarpus tree>. Furthermore, on another level of consciousness, in *phenomenological reflection*: the *noema* of my visual experience—the *sense* <this tall Podocarpus tree>, which presents the "tree-as-perceived"—is itself "constituted" thanks in part to the *higher-order sense* correlated with my reflective judgment that "I see this tall Podocarpus tree." And the noematic sense <this tall Podocarpus tree> is modified by the thetic character <I now here see> in the full noema <I now here see this tall Podocarpus tree>.

To elaborate: The tree itself is "constituted" through an indefinitely large manifold of meanings correlated with how the same tree is presented from many different perspectives, specifically, in perceptual experience: thanks to the "horizon" of "motivated" further potential experiences of the same tree. That is part of what the "reality" of the tree *means*, if you will. And now consider the ideal noematic sense <this tall Podocarpus tree>. That ideal meaning is itself "constituted" through a structure of ideal meanings that are "entertained" in our *reflective judgments* about the *intentional relations* between possible visual experiences and the actual tree. Those judgments define the phenomenological "*semantics*" of the ideal form of experiences that would "entertain" the given noema.

VI. Husserl's "zigzag" Between the World and Phenomenology

On several occasions, as noted, Husserl uses the metaphor of a "zigzag" (*Zickzack*) movement forward and backward between phenomenology and foundations of either logical or epistemological or social structures concerning the world. This methodological figure appears throughout Husserl's writings, as a kind of side note. Yet the figure suggests a meta-metaphysical principle at work. We take note of Husserl's use of the figure in both *Logical Investigations* (1900–01) and the *Crisis* (1935–38), as bookends to his implicit use of the figure in *Ideas* I: in the very practice of epoché.

In the *Crisis* Husserl ramifies his conception of phenomenology, with direct relevance to our question of a phenomenological metametaphysics. The running theme for Husserl in these writings of 1935–38 is how our "mathematized" theory of nature in physics relates to our everyday experience of things in nature. His larger concern, however, is how our physics—with an implicit eye on metaphysics—is grounded in our experience in the "life-world": a case study, as it were, in "constitutive realism."

In *Ideas* I (§§ 27–29), Husserl spoke of the *Umwelt*, the "surrounding world," whose actual existence is to be bracketed in epoché (§§ 30–32). In the *Crisis*, Husserl adapts the by-then-current term "*Lebenswelt*," or "life-world," which aligns with his prior talk of the *Umwelt* in *Ideas* I informed by his analyses of embodiment and intersubjectivity in *Ideas* II. For Husserl, the *life-world* is the surrounding world *within which* we live, the everyday world in which we see things, form judgments about things, and willingly act upon things. However, strictly, the life-world is the world *as we experience it* in everyday life. For Husserl, when we speak of things-as-intended, we are speaking of noematic sense. So the key problem of the *Crisis* is the relation between our *sense of things* in mathematical physics and our *sense of things* in everyday perception, thought, and action. Methodologically: we find ourselves living and acting in the surrounding *Lebenswelt*, a world of nature and culture, *on the ground of which* we pursue our study of nature in mathematical physics. Later philosophers would contrast the "scientific image" with the "manifest image" of the world, as formulated respectively in physics and in everyday language. (On Husserl's account of the life-world, and its relation to scientific theory as well as everyday experience, see Hyder and Rheinberger eds., 2010, including Smith 2010. On Husserl's account of intersubjectivity, see Kjosavik et al. 2019; including Smith 2019a.)

Briefly, our judgments in mathematical physics are *founded on* our judgments in everyday life, the latter beginning in everyday perception or observation. Not only do everyday perception and careful reasoning provide *Intuition* and *Evidenz* in epistemological support for scientific judgments in mathematical physics. Fundamentally, the *noematic contents* of everyday experience provide logical or semantic support for the very *propositions* (*Sätze*) which are the noematic contents of judgments formed in the practice of physics. Accordingly, in the *Crisis*, Husserl distinguishes two levels of epoché. First, we may bracket the mathematical-physical properties of things in nature, leaving still our *Lebenswelt* experience of natural "things." Second, we may bracket also the everyday properties of things in nature, in a full "Cartesian" epoché. Tracking the domains of *meaning* revealed in these two levels of "phenomenological reduction," we then see that our sense of things according to our life-world experience is *logically or semantically prior* to our sense of things according to our mathematical physics. Thus, our experience in physics expands on the "horizon" of our experience in everyday life. For the meaning of things according to relativity theory (and quantum mechanics) is understood as specifying further and more fundamental "mathematized" physical features of things in nature, including things as given in our everyday perceptions.

Here we see an extrapolation from Husserl's phenomenology of the "transcendence" of things in nature, beginning with the ways we experience them in everyday life. Indeed, Husserl's methodology prescribes our movement back and forth between our *Lebenswelt* sense of things and our mathematized sense of things. In particular, as the *Crisis* argues in detail, our now-familiar geometric concepts—noematic meanings such as <straight line>, etc.—carry with them a long historical genesis. "Genetic" phenomenology adds this dimension to "transcendental" phenomenological analysis of our noematic meanings regarding spatiotemporal things in the world around us. (Cf. the essays in Hyder and Rheinberger eds., 2010.)

Accordingly, in the *Crisis*, Husserl defines a "zigzag" methodology in the practice of transcendental phenomenology. Husserl writes (in § 9, "Galileo's mathematization of nature," in subsection l of § 9, "Characterization of the method of our exposition," p. 58 in the English translation):

> Thus we find ourselves in a sort of circle. The understanding of the beginnings [of our mathematized science of nature] is to be gained fully only by starting with science as given in its present-day form, looking back at its development. But in the absence of an understanding of the *beginnings* the development is mute as a *development of meaning*. Thus we have no other choice than to proceed forward and backward in a zigzag [*Zickzack*] pattern; the one must help the other in an interplay. Relative clarification on one side brings some elucidation on the other, which in turn casts light back on the former. In this sort of historical consideration and historical critique, then, which begins with Galileo (and immediately afterward with Descartes) and must follow the temporal order, we nevertheless have constantly to make *historical leaps* which are thus not digressions but necessities.

Already in the *Logical Investigations* Husserl used the "zigzag" metaphor in regard to "the phenomenological founding of logic." As we seek to clarify and develop a concept in logic and epistemology, Husserl writes:

> The phenomenological founding of logic involves the difficulty that we must, in our exposition, make use of all the concepts we are trying to clarify. This coincides with a certain wholly irremovable defect

> which affects the systematic course of our basic phenomenological and epistemological investigations. If a type of thought requires prior clarification, we should not make uncritical use of its terms or concepts in that clarification itself. But one should not expect that one should only be required to analyze such concepts uncritically, when the actual interconnection of one's logical materials has led up to them. Or, put differently, systematic [= phenomenological] clarification, whether in pure logic or any other discipline, would in itself seem to require a stepwise following out of the ordering of things, of the systematic interconnection in the science to be clarified. Our investigation can, however, only proceed securely, if it repeatedly breaks with such systematic sequence, if it removes conceptual obscurities which threaten the course of investigation *before* the natural sequence of subject-matters can lead up to such concepts. We search, as it were, in zigzag [*Zickzack*] fashion, a metaphor all the more apt since the close interdependence of our various epistemological concepts leads us back again and again to our original analyses, where the new confirms the old, and the old the new.

(Husserl 1900/2001, *Logical Investigations*, Volume 1, p. 175, in the English translation, 2001 edition, by Routledge: from Volume II, Part I of the German editions. Introduction, § 6.)

The foundational concepts for pure logic and thence for epistemology, in the *Logical Investigations*, will feature the concepts of intentionality and intentional content, and these phenomenological concepts themselves would continue development for Husserl in *Ideas* I, in terms of noematic content.

Indeed, Husserl observes a similar methodological movement in *Ideas* I. As noted earlier, in *Ideas* I, § 65, Husserl speaks of phenomenology's "relating back" (*Rückbeziehung*) to itself, where the "new science" of phenomenology relates back to its origin in everyday experience and empirical "psychology": all *within* the methodology of epoché that demarcates "pure" phenomenology. In the practice of epoché, Husserl emphasized (§ 32), "I *do not negate* this 'world,' as though I were a sophist [or idealist]; *I do not doubt its existence*, as though I were a skeptic." As I practice epoché, addressing "pure" consciousness (§§ 30–34), since I do not doubt or negate my surrounding world, "'The' world is as actuality [*Wirklichkeit*] always there" (§ 30). That is, even as I bracket the thesis of its actuality or being-there, the world remains in place as a "horizon of indeterminate actuality" for all my activities (§27). These activities include my practice of phenomenology itself (as unfolded in §§17–34). This practice entails a "zigzag" as I rely on my everyday experience *in the world* to which I may "relate back" even as I bracket the question of that world's actual existence, viz., as I turn to my "pure" experience as of things in the world, in my *Umwelt*.

From this "zigzag" methodology we may draw a distinctively Husserlian metametaphysics. In brief: We find ourselves in a surrounding world, wherein we interact with things in key forms of intentional "acts" of perception, thought, imagination, embodied action, and social relations including empathy. By Husserl's technique of epoché, we turn our regard to the ways in which we experience things in these acts of consciousness. In reflection, we analyze the structure of intentionality, and in that practice we abstract forms of noematic meaning. In this practice of phenomenology, over time and in intersubjective philosophical dialogue, we appraise the ways in which things are "constituted" in these formations of noematic meaning. We zigzag back and forth between our surrounding world and our experience of that world. In this pattern of activity, we weave together our metaphysics of *the world* and our phenomenology of *our consciousness of* the world.

VII. Conclusion: A Phenomenological Metametaphysics

The foundations of our metaphysics lie in our lived *intentional relations* with the world: here is the guiding principle of a phenomenological metametaphysics.

Accordingly, we find that: the world surrounding us is real and existent; our lived intentional experience is part of that world; things in that world are "constituted" through a manifold of relevant meanings entertained in our experiences; and those meanings we abstract from our experience by phenomenological reflection.

Following Husserl's detailed analyses, those meanings range from noematic contents in our *everyday experiences* of perception, thought, and action to noematic contents in our *reflective judgments* in the practice of phenomenology and metaphysics writ large. Accordingly, in Husserlian phenomenology we find a metaontology of "constitutive realism"—reforming "transcendental idealism," if you will. The very *structure of intentionality*, on Husserl's analysis, defines this constitutive realism: as we move between *the world* surrounding us and our lived *experience of* that world.

Addendum

Note 1: Following out the Husserlian picture of ideal noematic meaning, we would envision a *formal ontology* of noematic meaning. Down this road we may see Alfred Tarski's conception of formal semantics, a metalogical theory that grew into mathematical model theory. But, of course, model theory itself is a set-theoretic construction: as it were, a *mathematized* "model" of the relevant forms of intentionality. So we should see a "zigzag" relation between intentional consciousness itself and its mathematized model: quite in the spirit of the problem of mathematization

Husserl studied in the *Crisis*. On Husserl vis-à-vis Tarski, again, see Smith 2016a, 2016b.

Note 2: Rudolf Carnap's *Aufbau*, Carnap 1928/2003, develops a "logistic" theory of the "constitution" of things in cognition. Carnap thought of the ontology of his "constitution theory" (*Konstitutionstheorie*) as pragmatic and ultimately anti-metaphysical. A contemporary metametaphysics that is called Carnapian as opposed to Quinean offers an "easy" ontology of perfectly "real" ordinary objects that we experience and conceptualize in everyday ways: see Amie L. Thomasson, *Ontology Made Easy* (2015). There are Husserlian roots in Thomasson's metaontology, as there are in Carnap's original perspective.

A currently popular view sets Carnap's pragmatic, anti-metaphysical "logical empiricism" in opposition to W. V. Quine's logical-*cum*-epistemological "naturalism." Cf. Chalmers et al. 2009. Yet Quine's "ontological relativity" can be aligned with Husserl's "transcendental idealism," or in the current idiom, "constitutive realism." See Smith 1994, "How to Husserl a Quine—and a Heidegger too."

Note 3: In the context of twenty-first-century philosophy, we may seek to bring Husserlian phenomenology itself into relation with a neuroscientific model of intentionality and consciousness. A scientific metametaphysics for a properly Husserlian phenomenology, featuring the "constitution" of things in the natural world, would include an analysis of the neural substrate of consciousness and its intentionality. Such is the aim of cognitive neuroscience today, to recognize lived consciousness itself and to analyze its ontological dependence on appropriate neuronal activity in the subject's brain and body. Here we may see a "zigzag" methodology, and a zigzag metaphysics.

Yoshimi 2016 develops a system of Husserlian "constitutive phenomenology" integrated with a model of dynamical systems of neural activity correlated with correlative forms of intentional consciousness dependent on such neural activity. In Jeff Yoshimi's system, the *formal structure* of the horizon of an experience is correlated with the *formal structure* of neuronal activity, which itself is a "mathematized" form of intentionality. Here is a "neurophenomenology"—in Francisco Varela's idiom—consonant with a metametaphysics along the lines pursued earlier. If we bring in the "zigzag" form, we find ourselves moving back and forth between our lived experience and our neuroscience model of that experience. My thanks to Jeff Yoshimi for finding Husserl's varied uses of his "zigzag" model.

Note 4: As pointed out, Husserl's model of intentionality aligns with Tarski's set-theoretic model of truth: on the interpretation in D. W. Smith 2016a, 2016b. Without venturing too far from the present study, we may observe that the "zigzag" methodology can be seen as assuming a holistic structure of intentionality-in-the-world. The zigzag movement tracks the intentional relation by moving among act, noema, and object: all occurring in the world, the larger "horizon" or context within which phenomenology and ontology are fundamentally linked. Arguably, the linkage is assayed in different ways by Husserl and then Heidegger. Looking toward logical theory, where Husserl began, see Livingston 2017. Paul Livingston there pursues a holistic account of truth in the spirit of Donald Davidson's extrapolation of Tarski's theory of truth, working toward a "transcendental" logic of truth, drawing on Husserl, Frege, Tarski, and Heidegger. Livingston envisions a "metaformal realism" in the theory of truth underlying interpretation of language. In the present perspective, Livingston's interpretation of truth-in-language charts variant forms of "constitutive realism."

References

Beyer, Christian. *From Bolzano to Husserl: A Study of the Origin of the Phenomenological Theory of Meaning*. Phaenomenologica 139. Dordrecht: Kluwer, 1996. Reprint 2012.

Carnap, Rudolf. *The Logical Structure of the World and Pseudoproblems in Philosophy*. Translated by Rolf A. George. Chicago and LaSalle, IL: Open Court, 1928/2003. From the German original, *Der Logische Aufbau der Welt*, 1928. Known as the *Aufbau*.

Chalmers, David J., Manley, David, and Wasserman, Ryan (Eds.) *Metametaphysics: New Essay on the Foundations of Ontology*. Oxford and New York: Oxford University Press, 2009.

Dreyfus, Hubert L., and Harrison Hall. *Husserl, Intentionality and Cognitive Science*. Cambridge, MA: MIT Press, 1982.

Drummond, John J. "The Doctrine of the Noema and the Theory of Reason". In Andrea Straiti, ed., *Commentary on Husserl's "Ideas I"*. Berlin and Boston: Walter de Gruyter GmbH, 2015, 257–285.

Føllesdal, Dagfinn. "Husserl's Notion of Noema". In Dreyfus, ed. 1982. Reprinted from *The Journal of Philosophy*, 1969.

Føllesdal, Dagfinn. "The Emergence of Justification in Ethics." *European Review* 13, No 2 (May 2005), 169–182. published online April 28, 2005.

Føllesdal, Dagfinn. "Brentano and Intentionality". In Denis Fisette, Guillaume Fréchette and Friedrich Stadler, eds., *Brentano Centenary Conference. Vienna Circle Yearbook*. Dordrecht and Boston: Springer, 2019.

Husserl, Edmund. *Cartesian Meditations: An Introduction to Phenomenology*. Translated by Dorion Cairns. The Hague: Martinus Nijhoff. (Now: New York: Springer.), 1929/1960. First published in French in 1931, first German edition 1950. Original manuscript, in German, 1929.

Husserl, Edmund. *Formal and Transcendental Logic*. Translated by Dorion Cairns. The Hague: Martinus Nijhoff. (Now: New York: Springer.), 1929/1969. German original, 1929.

Husserl, Edmund. *The Crisis of European Sciences and Transcendental Phenomenology: An Introduction to Phenomenological Philosophy*. Translated by David Carr. Evanston, IL: Northwestern University Press, 1935–38/1970. Original German manuscripts written, 1935–1938. Posthumous German edition, first published 1954. (Cited as *Crisis*.)

Husserl, Edmund. *Ideas Pertaining to a Pure Phenomenology and a Phenomenological Philosophy, Third Book: Phenomenology and the Foundations of the Sciences*. Translated by Ted E. Dlein and William E. Pohl. The Hague and Boston: Martinus Nijhoff Publishers, 1912/1980. Original manuscript dating from 1912, posthumously published in German in 1971. (Cited as *Ideas* III.)

Husserl, Edmund. *Ideas Pertaining to a Pure Phenomenology and to a Phenomenological Philosophy, Second Book: Studies in the Phenomenology of Constitution*. Translated by Richard Rojcewicz and André Schuwer. Dordrecht and Boston: Kluwer Academic Publishers, 1912/1989. Original manuscript dating from 1912, posthumously published in German in 1952. (Cited as *Ideas* II.)

Husserl, Edmund. *Logical Investigations*. Volumes One and Two. 1900. Translated by J.N. Findlay. Edited and Revised by Dermot Moran. London and New York: Routledge, 2001. German original, first edition, 1900–01; second edition, 1913, 1920. English translation, first edition, 1970.

Husserl, Edmund. *Ideas for a Pure Phenomenology and Phenomenological Philosophy, First Book: General Introduction to Pure Phenomenology*. 2014. Translated by Daniel O. Dahlstrom. Indianapolis/Cambridge: Hackett Publishing Company, 2014. German original, 1913. Called *Ideas* I.

Hyder, David and Rheinberger, Hans-Jörg, eds., *Science and the Life-World: Essays on Husserl's Crisis of European Sciences*. Stanford, CA: Stanford University Press, 2010.

Kjosavik, Frode, Beyer, Christian, and Fricke, Christel (Eds.) *Husserl's Phenomenology of Intersubjectivity: Historical Interpretations and Contemporary Applications*. New York and London: Routledge, 2019.

Livingston, Paul M. *The Logic of Being: Realism, Truth, and Time*. Evanston, IL: Northwestern University Press, 2017.

Merleau-Ponty, Maurice. *Phenomenology of Perception*. Transl. by Donald A. Landes. London and New York: Routledge, 1945. English edition 2012.

Smith, David Woodruff, and McIntyre, Ronald. *Husserl and Intentionality: A Study of Mind, Meaning, and Language*. Dordrecht and Boston: D. Reidel Publishing Company (Now Springer), 1982.

Smith, David Woodruff. "How to Husserl a Quine—and a Heidegger Too." *Synthese* **98**, No. 1 (January 1994), 153–173.

Smith, David Woodruff. *Husserl*. New York and London: Routledge, 2013. 2nd edition, revised; 1st edition, 2007.

Smith, David Woodruff. "Science, Intentionality, and Historical Background." In Hyder, David and Rheinberger, Hans-Jörg, eds., *Science and the Life-World: Essays on Husserl's Crisis of European Sciences*. Stanford, CA: Stanford University Press, 2010, 1–26.

Smith, David Woodruff. "Husserl and Tarski: The Semantic Conception of Intentionality cum Truth". In Guillermo Rosado Haddock, ed., *Husserl and Analytic Philosophy*. Berlin: Walter de Gruyter, 2016a, 143–173.

Smith, David Woodruff. "Truth and Epoché: The Semantic Conception of Truth in Phenomenology." In Jeffrey A. Bell, Andrew Cutrofello, and Paul M. Livingston, eds., *Beyond the Analytic-Continental Divide: Pluralist Philosophy in the Twenty-First Century*. London and New York: Routledge, 2016a, 111–128.

Smith, David Woodruff. "Book Review: *Husserl's Legacy: Phenomenology, Metaphysics, and Transcendental Philosophy*, by Dan Zahavi, Oxford and New

York: Oxford University Press, 2017." *European Journal of Philosophy* 27, No 1 (March, 2019b), 284–290.

Smith, David Woodruff. "Intersubjectivity: In Virtue of Noema, Horizon, and Life-World." In Frode Kjosavik, Christian Beyer, and Christel Fricke, eds., *Husserl's Phenomenology of Intersubjectivity: Historical Interpretations and Contemporary Applications*. New York and London: Routledge, 2019, 114–141.

Thomasson, Amie L. *Ontology Made Easy*. Oxford and New York: Oxford University Press, 2015.

Tieszen, Richard. *Phenomenology, Logic, and the Philosophy of Mathematics*. Cambridge and New York: Cambridge University Press, 2005.

Tieszen, Richard. *After Gödel: Platonism and Rationalism in Mathematics and Logic*. Oxford and New York: Oxford University Press, 2011.

Walsh, Philip J. "Motivation and Horizon: Phenomenal Intentionality in Husserl." *Grazer Philosophische Studien* 94, No 3, 410–435. Leiden, The Netherlands: Brill/Rodopi, 2017.

Yoshimi, Jeffrey. *Husserlian Phenomenology: A Unifying Interpretation*. Dordrecht and Boston: Springer, 2016.

Yoshimi, Jeffrey, Tolley, Clinton, and Smith, David Woodruff. "California Phenomenology." In Michela Beatrice Ferri, ed., *The Reception of Husserlian Phenomenology in North America*. Dordrecht and Boston: Springer, 2019, pp. 365–387.

Zahavi, Dan. *Husserl's Legacy: Phenomenology, Metaphysics, and Transcendental Philosophy*. Oxford and New York: Routledge, 2017.

10 Husserl on 'Besinnung' and Formal Ontology[1]

Mirja Hartimo

I. Introduction

In his *Logical Investigations* (1900–1901) and *Ideas I* (§10), Husserl conceived mathematics as the source of formal ontology (for a standard view of Husserl's formal ontology, see Smith 1989). However, in his *Formal and Transcendental Logic* (1929, Hua XVII, henceforth FTL),[2] Husserl found the structuralism of mathematical theories insufficient to serve as the ontology for the real world and developed a new conception of formal ontology, based on the ontological commitments of logic as opposed to those of mathematics. Crucial to Husserl's development is his usage of the method of *Besinnung*, as Husserl explains in the introduction to FTL: Husserl first states that the purpose of the essay is to provide "*an intentional explication of the proper sense of formal logic*" (Hua XVII, 14/10). 'Intentional explication' refers to the philosopher's task of clarifying and renewing the "final sense" of logic towards which the scientists have always been aiming. It thus assumes that the scientists—*exact* scientists, for Husserl—have been striving for certain goals for centuries. In FTL, Husserl seeks to make these goals explicit, examine them, and possibly revise them. Furthermore, according to Husserl, this aim should be pursued by means of *Besinnung*. Using radical *Besinnung* as his method, Husserl claims, he arrived at the contents of FTL (Hua XVII, 14/10). He points out also that his views have importantly changed in comparison to the *Logical Investigations* (1900–1901). The novelties of FTL are (1) the threefold stratification of logic that he claims was not yet completely clear in *Logical Investigations*; (2) the radical clarification of the relationship between formal logic and formal mathematics; (3) the definitive clarification of the sense of pure formal mathematics; and, connected to this, (4) the genuine sense of *formal ontology* (Hua XVII, 15/11).

In what follows, my aim is to examine the last novelty, that is, Husserl's new notion of formal ontology. Explaining this, however, requires some understanding of the other novelties as well. I will start by explaining Husserl's method of radical *Besinnung* and its relationship to Husserl's view of the "intentional history" of logic and mathematics. Thanks to his reliance upon *Besinnung*, Husserl's approach is informed by the practices of formal sciences and the goals of the logicians (in his terms, by the "living intentions of logicians" (Hua XVII, 14/10)). This is important, because in FTL it leads Husserl to distinguish between mathematics and logic and thus to isolate the proper sense of formal mathematics from logic as a *theory of science*, i.e., as a theory of the conditions any scientific theory should seek to fulfill to count as a science (for more detail, see Smith 1989, 29–31). Formal mathematics is a universal and a priori discipline, and, hence, a *potential* candidate for offering a formal ontology. However, in FTL Husserl argues that since formal mathematics has nothing to do with questions of actual existence and truth, the genuine sense of formal ontology is subservient to the interests of logicians rather than to those of mathematicians. Thus, Husserl's method of *Besinnung* makes his view of formal ontology sensitive to the development of modern mathematics into an independent discipline. In FTL, Husserl realizes that modern mathematics as a structuralist enterprise offers too little to serve as the source of ontology. Ontology, in the proper sense of the term, should be related to intuitable objectivities—to something that actually might exist. According to the final sense of logic, Husserl formulates a judgment theory through which ontology is related to the actual world, so as to make up a universal but "wordly" ontology.

II. *Besinnung* as a Method

As already briefly indicated, the aim of FTL is what Husserl calls intentional explication of the proper sense of formal logic (Hua XVII, 14/10). Logic in turn is understood to be a theory of science (Hua XVII, 13/9), in particular, it is a study of pure essential *norms* of science (Hua XVII, 7/3). In other words, logic is about what (formal) sciences *ought to be*. Husserl believes that the scientists' understanding of what sciences ought to be like has been guiding scientists for centuries. For Husserl this normative ideal is an "intentive sense" of the scientific research.

Husserl further explains that sciences should be approached by means of *Besinnung*, which he defines as follows:

> *Besinnung* signifies nothing but the attempt actually to produce the sense 'itself, . . ., it is the attempt to convert the 'intentive sense' . . . the sense 'vaguely floating before us' in our unclear aiming, into the fulfilled, the clear, sense, and thus to procure for it the evidence of its clear possibility
>
> (Hua XVII, 13/9).

By means of *Besinnung*, the normative ideals of the sciences are made explicit. Furthermore, Husserl holds that it requires standing in, or entering, "a community of empathy with the scientists" [*Mit den Wissenschaftlern in Einfühlungsgemeinschaft stehend oder tretend*, . . .] (Hua

XVII, 13/9). The "intentive sense" of the scientific research is thus drawn from scientists' activities, not from a priori sources. This feature makes Husserl's view context-dependent and "mathematics-first"—indeed, reliant on a kind of naturalism about mathematics.[3] The mathematics-first view is a conception in which mathematics is approached as an autonomous discipline, "on its own," as opposed to the philosophy-first views, in which the practice of mathematics is found subservient to different kinds of philosophical demands. In accordance with such a mathematics-first view, in FTL Husserl begins by discussing the aims of the scientists, especially those of the formal scientists—aims which are typically implicit. Thus, he does not start with a theory of evidence and claim that the mathematicians should hold on to it, but the other way around—he seeks to clarify the evidences that are already used in mathematicians' practices.

Husserl also holds that *Besinnung* should be radical. To be radical in Husserl's phenomenology is to attempt to expose the tacit presuppositions held in the practices, whether theoretical or not. In FTL, this brings Husserl to ask transcendental questions about logic, that is, to engage in what he calls 'transcendental logic.' Through transcendental logic, formal logic and mathematics are seen to aim for certain kinds of evidence. Husserl also identifies several presuppositions that are made in the exact sciences.[4] Such transcendental questioning distinguishes his approach from any mathematical naturalism akin to the one that can be found in Maddy: While Husserl evaluates the practices in terms of their goals, as Maddy does, too, he also aims at revealing how these goals are constituted, and, thus, their conditions of possibility. This adds a further revisionary element to Husserl's approach: By means of a transcendental examination, Husserl hopes to revise confused senses and concepts, so as to make the practices *genuine* [*echt*] (Hua XVII, 14/10).

Assuming that Husserl indeed used this method, as he claimed he did, to obtain the results published in FTL, one is led to examine his "fellow mathematicians." The books in his private library, and especially his notes in them, suggest that in the early 1920s the fellow mathematicians were primarily David Hilbert, Hermann Weyl, and Oskar Becker. Husserl had markings in Hilbert (1922) as well as in Weyl (1925, 1926) (see further Hartimo 2018b). Of the people around Husserl in the 1920s, Oskar Becker was the most knowledgeable one in mathematics and physics.[5] Becker worked as Husserl's assistant from 1923 and stayed in Freiburg until 1931 (Mancosu 2010, 281). During that time, Becker wrote *Mathematische Existenz*, published in Husserl's *Jahrbuch* in 1927. Husserl had read at least the beginning of it.[6]

Based on his methodological considerations, Husserl's FTL should thus be read as an evaluation and renewal of the aims of mathematics discussed primarily by Hilbert, Weyl, and Becker in the 1920s.[7] These aims concerned the axiomatic approach to mathematics and different ways of providing it with intuitionistic, predicative, or proof-theoretical foundations—motivated by the discovery of the set-theoretical paradoxes. As we will soon see, Husserl accordingly isolates the pure sense of mathematics as axiomatics, and then, in his logical considerations, explores the ways in which the formal sciences relate to intuition and to the world. Husserl's method involves examination of what the fellow scientists, especially mathematicians, are seeking, seeing that as part of historical developments towards certain goals—their "final senses," and evaluating these goals critically. Consequently, FTL should be read as a clarification of the potentially overlapping and unclear goals of the approaches discussed by Hilbert, Weyl, and Becker.

III. Intentional History of Logic, Intentional History of Mathematics

Husserl's examination of the intentional history of the formal sciences takes place in two distinguishable progresses: On the one hand, there is the development of the theory of judgments, and, on the other, there is the development of formal mathematics. Common to these two fields is that they are both "interested specifically in certain derivative formations of anything-whatever" (Hua XVII, § 24), i.e., they are both formal. However, they differ in being guided by different ideals, i.e., intentive senses. The sense guiding formal mathematics is the Euclidean ideal, concretely captured by the notion of "definite manifold" (Hua XVII, § 31). The definite manifold is a structure derived from the Euclidean axiom system by means of "formalization." With it, a theory-form is obtained from Euclidean geometry understood as the theory of intuited world-space, so that "all the materially determinate What-contents of the concepts—in the case of geometry, all the specifically spatial contents—are converted into indeterminates, modes of the empty 'anything-whatever.'" (FTL, § 29). It is complete in the sense that it captures its domain exhaustively ("there is no truth about such a province that is not deductively included in the 'fundamental laws' of the corresponding nomological science"). According to Husserl, the Euclidean ideal captures Hilbert's intentions that led Hilbert to add the 'axiom of completeness' to his axiomatizations of geometry and arithmetic around the turn of the century. He also views his own formulations of the notion of definite manifold as attempts to give a concrete articulation to the Euclidean ideal (Hua XVII, §31). I have argued elsewhere that Husserl's view of completeness embraces both categoricity and syntactic completeness (Hartimo 2018a). Husserl's notion of formalization thus refers to an abstraction from a domain of an individual theory (system) to the domain of a categorical theory (structure). According to Husserl, the great advance of pure mathematics, particularly thanks to Riemann, does not stop at characterization of such pure structures, but taking such structures as mathematical objects themselves (FTL, § 30). This suggests that in Husserl's view the guiding goal of the mathematicians is increasing abstraction, and, hence, what captures the sense of pure modern mathematics in Husserl's view.

The guiding concept of logic (i.e., theory of judgment) is that of truth. A closer inspection shows that truth presupposes non-contradiction and grammaticality. Thus, logic can be divided into three goals and, accordingly, into three 'layers': grammar, logic of non-contradiction, and logic of truth. These are linked to three different kinds of evidences: the most general evidence, distinctness, and clarity, respectively. Husserl discusses first the development of "apophantic analytics," which is purely formal and consists only of grammar and logic of non-contradiction. To it belong, quoting Husserl, "not only the whole of syllogistics, so far as its essential content is concerned, but also (as we shall show) many other disciplines, namely those of formal-mathematical 'analysis'" (Hua XVII, §14). Apophantic analytics operates with what he calls 'apophantic senses' and relates, besides the most general evidence, to the evidence of distinctness. In Husserl's view, this apophantic analytics and formal mathematics are equivalent disciplines. Formal logic (i.e., theory of judgment) adds to them an interest in truth. In fact, formal logic and formal mathematics are in the end distinguished only by their final senses or goals that are revealed by *Besinnung* of the scientists' goals. Logic and mathematics are practices carried out with different kinds of *intentions*. In addition to the most general evidence related to grammaticality and the distinctness related to non-contradictoriness, logicians aim at truth and its evidence of clarity. Hence, Husserl writes:

> a *formal mathematics, reduced to the above described purity, has its own legitimacy* and that, for mathematics, there is in any case no necessity to go beyond that purity. At the same time, however, a great advance is made philosophically by the insight that such a restrictive reduction of *logical* mathesis (formal logic, when it has attained the completeness befitting its essence)—namely its reduction to a *pure analytics of non-contradiction*—is essentially its reduction to a science that has to do with nothing but apophantic senses, in respect of their own essential Apriori, and that in this manner the *proper sense of 'formal mathematics'*, the mathematics to which every properly logical intention (that is: every intention belonging to a theory of science) remains alien—the mathematics of mathematicians—at last becomes *fundamentally clarified*. Here lies the *sole legitimate distinction between formal logic and mere formal mathematics.*
>
> (Hua XVII, §52, 146/140–141; italics in original)

In other words, there is (necessarily) no difference between formal logic and formal mathematics when their theories are considered purely formally. But when one pays attention to the mathematicians' and logicians' intentive senses, one notices the logicians' interest in giving a true description of the actual world, in grammatical evidence, distinctness, as well clear [*klar*] evidence, analogous to the one had when perceiving middle-sized physical objects, whereas the mathematicians do not need to worry about the evidence of clarity.

IV. Formal Ontology

These two historical developments, one within mathematics, the other within logic, can both be considered as pertaining to formal ontology insofar as they are about something that is universal and a priori. Husserl first maintains that since formal mathematics is about formal objects, "it is natural to view this whole mathematics as an *ontology* (an a priori theory of objects), though a *formal* one, relating to the pure modes of anything-whatever" (Hua XVII, § 24). Such objects are completely indeterminate, "objects of thinking," that are determined

> exclusively by the form of the connexions ascribed to them. These connexions themselves are accordingly as little determined in respect of content as the Objects connected; only their form is determined, namely by the form of the elementary laws assumed to hold good for them.
>
> (Hua XVII, § 28, cited from the *Prolegomena to the Logical Investigations*, §70)

In other words, these objects are "pure positions" of structuralist ontology, determined only by the place they have in a structure. Indeed, Charles Parsons has pointed out that the most developed statement of structuralism before World War II is due to Husserl (Parsons 2008, 41).

Since the questions of truth and what actually might exist are excluded from formal mathematics, Husserl finds this structuralist view of formal ontology insufficient. The mathematical objects as conceived in structuralism are too abstract to have anything to do with truth and the substrates 'themselves.' Accordingly, Husserl thinks that proper formal ontology has to be carried out with the "logical interest."[8] Husserl writes that

> [l]ike the sciences themselves, *analytics as formal theory of science is directed to what exists* [*ontisch gerichtet*]; moreover, by virtue of its apriori universality, it is *ontological*. It is *formal ontology*. Its apriori truths state what holds good for *any objects whatever*, any object-provinces whatever, with formal universality, *in whatever forms they exist* or merely *can exist*—as objects of judgments [*urteilsmässig*], naturally: since, without exception, objects 'exist' only as objects of judgments and, for that very reason, exist only in categorial forms.
>
> (Hua XVII, 125/120)

According to Husserl, objects have being for us only as making their appearance in judgments (Hua XVII, § 25). Furthermore, logically considered, the arithmetic of cardinal numbers and the arithmetic of ordinal numbers and so forth have existence on their own (Hua XVII, § 33), even though they are instantiations of the same structure. The structuralist ontology suggested by formal mathematics in its detachment from the questions of truth and existence is thus not formal ontology in the proper sense of the term. Formal ontology should relate to what is judged in formal apophantics to be possibly true. Thus, Husserl concludes that

> *The aforesaid pure mathematics of non-contradiction*, in its detachment from logic as theory of science, *does not deserve to be called a formal ontology*. It is an ontology of pure judgments *as senses* and, more particularly, an ontology of the *forms* belonging to non-contradictory—and, in that sense, possible—senses: possible in distinct evidence.
>
> (Hua XVII, 150/144)

Structuralist ontology operates with *distinct* evidence that is the kind of evidence intended in the logic of non-contradiction, i.e., in formal mathematics. The proper formal ontology should relate to possible objects and theories given in addition in *clear* evidence obtained in an encounter with the world:

> *[F]or a . . . 'pure' formal mathematics, there can be no cognitional considerations [Erkenntnissorgen] other than those of 'non-contradiction'*, of immediate or mediate *analytic consequence or inconsistency*, which manifestly include all questions of *mathematical 'existence'*. It is otherwise, to be sure, for the *logician*: Being interested in a theory of science even when consistently broadening the traditional confines, he presses onward to *mathesis universalis* (as I myself did in the *Logische Untersuchungen*), he will not easily come upon the thought of making this reduction to an analytics of pure senses; and therefore he will acquire mathematics as only an *amplified logic*, which, as a logic, relates essentially to *possible* object-provinces and theories.
>
> (Hua XVII, 145–146/140; emphasis in original)

Husserl thus distinguishes between mathematical, structural existence characterized by "non-contradiction" and, connected to it, distinct evidence, and "possible actuality" or "the possible true being." For him the objects of formal mathematics, or mere positions in structures, as structuralism will have it, are too abstract to account for what is meant in judgments about objects. The objects of formal ontology should have a relationship to judgments about individuals and, hence, to what is given in intuition, in evidence of clarity. Accordingly, Husserl explains in the introduction to FTL that

> though it seemed obvious that a science relating with this universality to anything and everything—to everything possible, everything imaginable—deserves to be called a formal *ontology*, still, if it is to be one actually, then the *possibility* of objectivities belonging in its sphere must be established by intuition [*Anschauung*].
>
> (Hua XVII, 16/12)

Whereas formal mathematics offers us a merely possible formal ontology, an actual formal ontology has to establish the possibility of the objects by relating them to experiences in which objects themselves are given in clear evidence, as in perceiving them.

V. Transitional Link

The logical interest in truth requires givenness of the meant objectivities themselves, and, hence, clear evidence that has its source in the world, outside the non-contradictory formal theory:

> Here a truth signifies *a correct critically verified judgment*—verified by means of an adequation to the corresponding categorial objectivities 'themselves', as given in the evidential having of them themselves: given originaliter, that is, in the generating activity exercised on the basis of the experienced substrates 'themselves.'
>
> (Hua XVII, § 46, 132/127)

It seems that the "generating activity on the basis of the experienced substrates" can be understood in two ways: either it refers to material applications of the formal theories (e.g., geometry, mechanics) or else it refers to a judgment theory as "a transitional link" [*Übergangsglied*] between logic of non-contradiction (formal mathematics) and logic of truth. Husserl explains the former route in more detail in *Ideas I*. The material realizations of the formal theories form material ontologies (Hua III/1, § 10). The basic concepts of these disciplines are concepts of exact material essences that can be derived from the theory but can also be obtained from intuition through the method of eidetic seeing, which Husserl later develops into the method of eidetic variation (Hua III/1, §§ 4, 66, 72; EU, § 87a, 410–411/340). Highest universalities delimit regions of objects (e.g., the region consisting of material things, the region of animate organisms, and the psyche). In these regions they form hierarchies, ranging from the most general (e.g., any physical thing whatever, any sensory quality, any spatial shape, any mental process) to the most specific, from the highest genus to the infimae species, the eidetic singularities (Ideas I, §12, 31/25).

However, these material ontologies are regional, and, hence, not universal, as Husserl thinks formal ontology ought to be. This suggests that Husserl needs another way to connect the logic of non-contradiction to the world. This is provided by the "transitional link":

> In the first place, we require here an important supplementation of the pure logic of non-contradiction, a supplementation that, to be sure, goes beyond formal mathematics proper, but still does not belong to truth-logic. It is a matter, so to speak, of *a transitional link* between them (Hua XVII, 209–210/202).

The transitional link is a judgment-theory [*Urteilstheorie*], which is more explicit than the apophantic analytics, discussed in the beginning of FTL (esp. § 13). Crucially, it carries in it the information about the grammatical cores of the judgments, which seems to be the source for its normalizability. Within pure apophantic logic, one can construct complex judgments out of simple forms of judgment. Husserl explains that, for example, from the judgment 'S is p' one can construct the form 'Sp is q' and then '(Sp)q is r.' These judgments can be 'modified' so that they can occur as component parts in, e.g., a conjunction or a hypothetical form of judgments. Such construction is law-governed and reiterative. In addition to this, the judgment theory that provides the transitional link "normalizes" (not the term Husserl uses), so that "any actual or possible judgment leads back to ultimate cores when we follow up its syntaxes" (Hua XVII, § 82, 210/202–203). Or, as Husserl also characterizes it:

> the reduction signifies that, *purely by following up the meanings*, we reach *ultimate something-meanings*; first of all, then, as regards the meant or supposed judgment-objects, *supposed absolute objects-about-which* [*vermeinte absolute Gegenständeworüber*]. (Hua XVII, 211/203)

Husserl's brief description of the reduction thus suggests that it is mechanical or computable. The judgment theory envisioned by Husserl appears to have enough "computable" content in its forms of judgment to enable what one might call "strong normalization," that is, every judgment is mechanically reducible to elementary judgments.[9] The complex judgments of the theory can thereby be mechanically reduced into ultimate subjects, predicates, universalities, and relations:

> it can be seen a priori that *any actual or possible judgment leads back to ultimate cores* when we follow up its syntaxes; accordingly that it is a syntactical structure built ultimately, though perhaps far from immediately, out *of elementary cores, which no longer contain any syntaxes*. . . . And always it is clear that, by reduction, we

> reach a corresponding *ultimate*, that is: *ultimate substrates*—from the standpoint of formal logic, *absolute subjects* (subjects that are not nominalized predicates, relations, or the like), *ultimate predicates* (predicates that are not predicates of predicates, or the like), *ultimate universalities*, *ultimate relations*.
>
> (Hua XVII, 210–211/202–203)

Thus, the transitional link leads back to what Husserl calls *ultimate cores*, but what could also be called *canonical forms of expressions*.

The reduction takes place first on the level of senses, and then, analogously, on the level of truth:

> To the reduction of judgments to ultimate judgments with an ultimate sense, there corresponds a *reduction of truths*: of the truths belonging to a higher level to those belonging on the *lowest level*, that is: to truths that relate directly to their matters and material spheres, or (because the substrates play the leading role here) that relate directly *to individual objects* in their object spheres—individual objects, objects that therefore contain within themselves no judgment-syntaxes and that, in their experienceable factual being, are *prior to all judging*. That judgments (not judgment-senses) relate to objects signifies that, in the judgment itself, these objects are meant as substrates, as the objects about which something is stated; and reductive deliberation teaches, as an *Apriori*, that *every conceivable judgment ultimately* (and either definitely or indefinitely) *has relation to individual objects* (in an extremely broad sense, real objects), and therefore has *relation to a real universe*, a 'world' or a *world-province*, 'for which it holds good.'
>
> (Hua XVII, 212/204)

The judgment-theory ultimately establishes that the complex judgments can be reduced to judgments about individuals in the world. This addresses the set of problems Husserl promised to discuss in this section by relating logical truth to the objects in the world:

> Furthermore there is the set of problems offered by the *relation of predicational truth to objects-about-which* and, finally, to 'ultimate substrates', objects of possible 'experience.' These objects, the material [*das Sachliche*] in the ultimate sense, are in the opinion of traditional logic, something '*Objective*': Experience as such is *Objective experience*; truth as such is *Objective truth*. Truth is truth in itself concerning "Objects"—belonging to an *Objective world*.
>
> (Hua XVII, 208/201)

The judgment-theory aims to provide the connection between the abstract structuralist formal mathematics, or what Husserl also calls logic of non-contradiction, and the objects that possibly actually exist. Husserl points out that this is not something mathematicians need to care about, but it is something we need to do if we are interested in truth, and, hence, in formal ontology:

> For *mathesis universalis*, as formal mathematics, these ultimates have no particular interest. Quite the contrary for *truth-logic*: because ultimate substrate-objects are *individuals*, about which very much can be said in formal truth, and *back to which all truth ultimately relates*. If one keeps to the formal of pure analytics, if the evidence—the evidence serving this discipline—accordingly relates only to pure judgment-senses as distinct, one cannot establish this last proposition. To have insight into it, one must *make ultimate cores intuited*, one must draw fullness of adequation, not from evidence of the judgment-senses, but instead from evidence of the 'matters' or 'affairs' corresponding to them.
>
> (Hua XVII, 211/203)

The role of judgment theory is thus not to prove a certain part of mathematics consistent or otherwise to justify a body of mathematics. Rather, its role is to transfer intuition of objects to more complex formations and, presumably, ultimately to (at least part of) formal mathematics. Judgment theory preserves evidence, whether distinct or clear. The body of mathematics that can be normalized into basic forms of judgments about actually existing objects can thus be known with clarity.

Husserl's notion of evidence is thus more general than, for example, Charles Parsons's broadly Kantian view of intuition. For Parsons, mathematical intuition is one which gives objects that instantiate concepts that have a sharp and precise character (Parsons 2008, 165). His paradigm example for mathematical intuition is intuition of strokes, or strings of strokes, that are "quasi-concrete," so that by way of perceiving a token of a type, the type is intuited (ibid., 160). By contrast, in his discussion of the evidence of ideal objects Husserl merely claims that it is analogous to the evidence of ordinary perception. He writes that in it

> [t]he identity and, therefore, the *objectivity* of something ideal can be directly 'seen' . . . with the same originality as the identity of an object of experience in the usual sense—for example: an experienced object belonging to Nature or an experienced immanent object (any psychic Datum).
>
> (Hua XVII, 163–164/155)

Decisive for it is that the evidently given object has an identity and that it is given in itself, as if "in person." For Husserl the basic mode of such evident givenness is perception, but he considers also more complicated modes, such as recollection (Hua XVII, §§ 58–59). Whereas Parsons is worried about the vagueness of our spatial perception, Husserl takes it as a fact that we are able to individuate objects. Whereas Parsons, like Hilbert, searches for certainty in intuition, Husserl readily acknowledges that "[t]he *possibility of deception* is inherent in the evidence of experience and does not annul either its fundamental character or its effect" (Hua XVII, § 58).

For Parsons intuitive knowledge can be preserved by certain logical inferences, e.g., simple tautologies, addition and multiplication, but to him reiteration is not always able to preserve intuitive knowledge (2008, § 29). One may raise a question about how exactly Husserl's judgment theory preserves evidence, and, hence, intuitiveness of knowledge. One answer could be, indeed, in its use of reiteration, which in a "Brouwerian" manner could be thought of as the fundamental intuition of mathematics (as Mark van Atten has argued against Parsons, whose concept of intuition runs out at this point), and, hence, as what enables passing on intuitive knowledge (cf. Parsons 2008, 175, 235–262). Taking into account also Husserl's earlier approaches to the problem, I am inclined to claim that for Husserl the criteria for whether inferences preserve intuitive knowledge lie in strong normalization, that is, in mechanical reducibility of the judgment to elementary judgments suggested in his discussion of the transitional link. Around the turn of the century, Husserl advocated a similar approach that took place by means of equational reductions (for the detailed argument, see Hartimo and Okada 2016). In FTL, mechanical reducibility is based on the structure of the judgments that includes information about their original "cores." These cores thus provide the "computational" content to enable normalization.

Husserl explains that the "*reductive deliberations* [*reductive Überlegungen*]," as here explained, uncover "*hidden intentional implications* included in judging and in the judgment itself as the product of judging. *Judgments as senses accordingly have a sense-genesis* [*Sinnesgenesis*]" (Hua XVII, 215/207). Husserl's "transitional link" is what reveals the sense-genesis of the judgments. Curiously, Husserl thereby arrives at a rather systematic judgment-*theory* [*Urteilstheorie*] in his transcendental questioning concerning the constitution of the judgment senses (Hua XVII, § 86).[10] Husserl's judgement-theory, and "true" mathematics formulable by means of it (to be sure, Husserl does not explicitly articulate such "true" mathematics, but it seems to be implied in what he does in FTL), resembles Hilbert's formulation of real mathematics that has an intuitive basis in intuition of strokes and primitive recursive operations. Both Husserl and Hilbert thus seek to investigate the extent of intuitive knowledge in mathematics (for the way in which Hilbert does it, see Parsons 2008, § 28). Husserl thinks that, instead of Hilbertian strings of strokes, the paradigm case of evidence is perception of external, concrete objects. Furthermore, he distinguishes the evidence of clarity from the evidence of distinctness, and, hence, the search for non-contradiction from the search for truth. Furthermore, Husserl thinks that consistency can be established model-theoretically, whereas Hilbert created his proof-theory for this purpose. Husserl thinks that mathematicians do not need to seek any intuitive basis as long as their theories are consistent. Considered from Husserl's perspective, Hilbert simply confuses distinctness and clarity. Thus, it seems that regarding this particular issue, Husserl's radical *Besinnung* is an evaluation of especially Hilbert's attempt at providing mathematics with intuitive foundations. Husserl does not, however, approach Hilbert's view "philosophy-first," but engages in *Besinnung* of the various normative goals of mathematicians. Only after having examined the sense of mathematics as opposed to the sense of logic is he in the position not only to reformulate his view of formal ontology but also to suggest revisions to Hilbert's project.[11]

VI. Conclusion

Husserl arrived at formal ontology with a method he termed *Besinnung*. By means of *Besinnung*, he engaged in gleaning the intentive senses of his fellow mathematicians, especially those in Hilbert. By its means, Husserl formulates the proper sense of formal mathematics in contradistinction to that of formal logic. As Husserl sees it, different aspects of formal ontology have been sought in different ways in mathematics and in logic. Whereas in mathematics one has aimed for Euclidean manifolds and has thereby reached the notion of "any objectivity whatever," logic as a theory of science is concerned with truth and intuitability of objects. Husserl's initial formulation of formal ontology suggests that it consists of objects as conceived of in purely structural terms. This is too abstract to properly capture the objects as they exist and relate to truth. Hence, in his ultimate conception of formal ontology, Husserl substantiates his otherwise structuralist ontology with a constructive-intuitive judgment theory.

Structuralism has been criticized in the literature because of the *incompleteness* of its objects. Probably the best-known instance of this criticism is due to Paul Benacerraf (Benacerraf 1964, 291). The incompleteness objection runs as follows. It must be possible to individuate the abstract objects of mathematics independently of the role they play in a structure. Objects, as conceived of in structuralism, are "incomplete," because they can only be ascribed properties defined by a structure. Their existence is not sufficiently independent. This indeterminateness poses problems, e.g., for the applications of mathematics (Parsons 2008, 106, 151). Husserl appears to share these concerns in his claim that formal ontology acquired from formal mathematics does not deserve to be called 'ontology.' But Husserl's approach is "mathematics-first": he thinks that mathematicians should not worry about such philosophical concerns. These concerns are of interest only to those who share the logical interest in truth.

Husserl then formulates a judgment theory, putatively with a strong normalization property. By its means, Husserl examines and describes the way in which evidence can be mediated from a direct confrontation with the concrete, actual world to the higher flights of abstraction. The judgment theory then helps to single out one universal but mundane ontology that is shared by all material ontologies. I conclude with a quote from Husserl's own conclusion on this:

> this mundane ontology explicates the all embracing Apriori of any purely possible world whatever [*das universal Apriori einer in reinem Sinne möglichen Welt überhaupt*], the Apriori of the eidos *world*—an eidos that must arise concretely by virtue of the method of eidetic variation, which starts with the world that is given us in fact and takes it as the directive 'example'. This thought is the basis from which arise, at successive *levels, the great problems* pertaining to a *world-logic* [*Welt-Logik*] that is to be grounded radically, a genuine mundane ontology—some parts of which have already been indicated.
>
> (Hua XVII, 296/291)

Husserl's reference to the a priori of the eidos world anticipates his later analyses of the life-world and it's a priori structures. However, that lies beyond the scope of the present chapter and will be left for another occasion (I discuss it in Hartimo 2018d).

Notes

1. I greatly acknowledge the support from the Centre for Advanced Study in Oslo, Norway, which hosted our research project 'Disclosing the Fabric of Reality—The Possibility of Metaphysics in the Age of Science,' during the academic year 2015/16. The present article was conceived as part of the project.
2. In this chapter, I will refer to the text by means of section numbers and thereby simultaneously to all editions and translations of the same work. When references are to pages in the Introduction and Conclusion that do not have numbered sections, I will first give the page number in the German version and then that of the English translation. The same convention is used in citations. The italics are from the originals unless otherwise indicated.
3. In particular, as I have argued elsewhere, on this point Husserl's view is remarkably similar to that formulated by Penelope Maddy (1997, 2007, 2011). It is important to note that this kind of naturalism does not entail reductionist scientific naturalism but rather, when generalized to all disciplines, a kind of liberal naturalism.
4. I explain all this in detail in Hartimo (2018c).
5. Becker had a background in mathematics, but he wrote his *Habilitationsschrift* entitled "Beiträge zur phänomenologischen Begründung der Geometrie und ihre physikalischen Anwendungen" (1922) with Husserl. Husserl praised the work, writing to Weyl that: "It is nothing less than a synthesis of Einstein's and your discoveries with my phenomenological investigations on nature . . ." (Letter to Weyl, dated April 9, 1922, cited from Mancosu 2010, 282).

6. Husserl-Chronik reports that on March 1937,"H. *hat* größere Abschnitte gelesen (insbesondere zum ersten Mal auch [?] die zweite Hälfte) *von* Oskar Becker, Mathematische Existenz, 1927." (Schuhmann 1977, 484).
7. Zermelo was at the time in Freiburg as well. His role for Husserl's views is unknown.
8. Parsons, too, notes that for Husserl structuralism does not give a complete account of mathematical objects (Parsons 2008, 41).
9. For these reasons, Husserl's judgment theory appears to intend something like intuitionistic type theory, in which so-called 'type checking' makes the strong normalization possible (Dybjer and Palmgren 2016). Crosilla (2019) explains this to be the import of the Curry-Howard isomorphism, which makes set theory and logic "entangled" in Martin-Löf's type theory.
10. What is curious about this is that Husserl's transcendental phenomenology is *defined* as a study in which the usage of scientific theories, logic, and mathematics is put in brackets so that they cannot be used in transcendental constitutional analyses (obviously scientific theories, mathematics, and logic can themselves be transcendentally analyzed). But here, the judgment-theory serves as an aid for the properly transcendental analyses. This shows how the analyses carried out in natural and transcendental attitudes can be interrelated in Husserl's phenomenology.
11. For the relationship between Hilbert and Husserl, see Hartimo (2017).

References

Benacerraf, Paul. "What Numbers Could Not Be." In P. Benacerraf and H. Putnam, eds., *Philosophy of Mathematics: Selected Readings*. Cambridge, New York and Melbourne: Cambridge University Press, 1964, 272–294.

Crosilla, Laura. "The Entanglement of Logic and Set Theory, Constructively." *Inquiry* (2019), special issue: *Critical Views of Logic*. doi.org/10.1080/0020174X.2019.1651080

Dybjer, Peter and Palmgren Erik. "Intuitionistic Type Theory." *The Stanford Encyclopedia of Philosophy* (2016). Winter 2016 Edition. Edward N. Zalta. <https://plato.stanford.edu/archives/win2016/entries/type-theory-intuitionistic/>.

Hartimo, Mirja. "Husserl and Hilbert." In S. Centrone, ed., *Essays on Husserl's Logic and Philosophy of Mathematics*. Synthese Library. Dordrecht: Springer, 2017, 245–263.

Hartimo, Mirja. "Husserl on Completeness, Definitely." *Synthese* **195** (2018a), 1509–1527.

Hartimo, Mirja. "Husserl's Scientific Context 1917–1938, a Look into Husserl's Private Library." *The New Yearbook for Phenomenology and Phenomenological Philosophy* **XVI** (2018b), 317–336.

Hartimo, Mirja. "Radical *Besinnung* in *Formale und transzendentale Logik* (1929)." *Husserl Studies* 34 (2018c), 247–266.

Hartimo, Mirja. "On the Origins of Scientific Objectivity." In F. Kjosavik, C. Beyer, and C. Fricke, eds., *Husserl's Phenomenology of Intersubjectivity: Historical Interpretations and Contemporary Applications*. London: Routledge, 2018d, 302–321.

Hartimo, Mirja and Okada Mitsuhiro. "Syntactic Reduction in Husserl's Early Phenomenology of Arithmetic." *Synthese* **193** (2016), 937–969.

Hilbert, David. "Neubegründung der Mathematik: Erste Mitteilung." *Abhandlungen aus dem mathematischen Seminar der Universität Hamburg* 1 (1922), 157–177.

Husserl, Edmund. *Formale und transzendental Logik* [FTL]. Husserliana XVII. Ed. by P. Janssen. The Hague: Martinus Nijhoff, 1974. English translation: *Formal and Transcendental Logic*. Trans. by D. Cairns. The Hague: Martinus Nijhoff, 1969. (Cited as Hua XVII.)

Husserl, Edmund. *Ideen zu einer reinen Phänomenologie und phänomenologischen Philosophie. Erstes Buch: Allgemeine Einführung in die reine Phänomenologie* [*Ideas I*]. Husserliana III/1. Ed. by K. Schuhmann. The Hague: Martinus Nijhoff, 1976. English translation: *Ideas Pertaining to a Pure Phenomenology and to a Phenomenological Philosophy*. Trans. by F. Kersten. Dordrecht, Boston and London: Kluwer, 1982. (Cited as Hua III/1.)

Maddy, Penelope. *Naturalism in Mathematics*. Oxford: Oxford University Press, 1997.

Maddy, Penelope. *Second Philosophy: A Naturalistic Method*. Oxford and New York: Oxford University Press, 2007.

Maddy, Penelope. *Defending the Axioms: On the Philosophical Foundations of Set Theory*. Oxford: Oxford University Press, 2011.

Mancosu, Paolo. *The Adventure of Reason: Interplay between Philosophy of Mathematics and Mathematical Logic 1900–1940*. Oxford: Oxford University Press, 2010.

Parsons, Charles. *Mathematical Thought and Its Objects*. Cambridge: Cambridge University Press, 2008.

Schuhmann, Karl. *Husserl-Chronik, Denk- und Lebensweg Edmund Husserls*. The Hague: Martinus Nijhoff, 1977.

Smith, Barry. "Logic and Formal Ontology." In J.N. Mohanty and W. McKenna, eds., *Husserl's Phenomenology: A Textbook*. Lanham: University Press of America, 1989, 29–67.

Van Atten, Mark. *Intuition, Iteration, Induction* (Unpublished manuscript), https://arxiv.org/abs/1510.01094v1.

Weyl, Hermann. *Die heutige Erkenntnislage in der Mathematik*. Erlangen: Verlag der Philosophischen Akademie, 1925.

Weyl, Hermann. *Philosophie der Mathematik und Naturwissenschaft*. München: Oldenbourg, 1926.

Part III

Contemporary Metaphysics

The Role of Language and Science

11 Quine on Truth and Metaphysics*

Charles Parsons

—In memory of James Higginbotham, 1941–2014

W. V. Quine was a prolific philosophical writer, but he did not write a lot about truth. I think the reason for this is that he thought the story he had to tell was relatively simple and uncomplicated. Unlike in the case of some signature Quinean themes, such as the analytic-synthetic distinction and the indeterminacy of translation, he did not feel the need to restate his arguments to dispel misunderstandings of his views. It does turn out that the matter looks more complicated in the light of later statements of his. They also suggest some comparison of Quine's views of truth with those of Donald Davidson.

In the last part of this chapter, I will discuss briefly the question whether Quine was a metaphysician and speculate, mainly on the basis of one example, about what Quine might have thought about the analytic metaphysics of today.

I

One of the first questions one naturally asks about Quine on truth is this: He regarded the notion of meaning or sense and related intensional notions used in reflecting on language as questionable, not suitable for serious science. But he was quite happy to use the notion of truth, and with qualifications other notions belonging to what he called the theory of reference, such as denotation and extension, in spite of the thesis advanced in 1968 of the inscrutability or relativity of reference. Why is truth acceptable but notions like meaning or sense are not?

The key lies in the simple truth schema, familiar from Tarski's writings. Take the stock example

> 'Snow is white' is true if and only if snow is white.

In an early essay, "Notes on the theory of reference," (essay 7 in Quine 1953), Quine discusses the general paradigm of which this is an instance, as well as others concerning reference, such as

> 'Harry S. Truman' names Harry S. Truman and no one else.

He mentions the fact that the truth schema and the related one for truth *of* or satisfaction can generate semantical paradoxes. He gives an account of the paradoxes and of the idea of a definition of truth that largely follows Tarski. But in spite of the paradoxes, he ends the essay with remarks to the effect that these paradigms make notions of truth and reference much clearer than notions of the theory of meaning such as synonymy and analyticity:

> It is a striking fact that these paradigms, despite the paradoxes which we associate with them, are so very much less foggy and mysterious than the notions belonging to the theory of meaning. We have general paradigms (7)-(9) [of which the above are instances—CP] which, though they are not definitions, yet serve to endow 'true-in-L', 'true-in-L of', and 'names-in-L', with every bit as much clarity, in any particular application, as is enjoyed by the particular expressions of L to which we apply them. Application of truth, in particular, to 'snow is white', for example, is every bit as clear as attribution of whiteness to snow.
>
> (1953, 137–138)

Earlier in the essay Quine mentions using the schema to give the condition for the truth of a sentence in one language in another language, with the trivial example:

> 'Schnee ist weiss' is true-in-German if and only if snow is white.

He does not mention the complications that translation gives rise to. Of course he does not ignore the issue in later writings, as we will see.

There is an obvious objection to treating truth as a property of sentences, the fact that whether a sentence expresses a truth can depend on all sorts of contextual features, in particular who utters it and when. 'I have a headache' was all too often true, as uttered by me, when I was a young man and even into middle life, but there have been few occasions in old age in which I could have uttered it truly.

Quine meets this objection in a rather curious way, which so far as I know has been followed by virtually no one else. The obvious way would be to add parameters to cover the contextual factors, to yield something like:

> 'I have a headache' uttered by x at time t is true iff x has a headache at t.

But that is not how Quine proceeds, as one sees both in § 40 of *Word and Object* and in the late *Philosophy of Logic* and other writings that largely repeat it. Rather what he proposes is that the vehicles of truth should be "eternal sentences," sentences purged of elements whose reference depends on context. Suppose that at 10 p.m. on January 1, 1975, I said, "I have a headache." Quine would propose to render the truth condition roughly as follows:

> 'Charles Dacre Parsons has a headache at 10 p.m., US Eastern Standard time, on January 1, 1975' is true if and only if. . . .[1]

One would naturally object that this evades the problem of giving the truth-conditions for the natural-language utterance "I have a headache" as uttered at that particular time by that particular person, which in addition to containing the indexical 'I' is significantly in the present tense.

Why would Quine follow this apparently perverse procedure? In *Word and Object* what he aims at is to paraphrase sentences into a canonical language for science. This is a very different undertaking from "semantics of natural language." Donald Davidson, deeply influenced in many ways by Quine, proceeded in a way that is much more natural to us, giving disquotational truth-conditions where, in addition to something designating the sentence, there are parameters for speaker and time. There are several publications in which either Quine or Davidson explores issues on which they disagree. I don't believe that this issue was brought up in those papers.

However, Quine followed the same procedure in later writings such as *Philosophy of Logic*, at least without saying explicitly that the motivation for it was the same as it had been in *Word and Object*. He may have been influenced by his general preference for extensional locutions: the new expressions introduced have the same reference as those that are replaced had in the context of utterance. Another factor is that the Davidsonian version is not a strict instance of the Tarskian schema, since for theoretical purposes, the speaker and time need to be designated on the right in a more context-independent way.

II

Frege famously wrote. "One can indeed say, 'The thought that 5 is a prime number is true.' However, if one looks more closely, one notices that actually nothing more has been said than in the simple sentence, '5 is a prime number'" (1892, 34; translation from Frege 1984). Quine makes similar statements, but regards truth as a property of sentences. He writes of Tarski's stock example, "To ascribe truth to the sentence is to ascribe whiteness to snow" (Quine 1992, 80). A sentence later he says, "Ascription of truth just cancels the quotation marks. Truth is disquotation."

One might object to what Quine says by noting that "'Snow is white' is true" cannot mean the same as 'Snow is white,' because the former contains reference to a sentence while the latter does not.[2] Quine would probably reply that the existence of the sentence and that the sentence in quotes designates it are truisms. One can dress this up in a view of assertion by saying that to assert "'Snow is white' is true" does no more than to assert 'Snow is white' because that 'Snow is white' designates the sentence will be common ground in a conversation before a speaker says that the sentence is true.

Quine's emphasis on the disquotational feature of truth will remind us of the views of later deflationists. Sometimes he says things that make him seem more deflationist than I think he really is, for example in his emphasis on the disquotational character of attributions of truth to individual sentences. But he does not comment on deflationism, although it is commented on in Davidson's Dewey Lectures (Davidson 1990), published in the same year as *Pursuit of Truth*. I don't think the absence of comment tells us much of anything; Quine comments little on others' views about truth in general, except for Tarski's and traditional ones, particularly the correspondence idea. Some remarks are in harmony with what deflationists say, such as the previous remark about attributing truth to 'snow is white' and his statement that where attributing truth is needed (for more than pragmatic reasons) is to state generalizations. But he does not make the negative statements characteristic of deflationists, that the concept of truth does not have the importance traditionally attributed to it or that it is an uncontentious concept. And he does not say that the meaning of 'true' is given by some version or other of the schema expressing the equivalence of a statement with the attribution of truth of the statement itself (where 'statement' here might refer to a proposition or to a sentence or other linguistic item). Here his reservations about the notion of meaning may well have been at work. He also pays attention to a central feature of Tarski's procedure: that instances of the truth schema are in an account of truth for a given language derived from atomic instances (generally for satisfaction rather than truth proper) and satisfaction conditions for sentences containing logical connectives.

Still, his comments caused Davidson (1994) to express some worry about whether Quine was a deflationist in a sense he would object to, although he concluded that this was not so. Quine (1994) reassures him further in his reply.

Some deflationists are guilty of great naïveté about what I would call the logic of truth, the inquiry, now rather elaborate, inaugurated by Tarski's *The Concept of Truth in Formalized Languages* and perhaps made a mature subject by the work of Saul Kripke.[3] The semantic paradoxes and related circularities are at the center of this subject, and the different developed constructions are most likely to differ in their approach to these paradoxes. Quine, approaching the subject as a logician, and in general much influenced by Tarski, would not be naïve about paradoxes. However, he sticks closely to Tarski's own approach. He must have learned of some later work such as Kripke's, but he does not comment on it, still less adopt ideas from it.[4]

III

One of the most frequently quoted of Quine's statements about truth occurs at the end of the first chapter of *Word and Object*:

> Have we now so far lowered our sights as to settle for a relativistic doctrine of truth—rating the statements of each theory as true for that theory, and invoking no higher criticism? Not so. The saving consideration is that we continue to take seriously our own particular aggregate science, our own particular world-theory or loose total fabric of quasi-theories, whatever it may be. Unlike Descartes, we own and use our beliefs of the moment, even in the midst of philosophizing, until by what is vaguely called scientific method we change them here and there for the better. Within our own total evolving doctrine, we can judge truth as earnestly and absolutely as can be; subject to correction, but that goes without saying.
>
> (2013, 22)

That we can judge truth "within our own total evolving doctrine" may seem primarily to make an epistemological point: when we judge something true, we are using our own knowledge (although in many cases we will rely on authorities); we don't claim to have access to The Truth quite independently of that.[5] But I think Quine is also intimating a semantical point, which he expressed in later writings by saying that truth is immanent. Judging truth "within our own evolving doctrine" is judging truth in our own language, so that there is no question of translation or interpretation.

Quine is not bothered by the idea that truth is relative to a language. Of course it follows from his taking sentences to be the primary vehicles of truth. It also does not bother him that judgment of truth is relative to a theory:

> In being able to speak of the truth of a sentence only within some inclusive theory, one is not much hampered, for one is always working within some comfortably inclusive theory, however tentative.
>
> (2013, 68)[6]

Quine also says we can judge truth "absolutely." He does not elaborate on that point, but I believe he took it to be part of the grammar of 'true.' If we revise our theory and cease to accept a sentence that we formerly accepted, we do not say that it was previously true and is now no longer; rather, we say that what we previously took to be true is not and never was. Later he writes of this feature:

> Such is the idiom of realism, integral to the semantics of 'true'. Such is scientific method: interrogation of nature is a cosmic true-false test. Man proposes, nature disposes.
>
> (1994, 500)

That of course goes with the preference for eternal sentences as vehicles of truth; it would be natural in ordinary usage for me to say that 'My wife is asleep' was true at 2:30 in the afternoon of a certain day and is not true at 4:50 p.m. on the same day.

One may be worried by the fact that I judge truth on the basis of my evolving theory, and you will judge truth on the basis of yours. There may be no assurance that they will be the same. So are we condemned to relativism? There is no indication that Quine is concerned about this possibility, although it does arise in some late writings where he discusses the possibility that theories might be empirically equivalent. If the theories are not empirically equivalent, then there is some difference in what they imply about experience. But observation sentences are sentences about which speakers in close to the same position to observe will generally agree. So the situation is one of disagreement, which it is in principle possible to resolve.

In spite of the absoluteness arising from the grammar of 'true,' Quine does struggle in late years with an issue of relativism in his discussion of situations where two comprehensive theories are empirically equivalent. They may be intertranslatable in a natural way, in which case we can conclude that they are different formulations of the same theory. But on the theoretical level on which he discusses the issue, one can't count on this. Another possibility is that one theory is selected on the grounds of theoretical virtues, such as simplicity, elegance, or naturalness. But what if there is still a tie after these considerations are weighed?

In the passage quoted earlier from *Word and Object*, Quine says that we judge truth "within our own total evolving doctrine." Suppose we encounter another theory that is empirically equivalent to our own, and moreover cannot be transformed into our own by what Quine calls reconstrual of predicates. If we find it superior in the theoretical respects just mentioned, that should settle the question in favor of the second theory, so that our "evolving doctrine" will change accordingly. But suppose we do not find it so. The position of *Word and Object* would imply that we should still find the rival theory false. Quine says as much in an essay of 1975:

> Whatever we affirm, after all, we affirm as a statement within our aggregate theory of nature as we now see it; and to call a statement true is just to reaffirm it. Perhaps it is not true, and perhaps we shall

> find that out; but in any event there is no extra-theoretic truth, no higher truth than the truth we are claiming or aspiring to as we continue to tinker with our system of the world from within. If ours were one of those two rival best theories that we imagined a moment ago, it would be our place to insist on the truth of our laws and the falsity of the other theory where it conflicts.
>
> (2008, 242)

This is the response to his puzzle that he later calls "sectarian." He says it has the ring of cultural relativism. That he considers paradoxical. Someone who holds the rival theory will judge its truth just as absolutely as Quine envisages for the original theory in the passage just quoted.

Even in this essay, Quine does not rest there.[7] He suggests we might "settle for a frank dualism" and oscillate between the two theories. He ends by hinting at a position he proposes later:

> Where there is forever no basis for choosing, then, we may simply rest with both systems and discourse freely in both, using distinctive signs to indicate which game we are playing. The use of distinctive signs leaves us with two irreducible and unconflicting theories.
>
> (2008, 243)

This suggestion is made more explicit and less desperate-sounding in *Theories and Things* (Quine 1981, 29–30).

Quine's final published statement on this subject is an essay of 1994.[8] He supposes that we have two empirically equivalent theories, one of them our own, and considers three cases. In Case 1, the other theory is compatible with our own and is expressed "directly in our own terms." In that case, we can simply add the other theory to our own (maybe purging redundancy). In Case 2, "Again the other theory is logically compatible with ours, but it hinges on some theoretical terms not reducible to ours." This is the more difficult case. Case 3 is where the theories are logically incompatible. A sentence S that one theory implies and the other denies must contain (essentially for that implication, given the equivalence) some term that is "not firmly pinned down to observable criteria." The theories can be made compatible by replacing the relevant term in one sentence by a new one.[9] The result will make the theories compatible, thus reducing to case 2.

Assuming the two theories are equally simple and elegant, Quine proposes accepting both theories as true, as grasping the world in different ways. Because they are compatible, this does not in Quine's view give rise to relativism.

IV

In contexts where he emphasizes disquotation, Quine is treating truth as immanent, although the terminology arises only in later writings. That could not be the whole story about truth, even for Quine, because we assess for truth statements we encounter in other languages, even in the context of radical translation. Unsurprisingly, Quine acknowledges this:

> Of course the truth predicate carries over to other languages by translation. If this is transcendence, truth is indeed transcendent. But that much can be said even of reference, despite ontological relativity. It likewise carries over by translation.
>
> To call a sentence true, I said, is to include it in our science, but this is not to say that science fixes truth. It can prove wrong. We go on testing our scientific theory by prediction and experiment, and modifying it as needed, in quest of the truth. Truth thus looms as a haven that we keep steering for and correcting to. It is an ideal of pure reason, in Kant's phrase. Very well: immanent in those other respects, transcendent in this.
>
> (1995, 353)

This statement makes a transition between the preceding discussion and some very brief remarks I want to make about Davidson. An important difference between the two is that typically Davidson deploys the truth predicate in theoretical discussion where it is in Quine's terms transcendent. That is quite evident in Davidson's account of radical interpretation. Although Quine himself had not yet put things that way, the difference already existed in "Truth and meaning." Davidson writes as if linguistic understanding itself is interpretation in his sense, although I haven't studied the matter enough to be certain that that is his view. But one doesn't find locutions such as that of Quine when he speaks of being "at home in our own language."

Davidson also criticizes other views of truth, in particular the deflationism associated with Paul Horwich and Hartry Field. I see Horwich as a "purist" about truth, aiming to characterize what belongs to the concept of truth as independently as possible of other concepts. Davidson is an "anti-purist." In his Dewey Lectures, when he gives his own view of the content of the concept of truth, he emphasizes its interconnections with other concepts. He may go further than Quine does in saying that something is bound up with the concept, as opposed to being true or obviously true about it.

V

Was Quine a metaphysician? If so, in what sense? Some remarks on those questions will help to bring this chapter into connection with the project of this volume. Whatever Quine may have intended, should he be regarded as a major progenitor of the metaphysics pursued today, in particular "analytical metaphysics"? We might take as a text the following remark of Hilary Putnam. He says of some lectures of his on "Ethics without ontology" that they

> provided me with an opportunity to formulate and present in public something that I realized I had long wanted to say, namely that the renewed (and continuing) respectability of Ontology (the capital letter here is intentional!) following the publication of W. V. Quine's "On what there is" at the midpoint of the last century had disastrous consequences for just about every part of analytic philosophy.
>
> (Putnam 2004, 2)[10]

The remark suggests that the blame he is expressing belongs more to the reception of Quine's essay (and perhaps of later works of his) than to the essay itself. It may be that even if Putnam's historical judgment is quite right, the situation he deplores, insofar as it derives from Quine at all, rests more on misunderstandings of Quine's views than on a correct understanding of what Quine was advocating.

Two rather gross facts about Quine's philosophy point in opposite directions. The first is that Quine's first real philosophical mentor was Rudolf Carnap, who had been a full member of the Vienna Circle. At the time of Quine's memorable encounter with him in 1933, Carnap was certainly resolutely opposed to metaphysics as he and the Circle understood it and considered metaphysical questions to be meaningless. Commenting on Carnap's reaction to "On what there is," Quine wrote that Carnap would not like his own term 'ontology' since he would think it meaningless. In effect, Quine's answer was that he was then free to give it his own meaning (see Quine 1976, 203; from Quine 1951).

Secondly, Quine's dissent from the analytic-synthetic distinction meant that he could not accept a pillar of the Circle's and Carnap's view of mathematics, that mathematics "has no real content" and is in a sense empty. Burton Dreben has suggested that Quine was from the beginning not persuaded by this. Certainly, once he had developed views in which the analytic-synthetic distinction had no place, he would not make a distinction on the dimension of "real content" between mathematics and empirical science.

Both what we have said earlier about Quine's view of truth and the remarks just made go with the picture of Quine as at least a minimal realist. That is hardly enough to make him a metaphysician. A point that was emphasized by Quine in later statements and stressed by many commentators, and which separates him from many who would be called metaphysicians, is his naturalism. One aspect of this is his firm rejection of "first philosophy," philosophy that would proceed purely a priori and at least possibly legislate for the sciences. However, I don't think that that of itself is enough to deny him the title of metaphysician. Another, however, is his privileging of natural science.

"On what there is" is basically about what theories say that there is, and in that and other writings he argues that one cannot properly answer that question unless the language of the theory is regimented into the language of first-order logic. But does Quine offer an answer to the question what there *really* is? I would say, "Yes and no."

The "yes" answer is suggested by the fact that, according to Quine, philosophy is continuous with science, and it is science that gives the most objective account of how the world is. So if one asks Quine what there really is, it seems that the answer he should give is that it is what our best scientific theories are committed to, roughly at least in the sense of "On what there is." This would include the elementary physical particles, curved space-time, and more aggregative objects such as planets, stars, and galaxies. I won't go into Quine's view of the animal and human world, which would be a large subject and touch on matters that I am not sure I understand. But to the extent that there is a science of these matters, what there really is in this domain is whatever the science of it requires. However this turns out, Quine leaves no room for a distinctively metaphysical vocabulary. Recall that about the word 'ontology' he doesn't argue with the claim he attributes to Carnap that it is meaningless; rather, he claims the freedom to give it his own meaning, which has to do with the range of the variables of quantification in scientific language, suitably regimented.

There is, however, something that seems to throw a monkey wrench into the whole question of what there really is and suggests the "no" answer to our question. That is ontological relativity. Suppose we have a comprehensive theory, which may or may not talk of entities beyond physical science, but is still a physical theory. Its ontology is not fixed even by a determination of what sentences in the language of this theory are true. This conclusion can be reached by two routes, in his earlier writings by the indeterminacy of translation, and in later writings by proxy functions, which map the objects of a theory bijectively onto others, adjusting the understanding of the predicates so that truth-values are undisturbed. Quine even speculates about possibly eliminating physical objects altogether.

If we are "at home in our own language" and don't venture to interpret theories, in particular with respect to their ontologies, then we take our talk of "electrons, muons, and curved space-time" quite at face value. That doesn't remove ontological relativity; it just means that the questions that give rise to it are not raised. But the possibility of a more reflective stance means that we don't have a final answer to the question of our ontology. We can't be what Hilary Putnam calls metaphysical realists.

VI

What would Quine think of today's analytical metaphysics? I confess that my knowledge of that work is not sufficient for an informed answer.

I will, however, comment on two examples. I will begin with some rather general comments on the views of David Lewis. He would have to be viewed as a transitional figure. On the one hand he is counted as a father, maybe the principal father, of analytical metaphysics. On the other hand he was a student of Quine. He was certainly an original thinker and not a disciple. He wrote about a wide variety of topics, many of which Quine never took up. He embraced metaphysics much more wholeheartedly that Quine ever did.

However, one of his most distinctive views has a certain Quinean flavor. That is his realism about possible worlds. Lewis evidently did not share Quine's view that modality should not be part of the language of science, at least that language that is to "limn the true and ultimate structure of reality." He wanted for his own purposes the expressive power that a modal language provides; in fact, he argues that the usual language of modal logic is not expressive *enough*. In effect, what he does is to bring the possible worlds that are central to a model theory for modal languages into his object language. Although his ontology includes objects, namely possible worlds, that Quine would not countenance, to all appearances his way of talking about them is from there on extensional. The "counterpart theory" that he applies does not make room for an object existing in one possible world to recur *in propria persona* in another, although it can have counterparts whose properties and history provide a way of talking about how that object might have been if circumstances had been different. This doesn't have the literal truth that is claimed by the more mainstream view, exemplified forcefully by Saul Kripke.

Lewis's theory of possible worlds strikes me as more in the spirit of Quine than the embrace of modal logic by many of his contemporaries such as Kripke. The difference with Quine is a rather straightforward difference about whether it is desirable to postulate certain entities, possible worlds. I think that this aspect of Lewis's philosophy was one that Quine could at least make sense of, even if he did not accept it.

I will not pursue Lewis further, because he was a prolific writer and his philosophy is quite rich. I do not have a good enough grasp of it to zero in on some essential aspect of it that I could single out for comparatively brief discussion for the purposes of this chapter.

What I will do is to take a single example of contemporary analytic metaphysics and comment on it from what I will imagine to be Quine's point of view. The philosopher serving as my example is Theodore Sider, as represented by his book *Four-Dimensionalism* (2001). At the outset Sider signals what looks like a significant difference with Quine. He writes that "the reasons I provide for my conclusions are largely a priori" (2001, xiv). However, he believes that metaphysics should proceed without waiting for an epistemological theory to underpin metaphysical knowledge. It is not clear how much weight he intends to place on the a priori character of his investigation. He aims to make his conclusions consistent with existing scientific knowledge, and at one point argues at some length that his views are consistent with special relativity.[11] I would guess that he does not share Quine's rejection of the idea of a priori knowledge as found in earlier philosophy (even up to Carnap) but is not prepared to make a case to the effect that his claims could not be overthrown by discoveries in empirical science. It may be that he would think of his conclusions as a priori in the weak sense in which mathematics is uncontroversially a priori, in that results are not obtained by experimental reasoning or checked empirically by the refined methods of validating conclusions in the natural sciences. That observation does not fully rule out the possibility that some quite unforeseen change in natural science would lead to a change in accepted mathematics, or for Sider in his metaphysical conclusions.[12]

However, Sider's title already suggests a point of agreement with Quine. Quine also thought that the objective view of things would be in terms of four-dimensional space-time. Sider, however, does not simply take this as mandated by contemporary physics. Prior to that, I think Sider agrees with Quine's view that tense should not be part of the language of science. He devotes space to arguing against "three-dimensionalist" views and against tensed language as the philosophically appropriate way to talk about time. About time, he is what is called a "B-theorist," in that for him, even if (assuming relativity) one has to stay within a single frame of reference, temporal order is simply that of before and after, and "now" is just where one happens to be in that order, so that tense does not express anything fundamental about time. So far as I know, the issue between a view like Sider's and more tense-friendly views ("A-theories") is not addressed directly in Quine's writings, but he indicates that a canonical language for science would be tenseless.

Within the four-dimensional perspective, however, Sider has significant differences with Quine. One is that Sider accepts unrestricted mereological composition, so that for any plurality of objects (at least bounded in space-time), there is an object that is its sum, occupying any point of space-time occupied by one of the given objects.[13] I won't pursue this issue. Although I think Quine did not think that the language of science operated in that way, I don't know of his making an issue of the point. More significantly, Quine has no problem in viewing spatiotemporal objects as extending in time, "perduring" in the terminology of contemporary analytic metaphysics. We can view a spatiotemporal entity such as the astronaut John Glenn as a space-time worm. Sticking to an earthbound frame of reference, this worm never covers very much space at a fixed time, but at some times a segment of it is at enough distance from the earth's surface to be in orbit. In time, it extends from Glenn's birth through his service as astronaut to a later stretch as a U.S. Senator, to his retirement and eventual death.[14] Quine allows the singling out of "Glenn-stages," which are slices of the worm that are short temporal segments of it. In this way he can speak, as Sider does, of temporal parts. However, Sider points out that the worm theorist can accommodate the idea of instantaneous temporal parts in a way that commits him to very little, by simply taking them as pairs of the continuing objects and instants of time during their histories. What gives Sider's advocacy of temporal parts some metaphysical force is that he views the instantaneous stages as the fundamental realities and space-time worms as composed from them.

Why does this matter? The manner in which Sider defends and applies this point of view points to a respect in which Sider is not alone among contemporary analytic metaphysicians, which I believe would have troubled Quine. In chapter 5 of his book, Sider argues that it gives a more convincing view of some well-known puzzles, such as the ship of Theseus puzzle and the question of the relation of a statue to the lump of clay from which it was composed.

Sider seems concerned to avoid the notion of coincident objects, that is physical objects that at least for a time occupy the same region of space. Consider the well-known case of a lump of clay that is fashioned into a statue. We will simplify by assuming that all of the clay in the lump finds its way into the statue; none was superfluous. Once the statue is made, the lump and the statue occupy the same space, but by usual reckoning they are not identical; for example, the clay existed before the statue existed, and it might survive the statue. There is no violation of the substitutivity of identity. Sider's objection to coincidence in this sort of case is metaphysical, not logical.

Sider adopts a point about change expressed by David Lewis (1986, 202–204). Consider a certain leaf, say on a maple tree in northern New England. On September 15 it is green, but on October 15 it is red or yellow or some mixture of both. To simplify, let us say that it is yellow. It thus has the properties "green on September 15" and "yellow on October 15." That description, according to Lewis, represents the properties of being green or yellow as relations to times, whereas change is change in properties properly speaking. Lewis even suggests that on the view at issue these relations hold for entities of an "unchanging intrinsic nature." Consider Sider's example (already present in Lewis), of his being bent at one time (while sitting) and straight at another (while standing). Neither would be thought to be part of the intrinsic nature of Sider. I don't have firm intuitions either about intrinsicness or about essence or nature; however, being able to assume different positions (and thus different shapes) does seem plausible candidates for the nature (which I will assume to be intrinsic) of an animal such as a human.[15] And that feature may reasonably be considered unchanging, at least in the normal casc. In certain disabilities, one cannot change one's position by voluntary action.

Sider applies Lewis's view as part of an argument for the priority of stages (even instantaneous stages) over continuants, even if the latter are understood as space-time worms.

I believe that this issue did not move Quine. The distinction between predicates that express genuine properties and those that do not is not significant in his philosophy. In fact, the notion of property itself is not fundamental, given his nominalistic view of predication.[16] In particular, his classic text *Methods of Logic* (1950) does not give the notion of property the place that it has in the logical tradition. A more restricted notion of property could gain admission via Quine's later essay "Natural kinds" (in Quine 1969).

There is a broader feature of Sider's concerns and arguments that I think would have troubled Quine. That is a sort of mixing of what, following Wilfrid Sellars, is called the manifest and scientific images. The puzzles that underlie his arguments for the "stage view" rather than the "worm view" of a four-dimensional world are puzzles about ordinary objects: ships, statues, lumps of clay. One might be skeptical about the possibility of answering the question what is the fundamental reality underlying such objects without departing from the general framework. I will comment briefly on the ship of Theseus problem, which Sider mentions as the sort of puzzle that concerns him, although he does not go into it in detail in *Four-Dimensionalism*. A possible answer to the problem might be that it is a matter of convention or decision whether the ship resulting from replacement of planks one at a time or the ship reassembled from the planks discarded in that process is identical with the original ship. (The view that neither is would be a third competitor.) Quine's answer may well have been that the language of science is silent on the question, so that there is no compulsion to adopt one or another of the positions. He might opt for the result of replacing the planks one by one on the ground that there is a continuous ship there, whose existence is not interrupted. That way of putting the matter is tendentious; it seems to beg the question whether the ship finally resulting is the same ship as the one before the replacement starts. But this ship does have the advantage that if we accept it as one ship, it has an uninterrupted existence, whereas the reassembled ship seems to come into existence in the rebuilding; if we do regard it as the same ship as the original, we do have to admit that there was a time when the existence of the ship was interrupted.[17]

Can Sider maintain that taking the piece-by-piece replacement as maintaining identity of the ship, while reassembly of the discarded planks does not, come from a conception that "carves nature at the joints"?[18] I think it might be argued on the basis just indicated, that physical things do not have discontinuous existence. That may, to Sider's taste, be too much like an appeal to "our conceptual scheme." But it may be that we cannot emancipate ourselves from that as fully as Sider seems to claim. But as regards the main theme of this chapter, the issue as regards the identity of ordinary macroscopic objects is remote from those that drove Quine.

Notes

* An early version of this chapter was presented to the Semantics Workshop at Rutgers University, October 3, 2015, commemorating James Higginbotham. Obviously, that is the reason for dedicating it to Jim's memory. I thought it appropriate because he had written his dissertation with Sidney Morgenbesser and me on a Quinean theme, and unlike many who worked on semantics on the paradigm of generative linguistics, he continued to take Quine's philosophy seriously.

I am greatly indebted to Camilla Serck-Hanssen and Frode Kjosavik, leaders of the project that brought me to the Centre for Advanced Study of the Norwegian Academy of Science and Letters, for inviting me to participate. An intermediate version of the chapter, with only very sketchy remarks about metaphysics, was presented at the project's conference at Kleivstua. Thanks to the audience for comments.

1. There are many people named Charles Parsons, but I think it a reasonable guess that 'Charles Dacre Parsons' identifies me uniquely. I could be wrong: I recall that Quine once proposed in a lecture that 'W. V. Quine' is an eternal singular term. However, now it could designate his grand-nephew William van Orman Quine, who was, however, not yet born at the time.
2. Of course a similar objection could be made to Frege's remark.
3. Horwich (1990) is for me the paradigm example. I comment (all too briefly) on his relevant views in Parsons (2016).
4. The same could be said of Davidson; see my following comments. In my book *Mathematics in Philosophy* (1983), my 1974 paper "The liar paradox" was reprinted with a postscript commenting on later work, including Kripke's. Quine reviewed the book (Quine 1984), but his comments on that paper are sparse.
5. If Quine had been inclined to say more about Descartes in this context, he might have pointed out that Descartes found it necessary to appeal to God to resolve the doubts of the beginning of the *Meditations*. However, Quine was not only not a believer, he was not really interested in theology as a theoretical inquiry.
6. He goes on to say that "the parameters of truth stay fixed most of the time" and contrasts this with the case of translation.
7. Still, Quine briefly embraces the sectarian view (1981, 21–22). After considering alternatives, he returns to it in (1986, 157). As noted shortly, that is not Quine's last word on the subject.
8. "Truth," in Quine 2008, 434–437. It mostly repeats what was said a few years earlier in *Pursuit of Truth*.
9. Although Quine does not remark on this, this most likely has to be done a number of times. He credits the idea for the reduction to Davidson.
10. Cf. Quine 1948, 21–38, reprinted in Quine 1953.
11. He states without argument that general relativity would present no new problems.
12. About mathematics, I have argued elsewhere that this is unlikely; what it would lead to is rather a change in what mathematical theories are applied. See Parsons (1983, 195–197).
13. Hirsch (2004) focuses on this view of Sider's in his criticism.
14. Glenn was born in 1921 and died in 2016. Thus, as human space-time worms go, his extension in (earth) time was unusually long.
15. Some animals, such as many insects, are less flexible in this respect than humans.

16. I discuss this at length in Parsons (2011).
17. The idea that a physical object like a ship might have an interrupted existence strikes me as odd. The religious idea of the resurrection of the dead may also offer room for interrupted existence. But neither Quine nor Sider engages with religious conceptions. A less problematic example is institutions. My family has for a long time patronized a restaurant, Peter Christian's Tavern, in New London, New Hampshire. But some years ago it went out of business, apparently because of bankruptcy. However, after a time it was revived, and the decor and menu hardly differed from the original, although the management is different. Now it is again closed, because of the rebuilding of the building where it is located. But it is anticipated that it will reopen again. (As of 2019, this has happened.)
18. Whether concepts carve nature at the joints is a major concern of his later book *Writing the Book of the World* (2011). I have not ventured to comment on this work.

References

Davidson, Donald. "The Structure and Content of Truth." *The Journal of Philosophy* 87 (1990), 289–328.

Davidson, Donald. "What Is Quine's View of Truth?" *Inquiry* 37 (1994), 437–440.

Frege, Gottlob. "Ueber Sinn und Bedeutung." *Zeitschrift für Philosophie und philosophische Kritik* 100 (1892), 25–50.

Frege, Gottlob. *Collected Papers on Mathematics, Logic, and Philosophy*. Edited by Brian McGuinness. Oxford: Blackewell, 1984.

Hirsch, Eli. "Comments on Theodore Sider's *Four-Dimensionalism*." *Philosophy and Phenomenological Research* 68 (2004), 658–664.

Horwich, Paul. *Truth*. Oxford: Blackwell, 1990.

Lewis, David. *On the Plurality of Worlds*. Oxford: Blackwell, 1986.

Parsons, Charles. *Mathematics in Philosophy*. Ithaca, NY: Cornell University Press, 1983.

Parsons, Charles. "Quine's Nominalism." *American Philosophical Quarterly* 48 (2011), 213–228.

Parsons, Charles. "Theories of Truth and the Value of Truth." *Diotima* 44 (2016), 82–100.

Putnam, Hilary. *Ethics without Ontology*. Cambridge, MA: Harvard University Press, 2004.

Quine, Willard van Orman. "On What there Is." *Review of Metaphysics* 2 (1948), 21–38.

Quine, Willard van Orman. "On Carnap's Views on Ontology." *Philosophical Studies* 2 (1951), 65–72.

Quine, Willard van Orman. *From a Logical Point of View*. Cambridge: MA: Harvard University Press, 1953.

Quine, Willard van Orman. *Ontological Relativity and Other Essays*. New York: Columbia University Press, 1969.

Quine, Willard van Orman. *The Ways of Paradox and Other Essays*. 2d ed. Cambridge, MA: Harvard University Press, 1976.

Quine, Willard van Orman. *Theories and Things*. Cambridge, MA: Harvard University Press, 1981.

Quine, Willard van Orman. "*Mathematics in Philosophy* by Charles Parsons (Review)." *The Journal of Philosophy* 81 (1984), 783–794.

Quine, Willard van Orman. "Reply to Roger F. Gibson, Jr." In L.E. Hahn and P.A. Schilpp, eds., *The Philosophy of W. V. Quine*. Chicago and La Salle, IL: Open Court, 1986, 155–158.

Quine, Willard van Orman. *Pursuit of Truth*. 2nd ed. (1st edition 1990). Cambridge, MA: Harvard University Press, 1992.

Quine, Willard van Orman. "Responses." *Inquiry* 37 (1994), 495–505.

Quine, Willard van Orman. "Reactions." In Paolo Leonardi and Marco Santambrogio, eds., *On Quine: New Essays*. Cambridge: Cambridge University Press, 1995, 347–361.

Quine, Willard van Orman. *Confessions of a Confirmed Extensionalist and Other Essays*. Edited by Dagfinn Føllesdal and Douglas B. Quine. Cambridge, MA: Harvard University Press, 2008.

Quine, Willard van Orman. *Word and Object*. (1st Edition 1960). Cambridge, MA: MIT Press, 2013.

Sider, Theodore. *Four-Dimensionalism*. Oxford: Oxford University Press, 2001.

Sider, Theodore. *Writing the Book of the World*. Oxford: Oxford University Press, 2011.

12 The Paradox of the Largest Number

From Aristotle to Cantor

Øystein Linnebo

Is there a largest number? Suppose a young child asks the question. Most parents would answer 'no.' They might add the following explanation. Suppose a gazillion was the largest number. Then a gazillion and one would be an even larger number. So there cannot be a largest number. The child might, if she is sufficiently smart and quarrelsome, respond by asking whether infinity isn't a largest number. She might reason as follows. There are infinitely many natural numbers. If we added one more thing, we would still have infinitely many things. So there cannot be a number greater than infinity. And this means that infinity is the largest number.

This is a very simple version of *the paradox of the largest number*. While admittedly somewhat naive, this reasoning should not be dismissed as just childish confusion. On the contrary, I shall argue, there are important lessons to be learnt from pondering some more subtle versions of the paradox. These versions force us to rethink the relation between mathematics and metaphysics, especially the metaphysics of modality.

To defend this claim, I first show how an interesting version of the paradox arises in the context of the ancient conception of infinity as the property of unboundedness. Next, I explain how an elegant solution is provided by the Aristotelian conception of potential infinity. This remained the dominant conception of infinity in mathematics and philosophy for more than two millennia, until the Cantorian revolution in the second half of the nineteenth century, which ushered in the now-standard view. Although the Cantorian revolution solves many of the traditional paradoxes of infinity, I show that a version of the paradox of the largest number remains. To improve on the situation, I articulate a "successor concept" to the ancient concept of potential infinity which, I argue, enables us to resolve the paradox. If I am right, the concept of potential infinity is not only of great historical importance but also of profound significance for contemporary metaphysics and philosophy of mathematics. This illustrates how the history of philosophy (in our case, Aristotle and Kant) can serve as an inspiration and source of ideas for the contemporary philosophical debate.

I. The Ancient Conception of Infinity

In the ancient period, infinity was understood first and foremost as *apeiron*, that is, the property of being without limit or bound. Suppose, for example, that the world will never come to an end. Then the days yet to come are infinite in this ancient sense of being unlimited: there is no number large enough to count, or provide an upper bound on, all the days ahead. In fact, this conception of the infinite is implicit in the word that is still used. When 'infinite' is understood in its etymologically correct sense, to be infinite is precisely to be unlimited or unbounded.

An interesting variant of this ancient conception of infinity is found in Kant, who writes:

> The infinite is a magnitude for which no determinate measure can be specified (Ak. 28: 568).

How is this notion of lacking a "determinate measure" related to the notion of being unlimited or unbounded? I think a strong case can be made for their equivalence: for a bound on some quantity or magnitude would serve as a determinate measure and vice versa.[1] Of course, this equivalence assumes that the possible bounds coincide with the determinate measures that are available. But this assumption seems plausible. The result is a conception of the infinite as the property of being "off the scale" that is provided by the number line. None of the determinate measures that are found on this scale suffices to characterize the magnitude in question.

To better understand the property of being off the scale of determinate measures, let us take a closer look at the scale itself. What are the determinate measures that make up the scale, and what determinate measures are there? In other words, what are numbers, and what numbers are there? For present purposes, a number can be taken to be a cardinal number, that is, an answer to a "how many" question, such as: How many are these objects? or: How many members does this collection have? Thus, what numbers there are will turn on how many objects there are—or at least could be.

Aristotle's view—which was reaffirmed by much of the mathematical and philosophical tradition right up until Cantor—is that there could be any (positive) natural number of objects—but no more.[2] The positive half of this view is straightforward enough. Clearly, there is at least one object. And plausibly, for any objects, there could be all these objects and one more. The negative half of the view is far more controversial, especially by the lights of contemporary mathematics. Why deny the possibility of an infinite collection of objects—and hence also the existence of an infinite number? In the next section, we shall consider two arguments for this (by our lights) surprising view.

First a note on terminology is required. Throughout this chapter, I use the word 'collection' in a loose and informal sense. One of the main lessons to emerge will be that this informal notion needs to be disambiguated. There are two very different conceptions of a collection. On the *combinatorial* conception, a collection is individuated in terms of its members. A collection of this sort therefore presupposes each of its members. Pluralities and sets are examples. By contrast, a *conceptual* collection is individuated in terms of its membership condition and therefore does not in general presuppose its members. Properties and Fregean extensions are examples.[3]

II. Two Paradoxes of Infinity

Apparent paradoxes of infinity are a dime a dozen. Let me describe two paradoxes which I find particularly interesting and believe pre-Cantorian thinkers were right to take seriously. Both paradoxes arise in connection with infinite collections of objects and can thus be understood as arguments against the existence—or indeed even possibility—of infinite collections. While I believe the first paradox was conclusively solved by Cantor, I shall argue that the second is far more recalcitrant.

The first paradox was made famous by Galileo's *Two new sciences* (1638), although it was actually formulated several centuries earlier.[4] So the paradox is now often known as Galileo's paradox. It is based on two *prima facie* plausible principles about numbers:

Euclid's Principle

Every whole is larger than any of its (proper) parts.

Bijection Principle

The number of any collection is preserved under any one-to-one correspondence (also called a 'bijection').

If infinite collections are possible, Galileo observed, then these two principles will come into conflict. We are therefore told to conclude that infinite collections are *not* possible.

To understand how the conflict arises, it is useful to consider the famous thought experiment of "Hilbert's hotel." Suppose there is a hotel with infinitely many rooms, each labelled with a distinct natural number. It's a busy night at the hotel: every room is occupied. Then another guest shows up. Fortunately, the receptionist has a brilliant idea. What if the guest in room one moves to room two, the guest in room two moves to room three, and so on? Then every current guest will have a room to herself, while room one has been made available for the new guest. This is puzzling! By reassigning rooms in an entirely full hotel, we can free up a room for the newly arrived guest! Obviously, no such thing is possible in an ordinary hotel with "only" finitely many rooms.

Let us try to understand precisely why Hilbert's hotel is so puzzling. A good diagnosis is that Euclid's Principle comes into conflict with the Bijection Principle. By the first principle, the hotel has more rooms altogether than it has rooms numbered two and up. But by the second, these two collections of rooms have the same number. Since this is a contradiction, many thinkers have concluded that Hilbert's hotel is *not* possible. And since Hilbert's hotel is just a particularly vivid illustration of a phenomenon that arises for all infinite collections, we can generalize and conclude that no infinite collection is possible.

The second paradox of infinity goes as follows. Assume, for contradiction, that there could be infinitely many objects, say, an infinite collection of pebbles. So the number of this collection is ∞. However, there could have been an even more numerous collection; for example, there could have been all of the mentioned pebbles and one more. Let N be the number of this extended collection. Thus, there is a number $N > \infty$. But this is unacceptable! The existence of a number larger than infinity contradicts our concept of infinity as being unbounded or lacking a determinate measure.

In short, if it was possible for there to be an infinite collection, then, by our concept of infinity, there would be a largest number. But there cannot be a largest number because any collection can be extended to an even more numerous collection. This is an interesting version of the paradox of the largest number. To avoid the paradox, some thinkers claim, we must deny the possibility of infinite collections.[5]

Let us spell out the argument. I begin by making explicit two background assumptions.

First:

Existence of Numbers

Necessarily, every collection has a number.

(As announced, the crucial word 'collection' is used in a loose and informal way, but will be clarified in due course.) For any collection X, we may therefore write '$\#X$' for its number.

Next:

Cardinality Is Intrinsic

If a collection X has a number N, then necessarily, if X exists, then X has the number N.

This second assumption ensures that the number term '$\#X$' is a rigid designator; that is, at every possible world where the term refers at all, it refers to the same number.[6] This means that our number terms can freely be used in modal contexts.

Both background assumptions are plausible. General nominalist worries can safely be set aside as orthogonal to our present concerns. When this is done, it is plausible to regard [Existence of Numbers] as implicit in our mathematical practice. Notice also that this assumption is weak enough to be compatible with the Aristotelian view that numbers have to be instantiated in order to exist. Next, [Cardinality is Intrinsic] is also plausible, at least when the collections in question are understood as pluralities of objects. For example, the fact that some objects are 17 in number is intrinsic to these objects: There is no need to "look beyond" these objects to determine that the number ascription is true.

We are now ready to present the heart of the argument. Assume there is an infinite collection. Based on this assumption, the argument seeks to prove both a thesis and a contradictory antithesis, thus showing that we must reject the assumption that there is—or even could be—an infinite collection.[7] The reasoning is straightforward. First, recall from our analysis of the ancient concept of infinity that we have:

Infinity as Unboundedness. Necessarily, there is no number $N > \infty$.

Let us say that a collection is *unsurpassable* just in case it is impossible for there to be a collection that is strictly more numerous. In symbols, X is unsurpassable just in case:

$$\Box\forall Y(\#Y \leq \#X)$$

When we combine the conceptual truth [Infinity as Unboundedness] with our assumption of an infinite collection, we obtain:

Thesis: There is an unsurpassable collection.

Next, it is plausible to accept:

Extendability. For any collection X, it is possible for there to be an even more numerous collection that extends X. That is:

$$\Box\forall X\Diamond\exists Y(\#X < \#Y)$$

This principle is particularly plausible when the collections in question are understood as pluralities of objects.[8] Given any collection of pebbles, for example, there could be all those pebbles and one more. Finally, [Extendability] obviously entails:

Antithesis: Necessarily, every collection is surpassable.

As mentioned, this contradicts the Thesis. We are therefore told to reject the assumption that there could be an infinite collection.

III. Potential Infinity as a Solution

Aristotle, we recall, denies the possibility of infinite collections or magnitudes. Any such collection or magnitude would lead to paradox, he argues. His view thus avoids the two paradoxes described in the previous section. However, the Aristotelian view immediately gives rise to an objection. Surely, there many examples of infinite totalities, for instance, space, time, and the natural numbers!

Aristotle offers a brilliant reply: These apparent counterexamples are merely *potential* infinities; what he rejects are *actual* infinities.[9] To explain the distinction between potential and actual infinities, it is instructive to consider some examples. According to Aristotle, a stick s is infinitely *divisible*; that is:

(1) Necessarily, for any part x of s, possibly x has a proper part.

But Aristotle denies that s is, or even could be, infinitely *divided*; that is:

(2) For any part x of s, x has a proper part.

Thus, while he affirms the potential infinity of the parts of the stick, he denies their actual infinity.

Aristotle takes an analogous view of the natural numbers, affirming their potential infinity while denying even the possibility of their actual infinity:

(3) $\Box\forall m\Diamond\exists n\text{Successor}(m,n)$

(4) $\neg\Diamond\forall m\exists n\text{Successor}(m,n)$

However large a number m has been instantiated (say, by means of some parts of the stick), it is always possible to instantiate an even larger number n (say, by cutting one of these parts in two). But it is impossible for every number *in fact* to have a successor, since the instantiation of all these numbers would require an infinite collection of objects.

On this view, there is a fundamental difference between each natural number, on the one hand, and potential infinity, on the other, having to do with what these measures of cardinality are ascribed to. Each natural number specifies that cardinality of certain pluralities of objects. A particular plurality of pebbles, for example, can be 1001 in number. By contrast, potential infinity is ascribed, not to a particular plurality, but to *a type* of object. It makes no sense to say of a particular plurality of pebbles that it is potentially infinite. Any particular plurality has some particular number of members, which, according to Aristotle, has to be finite. Rather, potential infinity is ascribed to *a type* of object, say, the type *part of this stick* or *pebble that God might create*.[10]

Let me clarify this last claim. To say that objects of some type F are potentially infinite is to say that necessarily, for any plurality xx of Fs, it is possible for there to be an even more numerous plurality of Fs that extends xx.[11] Let '$\#F^{\Diamond}$' (to be read as "the largest possible number of Fs") denote the upper bound on the cardinality of possible pluralities of Fs, if there is such a bound, and ∞, if not. For example, suppose 'P' means "passenger in this elevator." If the elevator can accommodate a maximum of ten people, then $\#P^{\Diamond} = 10$. It is important to notice that these number terms too are rigid designators. The reason is that the possible Fs, as seen from one possible world, are the same as the possible Fs, as seen from any other possible world.[12]

Our analysis has an important upshot. *There is a qualitative, not merely a quantitative, difference between the finite and the infinite.* Each natural number is a determinate measure of the cardinality of certain pluralities. Each of these measures is, or at least could be, realized by some plurality. By contrast, infinity—which the Aristotelian understands as *potential* infinity—isn't a determinate measure at all but a property of some type of object, namely, the property that no determinate measure is large enough to provide an upper bound on the cardinality of every possible plurality of objects of this type. In short, while each natural number is a point on the scale by which we measure cardinality, potential infinity is the property of being off this scale altogether.

Let us return to the paradox of the largest number. I claim that the Aristotelian view—even if ultimately far too restrictive—provides an elegant solution, namely that the paradox conflates the two different notions of collection and number. First, suppose 'collection' and 'number' are used to mean plurality and cardinality thereof, respectively. On this interpretation, the Antithesis is true, namely:

$$\Box\forall xx\Diamond\exists yy(\#xx < \#yy)$$

By contrast, the Thesis now loses its attraction: there is no unsurpassable plurality. There is, for example, no plurality of all possible pebbles. For a plurality exists at some possible world only if each of its members exists at this world. And there is no possible world at which all possible pebbles exist: for however many there are, there could be even more.

Second, suppose 'collection' and 'number' are understood in a looser sense that includes a type of object and the property of there being no upper bound on the cardinality of possible pluralities of objects of this type. On this alternative interpretation, the Thesis is true; for example, we have:

$$\Box\forall F(\#F^{\Diamond} \leq \#\text{Pebble}^{\Diamond})$$

By contrast, the Antithesis loses its attraction. When 'collection' and 'number' are understood in this loose sense, a collection need not be surpassable. Consider the number ascription $\#\text{Pebble}^{\Diamond} = \infty$. This ascription takes into account not only what pebbles there *are* but also what pebbles there *could be*. Every possible increase is thus *already* taken into account. So there is no room for any *further* increase by considering alternative possible worlds.

The upshot is that the paradox dissolves once we distinguish clearly between the two different senses of 'collection' and 'number.'

It is less obvious how an Aristotelian should respond to Galileo's paradox. One plausible option is to restrict Euclid's Principle and the Bijection Principle to extensional collections (say, to pluralities).[13] For only collections in this sense have a number in the strict sense of a determinate measure. And thus understood, both principles are valid on the Aristotelian view, which prohibits infinite pluralities.

IV. Cantor's Paradise

Cantor's response to Galileo's paradox is very different. He wholeheartedly accepts the Bijection Principle but severely restricts Euclid's Principle. Although the latter principle holds for all finite collections, its extrapolation into the infinite is unjustified, he claims. In this way, Cantor develops what we may call a *non-Euclidean* theory of the infinite.

According to this theory, actual infinities, such as that of Hilbert's hotel, are perfectly coherent. The fact that the collection of rooms in Hilbert's hotel can be put in one-to-one correspondence with a proper subcollection is taken to be just a defining feature of the infinite—initially surprising but ultimately unproblematic. Likewise, Cantor quickly persuaded other mathematicians that it is permissible to postulate various actual infinities (or "completed totalities"), for example, the set $\mathbb{N}$ of all natural numbers and the set $\mathbb{R}$ of all reals.

Investigating these infinite sets, Cantor proved the famous theorem now bearing his name.

Cantor's theorem

Let S be any set (which may be actually infinite). Then S has more subsets than it has elements. That is, the powerset of S is strictly larger than S:[14]

$$\text{card}(S) < \text{card}(\wp(S))$$

This means that the number line extends far beyond the natural numbers:[15]

$$0, 1, 2, \ldots, \aleph_0 = \text{card}(\mathbb{N}), \ldots, 2^{\aleph_0}, \ldots, 2^{2^{\aleph_0}}, \ldots,$$

In short, the number line turns out to be far more extensive than we had thought.

We just observed that Cantor solves Galileo's paradox. What about the paradox of the largest number? The version of the paradox developed in section II turns on the ancient conception of infinity as unboundedness. But this conception has now been superseded by the Cantorian conception, on which every transfinite number is surpassed by even greater such numbers. So this paradox too appears to be solved. But this appearance is deceptive, I claim; all we have done is shift the bump in the carpet.

To defend this claim, let us take a closer look at what becomes of the ancient conception of infinity as unboundedness, given Cantor's discoveries. It is essential to observe that the distinction between the bounded and the unbounded is relative to a system of measures that can serve as bounds. An object may be too large to be bounded by any measure available on one measuring stick, while still being bounded by a larger measure available on a larger measuring stick. Cantor's discovery makes this observation particularly important. For the familiar measures provided by the natural numbers are now supplemented with the much larger measures provided by Cantor's new transfinite numbers.

In light of this development, what is the more natural continuation of the ancient concept of infinity? There are two options. One is to regard a magnitude as infinite just in case it is unbounded by any of the *ordinary* measures, namely the natural numbers. Another is to regard something as infinite just in case it is unbounded by *any* determinate measure, including Cantor's new and extraordinary ones. In fact, the former option prevailed. To be infinite is now defined as the property of being so large that no natural number suffices to specify the cardinality. The resulting picture is the one described previously: the natural numbers are followed by an unbounded sequence of "infinite numbers."

In my opinion, the second option would have been a far more appropriate development. Since 'finite' originally meant 'bounded' or 'limited,' the result of discovering a new system of larger measures should be to regard more things as finite—albeit in a generalized sense. On this alternative conceptual development, Cantor's new numbers would have been regarded not as infinite but as a generalization of the finite.[16] On the conceptual development that in fact prevailed, the word 'infinite' shifted its meaning from something unbounded or unlimited to something that isn't bounded *by any natural number*.

Despite my misgivings about this conceptual revolution, I shall of course continue to use standard terminology and thus classify as infinite anything that cannot be bounded by a natural number.

Even though the word 'infinite' is defined in this way, it is interesting to examine the proper heir to the ancient notion of infinity, namely the property of having no bound whatsoever. Following Cantor, let us call this *absolute infinity*. This is an important and puzzling property. As observed earlier, the paradox of the largest number poses no threat to the now-standard notion of infinity. But I shall now argue that this paradox remains a serious problem for the Cantorian notion of absolute infinity.

V. Trouble in Paradise

Cantor convinced us that the notion of an infinite collection is coherent and mathematically fruitful. But an important question remains. Are there collections that are not only infinite but *absolutely* infinite? Two natural candidates are the collections of all ordinal numbers and of all sets.

While Cantor admits that there are such collections, he regards them as "inconsistent multiplicities," fundamentally different from sets, which are "consistent multiplicities." In a famous 1899 letter to the mathematician Richard Dedekind, he writes:

> [A] multiplicity can be such that the assumption that *all* of its elements 'are together' leads to a contradiction, so that it is impossible to conceive of the multiplicity as a unity, as 'one finished thing'. Such multiplicities I call *absolutely infinite* or *inconsistent multiplicities*. . . . If on the other hand the totality of the elements of a multiplicity can be thought of without contradiction as 'being together', so that they can be gathered together into '*one* thing', I call it a *consistent multiplicity* or a 'set'. (Ewald 1996, 931–932)

Cantor thus postulates a qualitative difference between the finite and transfinite, on the one hand, and the absolutely infinite, on the other. Finite and transfinite collections are consistent multiplicities, all of whose members can coexist, whereas absolutely infinite collections are inconsistent multiplicities, whose members accordingly cannot all coexist.

It is not obvious how Cantor's view should be understood, however. The modal and temporal language he uses has since been banished from the official language of mathematics. So the dominant view today is probably a more straightforward acceptance of absolutely infinite collections, such as the collections of all ordinals and of all sets.[17] On this view, there is only a quantitative difference between absolute infinity and the sizes represented by Cantor's transfinite cardinal numbers. Just as there are some objects that are all the natural numbers, so there are some objects that are all the ordinal numbers. It is just that the latter objects are far more numerous than the former.

I contend that this post-Cantorian view is prone to antinomies. My argument is based on a version of the paradox of the largest number, only this time applied to what Cantor calls absolute infinity rather than to infinity in the now-standard Cantorian sense. In a nutshell, the argument goes as follows. Assume there could be an absolutely infinite collection. If so, there could be even more numerous collections as well. But this would entail the existence of numbers that exceed absolute infinity, which is unacceptable.

Let me spell things out. Assume there is an absolutely infinite collection, say, the collection *On* of all ordinal numbers, which is as good a candidate as any. I claim that this collection is unsurpassable, in our earlier sense that there could not be a more numerous collection.

Assuming this claim is correct, it follows that:

Thesis: There is an unsurpassable collection.

It remains to establish the claim. Let me begin with an informal sketch of my argument.

Assume, for contradiction, that there could have been a collection *X* that is more numerous than *On*. If so, we would need more ordinals than are available in *On* in order to represent the various additional well-order types that can be based on the more numerous collection *X* and that are suitable for defining ordinals. But this contradicts our assumption that *On* contains all ordinals.

It might be objected that this argument is just an instance of the familiar Burali-Forti paradox of the well-order type of all the ordinals. Aren't we making the illicit assumption that every well-ordering corresponds to an ordinal? To see that we are not, let us adopt a standard response to the Burali-Forti paradox, namely that precisely those well-orderings that involve fewer objects than the universe as a whole define an ordinal; in particular, the well-ordering of all the ordinals does not define an ordinal because the ordinals are equinumerous with the entire universe. Even so, the argument for the Thesis goes through (indeed, in a way that is easily rendered formally precise). We have assumed that it is possible for there to be a collection more numerous than *On*. Thus, there is a possible world at which *On* is less numerous than the entire universe. So at this world, the natural well-ordering of *On* does define an ordinal. Thus, it is possible for *On* to define an ordinal. Finally, since every ordinal exists by necessity, it follows that *On* actually defines an ordinal. But once again, this contradicts our assumption that *On* contains all ordinals.

So far, so good. The problem arises because we can also defend a contradictory claim, namely:

Antithesis: Necessarily, every collection is surpassable.[18]

How is the Antithesis defended? Let us assume that necessarily, for every collection of objects, there could be an equally numerous collection of angels. This assumption is plausible because angels, as incorporeal objects, impose no constraints on space or time: as has long been known, any number of angels can be co-located on the point of a needle.[19] It therefore suffices to consider a collection *A* of angels, which may be exceedingly numerous. There are compelling reasons, I claim, to think that even more numerous collections are possible.

One strategy for defending this claim is to invoke mereology, as suggested by Hawthorne and Uzquiano (2011). Angels can be assumed to be mereological atoms, that is, objects with no proper parts. (If you have problems with this theological assumption, just modify the example.) Let us now apply classical mereology to all these angels. By a generalization of Cantor's theorem, it follows that there are more mereological sums of angels than there are angels.[20]

Another strategy is to argue modally. If there could have been the mentioned collection *A* of angels, then there could have been even more. God might, for example, have been concerned to avoid strife among the angels in *A*. He might therefore have created even more angels so as to ensure that each subcollection of *A* has its own designated angel to oversee the peaceful coexistence of precisely the angels from this subcollection. Again, a generalization of Cantor's theorem shows that this second scenario contains even more angels than the first.

Either way, we obtain a defense of the Antithesis, which contradicts the Thesis. So a version of the paradox of the largest number still poses a threat.

VI. Absolute Infinite as a Solution

How should we respond to this paradox? I propose we seek inspiration from the elegant Aristotelian solution to an earlier manifestation of the paradox. This will involve transforming the Aristotelian view from a view about infinity in the ancient sense to a view about absolute infinity in Cantor's sense. If we succeed, this will firmly establish absolute infinity as a successor concept to the ancient Aristotelian concept of potential infinity. Some Aristotelian and Kantian ideas will thus be put to use to resolve a paradox at the intersection of contemporary mathematics and modal metaphysics.

I claim that there is a qualitative difference between the finite and the transfinite, on the one hand, and the absolute infinite, on the other. Again, the difference has to do with the kinds of collection to which these measures of cardinality are ascribed. Each finite and transfinite number specifies that cardinality of collections understood in a combinatorial sense, say, as just a plurality of objects. A particular plurality of angels, for example, can be $\aleph_{17}$ in number. By contrast, absolute infinity is ascribed, not to a particular plurality, but to *a type* of object. It makes no sense to say of a particular plurality of angels that it is absolutely infinite. Any particular plurality has some particular number of members, which is either finite or transfinite. Rather, absolute infinity is ascribed to *a type* of object, say, to the type *angel that God might create*.

Let me be more precise. To say that objects of some type *F* are absolutely infinite is to say that necessarily, for any plurality *xx* of *F*s, it is possible for there to be an even more numerous plurality of *F*s. Recall that '$\#F^{\Diamond}$' denotes the upper bound on the cardinality of possible pluralities of *F*s, if there is such a bound, and ∞, if not. As observed, these terms too are rigid designators. For the possible *F*s, as seen from one world, are the same as the possible *F*s, as seen from any other world.

On the view that emerges, each finite and transfinite number is a determinate measure of the cardinality of certain pluralities. Each of these measures is, or could be, realized by some plurality. By contrast, absolute infinity isn't a determinate measure at all but a property of some *type* of object, namely, the property that no determinate measure is large enough to provide an upper bound on the cardinality of every possible plurality of objects of this type. In short, while each finite and transfinite number is a point on the scale by which we measure cardinality, absolute infinity is the property of being off this scale altogether.

This view solves Hawthorne and Uzquiano (2011)'s puzzle about how many angels there could be. For any number on Cantor's extended number line, there could be that many angels. But it makes no sense to consider an absolutely infinite plurality of angels. Absolute infinity is the property of being "off the scale," which can only be ascribed to a type of object, not to a plurality of objects.

The view solves the paradox of the largest number as well, or so I claim. As before, the solution is that the paradox conflates the two different notions of collection and number. First, suppose 'collection' and 'number' mean plurality and cardinality thereof, respectively. On this interpretation, the Antithesis is true, namely:

(5) $\Box\forall xx\Diamond\exists yy(\#xx < \#yy)$

We defend this claim as follows. Since pluralities (and other combinatorial collections, such as sets) are individuated in terms of their members, they are tracked across possible worlds in terms of their members. This rigid tracking from possible world to possible world ensures that any plurality from one possible world can be surpassed at some other possible world, namely by letting the latter world contain sufficiently many new objects *that lie outside of the plurality*. In short, the rigid tracking of pluralities, combined with a sufficiently liberal conception of possibility, ensures that (5) is true. The rigidity of pluralities underpins a distinction between being inside or outside the plurality, and the liberal conception of possibility ensures that it is possible for the outside to be populated.

Second, suppose instead we use 'collection' and 'number' in a looser sense that includes a type of object and the property of there being no upper bound on the cardinality of possible pluralities of objects of this type. On this alternative interpretation, the Thesis is true; for example, the collection of possible angels is unsurpassable:

$$\Box\forall F(\#F^{\Diamond} \leq \#\text{ANGEL}^{\Diamond})$$

We see this as follows. The collection of possible angels is conceptually individuated in a way that allows it to be "spread out" across the modal dimension. And because the collection is "spread out" in this way, it cannot be surpassed by any further modalizing—no matter how liberal our conception of possibility.

In short, the paradox dissolves once we distinguish clearly between the two kinds of collection and cardinality ascription.

VII. Inconsistent Multiplicities

In the previous section, we discussed collections of non-mathematical objects, such as (possible) angels, and cardinality ascriptions made to such collections. I wish to end by briefly discussing collections of pure mathematical objects, such as the collection of all ordinals or of all pure sets.

Just as the conceptually individuated collection of possible angels, these two collections are absolutely infinite. However, the latter two collections differ from the former in an important respect. It is easy enough to see why there cannot be a plurality of all possible angels: there is no possible world at which all possible angels exist. However many angels there are, there could always be more. By contrast, ordinals numbers and pure sets are assumed to exist at every possible world. Thus, while it is clear enough why there cannot be a plurality of all possible angels, there seems to be no obstacle to the existence of a plurality of all ordinals or all pure sets. Why, then, should there not be such pluralities? But now we are headed for a clash with my view that combinatorial collections such as pluralities cannot be absolutely infinite.

The underlying problem is this. My response to the paradox is based on a contrast between two kinds of collections: those that are individuated combinatorially and those that are individuated conceptually. This contrast surfaces clearly in modal contexts. But since pure mathematical objects exist necessarily, the needed contrast appears to evaporate, thus threatening to undermine my response.

Let me sketch a solution based on a generative conception of mathematical objects. According to this conception, any given pool of objects suffices to generate even more objects. The famous iterative conception of sets provides an illustration. Given some pool of objects, we can use the 'set of' operator to generate a set $\{xx\}$ of any objects xx drawn from this pool. Let me now make a crucial observation. On this generative conception too the distinction between combinatorially and conceptually individuated collections surfaces and is theoretically important. Since a combinatorial collection depends on each of its members, such a collection can only exist at a stage of the generative process where all of its members have already been generated. Every combinatorial collection is therefore *complete*, in the sense that, by the stage where the collection is generated, all of its members too have been generated. By contrast, a conceptually individuated collection need not be complete. Since such a collection is individuated in terms of its membership condition, it does not depend on its members and can therefore be generated prior to the generation of all of its members.

The central insight of the previous section is that the rigid tracking of pluralities (and other combinatorial collections), combined with a sufficiently liberal conception of possibility, ensures that for every plurality it is possible for there to be a more numerous plurality. An analogous insight holds when we consider, not metaphysical possibility, but the generation of mathematical objects. The rigid tracking of pluralities, combined with a sufficiently liberal conception of generative possibility, ensures that for every plurality, it is possible to generate a more numerous plurality. The reason is the same as before. The rigidity underpins a distinction between being inside or outside the plurality, and the liberal conception ensures that it is possible to generate additional objects located on the outside.[21]

It follows that there can be no plurality, or other combinatorial collection, of all ordinals. For if there were such a plurality, it could be used to generate even more ordinals, thus contradicting our assumption that we started with all ordinals.

This completes my response, inspired by Aristotle and Kant, to the last and most serious version of the paradox of the largest number. Cantor's extended number line contains all determinate measures, which suffice to specify the cardinality of every plurality (or other combinatorial collection), whether actual or merely possible. Each of these measures is surpassed by an even greater measure, just as every plurality is, or could be, surpassed by an even more numerous plurality. In addition to these determinate measures, there is a single indeterminate measure, namely the property of being off the scale of determinate measures altogether. This single indeterminate measure is absolute infinity, which can only be correctly ascribed to conceptually individuated collections that are "spread out" across the modal or generative dimension. Because these collections are "spread out" in this way, they cannot be surpassed; and hence, nor can absolute infinity.[22]

Notes

1. At least this is so when the measure in question is the cardinality of a collection of objects. For other kinds of measure, the situation is more subtle; for example, the surface of a sphere is (at least in one sense) unbounded yet has a finite area. See Parsons (1992) for discussion.
2. The parenthetical qualification is required if we follow contemporary mathematics and include zero as a natural number.
3. The distinction between these two kinds of collection is related to Kant's distinction between two kinds of totality, namely a *totum syntheticum*, which is composed of independently existing parts, and a *totum analyticum*, which is metaphysically prior to its parts. (See Allison (1983, 43), who in turn cites Kant's *Reflexion* 393, according to the Erdmann numbering.) Kant famously took space to be a *totum analyticum*.
4. As Moore (1990) explains, the paradox was formulated by Gregory of Rimini in the fourteenth century. A geometric version of the paradox can be found in Duns Scotus a century earlier.
5. The argument is clearly formulated in Kant (1998), B458. Earlier formulations may well exist as well.
6. Throughout this chapter, I sometimes talk about possible worlds for heuristic purposes. This talk can always be eliminated in favor of modal claims about necessity and possibility.
7. This argumentative structure mirrors that of Kant's antinomies of pure reason. Indeed, my argument is related to the first two antinomies.
8. See Williamson (2013) for an alternative view, which I cannot discuss here (though see footnote 19).
9. See Linnebo and Shapiro (2019) for an analysis of the distinction between actual and potential infinity.
10. An analogous point applies to Brouwerian choice sequences, which are also supposed to be potentially infinite. It is not a sequence *in extension* (specified, say, by a plurality of ordered pairs) that is potentially infinite but rather a sequence *in intension*, which can thus give rise to ever longer (but always finite) sequences in extension.
11. Here I use plural logic, which extends the expressive resources of ordinary singular logic by adding plural variables such as '*xx*,' '*yy*,' etc., and plural quantifiers binding such variables. See Linnebo (2017) for an introduction.
12. We are here making the usual assumption that the accessibility relation between possible worlds is universal; that is, that any possible world has access to any other possible world (and thus deems possible anything that obtains at the other possible world). (In our official modal language, which eschews quantification over possible worlds in favor of modal operators, the mentioned assumption corresponds to the claim that the modal logic contains the system S5; cf. fn. 6.)
13. See Levey (2015) for an interesting discussion of some early modern responses to Galileo's paradox.
14. The powerset of a set S is defined as the set of subsets of S, i.e. the set $\wp(S)$ defined as $\{x : x \subseteq S\}$.

15. Where c is a cardinal number, 2^c is the number of subsets of a set of cardinality c. Thus, in particular $2^{\aleph_0}$ is the number of subsets of which in turn is fairly easily seen to be equivalent to the number of real numbers.
16. Cf. Mayberry (2000).
17. See e.g. Boolos (1984) and Uzquiano (2003).
18. The conflict between the Thesis and Antithesis is a Cantorian echo of Kant's antinomies; cf. fn. 7. See also Zermelo (1930, 1233) for a related observation.
19. Cf. Hawthorne and Uzquiano (2011). (To be absolutely clear, Hawthorne and Uzquiano also explore whether it might help to reject the assumption, plausible though it may be. They show that, for some influential views in the metaphysics of modality (including Williamson (2013)), this rejection won't help.)
20. See e.g. Florio (2014, 329).
21. This argument is spelled out in Linnebo (2010).
22. Thanks to Salvatore Florio, Peter Fritz, Frode Kjosavik, Gabriel Uzquiano, the members of a research group at the Centre for Advanced Study (2015–16) at the Norwegian Academy of Science and Letters, and audiences in Leeds and Oslo for comments and discussion. I am grateful to CAS for a research leave and an intellectual environment which made this chapter possible.

References

Allison, Henry. *Kant's Transcendental Idealism*. New Haven, CT: Yale University Press, 1983.

Boolos, George. "To Be Is to Be a Value of a Variable (Or to Be Some Values of Some Variables)." *Journal of Philosophy* **81** (1984), 430–449.

Ewald, William Bragg. *From Kant to Hilbert: A Source Book in the Foundations of Mathematics*. Oxford: Oxford University Press, Vol. 2, 1996.

Florio, Salvatore. "Untyped Pluralism." *Mind* **123** (2014), 317–337.

Hawthorne, John and Uzquiano, Gabriel. "How Many Angels Can Dance on the Point of a Needle? Transcendental Theology Meets Modal Metaphysics." *Mind* **120** (2011), 53–81.

Kant, Immanuel. *Metaphysik L2*. Berlin: Academy Edition. Vol. 28/2.1, 525–610. Ed. by G. Lehmann. Berlin and Leipzig: Walter de Gruyter, 1970. (Cited as Ak. 28.)

Kant, Immanuel. *The Critique of Pure Reason*. Translated and edited by P. Guyer and A. Wood. Cambridge: Cambridge University Press, 1998.

Levey, Samuel. "Comparability of Infinities and Infinite Multitude in Galileo and Leibniz." In D. Rabouin, P. Beeley, and N.B. Goethe, eds., *G.W. Leibniz, Interrelations Between Mathematics and Philosophy*. Berlin: Springer, 2015, 157–187.

Linnebo, Øystein. "Pluralities and Sets." *Journal of Philosophy* **107** (2010), 144–164.

Linnebo, Øystein. "Plural Quantification." In E.N. Zalta, ed., *The Stanford Encyclopedia of Philosophy*. Stanford: Metaphysics Research Lab, Stanford University, Summer 2017 edition.

Linnebo, Øystein and Shapiro, Stewart. "Actual and Potential Infinity." *Noûs* **53** (2019), 160–191.

Mayberry, John P. *The Foundations of Mathematics in the Theory of Sets*. Cambridge: Cambridge University Press, 2000.

Moore, Adrian W. *The Infinite*. London: Routledge, 1990.

Parsons, Charles. "The Transcendental Aesthetic." In P. Guyer, ed., *The Cambridge Companion to Kant*. Cambridge: Cambridge University Press, 1992, 3–62.

Uzquiano, Gabriel. "Plural Quantification and Classes." *Philosophia Mathematica* **11** (2003), 67–81.

Williamson, Timothy. *Modal Logic as Metaphysics*. Oxford: Oxford University Press, 2013.

Zermelo, Ernst. "Über Grenzzahlen und Mengenbereiche." *Fundamenta Mathematicae* **16** (1930), 29–47. Translated in Ewald (1996).

13 Symbolic Construction From the "Purely Infinitesimal"

Gauge Invariance, Lie Algebras, and Metaphysics chez Hermann Weyl

Thomas Ryckman

"(T)here is no shame in doing metaphysics as long as the activity is informed by scientific practice."

—(Earman 2004, 1227–1228)

"Only in the infinitely small may we expect to encounter elementary and uniform laws; hence the world must be comprehended through its behavior in the infinitely small."

—(Weyl 1949, 86)

I. Introduction

In Huw Price's 2009 philosophical fantasy, Carnap* mysteriously falls asleep somewhere in New Jersey in 1950, to awaken (in New Brunswick?) some 60 years later and to find, with dismay, that metaphysics is again respectable in what otherwise is recognized as "scientific philosophy." Whether Price (following Putnam) is correct to view the present flourishing state of the positivists' nemesis as originating in Quine's 1948 paper "On What There Is" or whether pragmatism is the appropriate antidote to the contagion (as Price maintains) is not a concern here. It will suffice to note that contemporary philosophers of physics are indeed far from excoriating metaphysics as "word music" or as "language on holiday." Many, perhaps most, are of Earman's persuasion; they evince little epistemological compunction in forwarding claims regarding "the nature of reality" based on fundamental physical theories. Of course, the genus of prevailing metaphysics is ontological naturalism and physicalism. Even so, the species differ widely in scope and variety: "ontic structural realism," "wave function realism," "many worlds," even Whiteheadian "potentialities" are only a few of the dishes on the contemporary menu. With the exception of a small group of mind-matter dualists who regard the mental state of the observer as pivotal in bringing about definite outcomes of quantum measurement,[1] non-naturalistic, more specifically transcendental, approaches to metaphysics informed by scientific practice are decidedly fewer in number and not widely known.

The current naturalistic monoculture presents all the more reason to revisit the case of Hermann Weyl (1885–1955). A preeminent mathematician of the twentieth century, Weyl also made seminal contributions to the twin pillars of fundamental physical theory, general relativity and quantum mechanics. A recent tribute of Fields medalist Sir Michael Atiyah concluded

> No other mathematician could claim to have initiated more of the theories that are now being exploited. His vision has stood the test of time.
>
> (Atiyah 2002, 13)

In mathematics, Atiyah pointed in particular to Weyl's work on the theory of Lie groups and algebras (1925/26). In regard to physical theory, Atiyah alludes to Weyl's idea of gauge invariance, subsequently to become the unifying framework of the current Standard Model of elementary interactions. Weyl also authored a handful of philosophical works giving expression to reflective musings on his practice as an innovative mathematician and physical theorist. Even so, these texts are somewhat difficult to read and understand, in part because they draw upon figures and traditions largely unknown to contemporary philosophers of science (post-Kantian German idealism including Fichte and Husserl, but also Nicholas of Cusa, and even the medieval mystic Meister Eckhart), in part because they are highly personal, closely intertwined with Weyl's intricate interests and achievements, and finally in part because his conclusions are somewhat fragmentary, even purposefully hesitant. After all,

> an epistemological conscience [*Erkenntnisgewissen*], sharpened by work in the exact sciences, does not make it easy . . . to find the courage for philosophical statement. One cannot get by entirely without compromise.
>
> (Weyl 1954b, 648)

In this chapter, we give an overview of how a non-naturalistic metaphysics of transcendental subjectivity proved an extremely fruitful heuristic in two of Weyl's central achievements, in the origin of the idea of "gauge invariance" in his 1918 "purely infinitesimal geometry," and in his 1925–6 purely mathematical work on Lie theory (on representations of semi-simple Lie groups and Lie algebras, see further ahead), attaining results that Weyl himself regarded as his greatest mathematical achievement. The two are not unrelated: Weyl would coin the term "Lie algebra" for the infinitesimal group structure of a Lie group; he also showed that this infinitesimal structure is a real-valued linear space, in fact, a vector space, a concept first defined in Weyl (1913). In this guise, Lie algebras play an important role in the contemporary gauge theories of the Standard Model. We argue that both achievements are heuristically motivated by Weyl's injunction to attempt to comprehend the world from its behavior in the infinitely small, an injunction implicating a metaphysics of transcendental subjectivity closely related, perhaps not identical, to transcendental phenomenological idealism. Our "revisit" may have more than mere historical interest in that Weyl's injunction may shed light on the meaning of unphysical degrees of freedom in gauge theories, considered by some as the most important contemporary problem of philosophy of physics.[2]

II. Transcendental-Phenomenological Idealism

The metaphysics undergirding Weyl's injunction is an idealism, not a realism, since

> Science concedes to idealism that its objective reality is not given but to be constructed [*nicht gegeben, sondern aufgegeben*], and that it cannot be constructed absolutely but only in relation to an arbitrarily assumed coordinate system, and in mere symbols.
>
> (Weyl 1927, 83, 1949, 117)

Readers of Kant's Transcendental Dialectic (1781/1787; A647/B675) will recognize the italicized phrase as an avowal of transcendental idealism. Weyl's idealism is also a transcendental-phenomenological idealism insofar as symbolic construction of the "objective reality" of the purportedly mind-independent objects of physics is, *per* Husserl, a constitution of the *sense* of such objects as having "the sense of 'existing in themselves'"[3] (emphasis in original). Weyl himself expressed this understanding of sense-constitution in a densely exposited account of the phenomenology of perception in the "Introduction" to all five editions of his masterful text on general relativity, *Raum-Zeit-Materie*.

> Upon general in-principle grounds: The real world, in each of its components and all their determinations, is and can only be, given as intentional objects of acts of consciousness. Given, purely and simply, are the conscious experiences that I have—as I have them. Certainly, in no way do they consist, as positivists perhaps maintain, only of the mere stuff of sensation. Rather a perception, for example, an object standing bodily there before me, each experience of which is known to everyone but not more exactly describable, is taken up in a completely characteristic manner to be designated, with Brentano, through the expression "intentional object." While I am perceiving, as in seeing this chair, I am thoroughly directed to it. I "have" the perception, but only when I make this perception itself into the intentional object of a new, inner perception (of which I am capable in a free act of reflection), do I "know" something about it (and not merely about the chair). . . . In this second act

> the intentional object is immanent like the act itself; it is an actual component of my stream of experience; but in the primary perceptual act, the object is *transcendent*, i.e., actually given in a conscious experience but not an actual component. The immanent is absolute, that is, exactly what it is as I have it and am able to bring its essence [*Wesen*] to givenness [*Gegebenheit*] before me in acts of reflection. . . . The given-to-consciousness [*Bewußtseins-Gegebene*] is the starting point at which we must place ourselves in order to comprehend the sense and the justification of the posit of reality [*Wirklichkeitsetzung*].
>
> (Weyl 1918a, 3–4, 1923a, 3–4)

Readers cognizant of phenomenology find the Husserlian resonances unmistakable; the book's first endnote states that the "precise wording" is "closely modeled" upon Husserl's *Ideen* (1913). This passage, utterly remarkable in a monograph largely establishing the modern mathematical machinery of general relativity, is not idle philosophical window dressing. It is crucial to Weyl's *transcendental* philosophy of natural science, setting out the central phenomenological distinction between "objects" beyond (*transcendent* to) experience and those immanent within experience, i.e., "intentional objects" produced in acts of reflection upon experience. The former is the realm of mind-independent objects, structures, etc., the subject of physical theory, the latter are the idealized mathematical surrogates of physical theory. Scholars have noted that at various times, mostly after the mid-1920s, Weyl expressed misgivings regarding phenomenology.[4] Yet evidence is abundant that even in later years Weyl's philosophical orientation is to something quite similar to it. A prime example is an affirmation from his 1930 Rouse Ball lecture in Cambridge, already some years after the most intense period (1918–25) of rather explicit immersion in Husserlian phenomenology,

> Reality [*Wirklichkeit*] is not a being-in-itself [*Sein an sich*] but rather is constituted *for* a consciousness.[5]

Further expressions of an affinity to phenomenology occur as late as the year before Weyl's death in 1955.[6]

III. An "idealism in the infinitesimal"

Following the apt term of Julien Bernard (2015), Weyl's transcendental metaphysics is an "idealism in the infinitesimal." It is a modern descendant of Leibniz's *principle of continuity* ("*natura non facit saltus*"), i.e., that all finite changes are to be comprehended as arising through infinitesimal increments acting in sequence. Weyl gave it a mathematical setting, inspired by two titans of mathematics in the second half of the nineteenth century, Bernhard Riemann and Sophus Lie:

> The productivity shown by the differential calculus, by contiguous action [field] physics [*Nahewirkungsphysik*], and by Riemannian geometry certainly rests upon the principle: To understand the world, according to its form and content, by its behavior in the infinitely small, clearly because *all problems can be linearized in passing to the infinitely small.*[7]

Weyl elsewhere observed that Riemann, in geometry, took the step that Faraday and Maxwell had taken in physics, according to "the principle: To understand the world from its behavior in the infinitely small."[8] Mathematically, the "purely infinitesimal" is the tangent space T_P to each point P of a differential manifold. As readers may know, the passage alludes to the fact that just as in elementary differential calculus, functions differentiable at a point P on the function's graph are locally linear there, so in a Riemannian manifold M mathematical relations are linear within the tangent space T_P to each point P of M. Riemann noted that the manifolds subsequently bearing his name exhibited "planeness in their smallest parts," where only linear relations are required.[9] In a Lie group, the group elements are points in a manifold, and are parameterized by continuous real variables. The points are combined by an operation obeying the group axioms; they compose to form a 'space.' Yet a core idea of Lie's theory, one that Weyl would exploit, is that of "Descent to the infinitely small": substitution of an infinitesimal group in place of a finite group.[10] Though generically non-linear manifolds, Lie groups too can be linearized in passing to an infinitesimal group acting in the tangent space of the group identity. In doing so, most (not all) information about the finite group can be obtained, a crucial mathematical fact on which the modern structure theory of Lie groups is based, yielding a precise classification of all Lie groups and their respective algebras.

In the application of both Riemannian geometry and Lie theory to physics, Weyl's "idealism in the infinitesimal" is an epistemological mandate that *comprehensibility* of the physical world, i.e., phenomenological "sense-constitution," is to be gained by bottom-up symbolic construction starting from mathematical relations in the infinitely small. With both Riemannian manifolds and Lie groups in mind, Weyl would regard the tangent space T_P as the legitimate epistemic reach of "eidetic vision" or "insight" or "*Evidenz*" (or whatever one will call it) of the cognizing, constructing subject, an "ego-center" whose range of vision is mathematically understood as the "infinitesimally small" bounded linear region surrounding each point.[11] This is the entry point of transcendental-phenomenological sense-constitution and a core assumption of "idealism in the infinitesimal":

> Only the spatio-temporally coinciding and the immediate spatial-temporal neighborhood has a directly clear meaning exhibited in intuition. . . . The philosophers may have been correct that our space

> of intuition bears a Euclidean structure, regardless of what physical experience says. I only insist, though, that to this space of intuition belongs the ego-center [*Ich Zentrum*] and that . . . the relations of the space of intuition to that of physics, becomes vaguer the further the distance from the ego-center.
>
> (Weyl 1931c, 49)

Phenomenological requirements on evidence, on what is given to consciousness and what can be evidently constituted (in Weyl's mathematically expanded notion of *Wesensschau*) on that basis, are expressly tied to the infinitely small: only in this limited region can a cognizing consciousness impose *evident* elementary and uniform laws. Other mathematical resources may be required for manifolds as a whole; their evidential basis is accordingly less direct. The injunction to comprehend the world from "its behavior in the infinitely small" is a recurrent theme running through Weyl's writings from 1918 to at least 1949 (cf. Weyl 1918a, 82, 1949, 86).

IV. Transcendental Phenomenological Origins of Gauge Invariance

We have previously argued that Weyl's 1918 reformulation of Einstein's general relativity (GR) within a "purely infinitesimal geometry" stemmed from his philosophical orientation to transcendental-phenomenological idealism (Ryckman 2005). As noted, the "purely infinitesimal" is the locus of origin of sense-constitution arrived at through phenomenological reflection upon the levels of sedimented mathematical structures defined on differential manifolds, a reflection that reveals

> a surreptitious substitution of the mathematically substructed world of idealities for the only real world, the one that is actually given through perception, that is every actually experienced, and experienceable.
>
> (Husserl 1954, 48–49)

The "purely infinitesimal" then bounds the immediate evidential reach of a situated ego, the constituting transcendental subject that, through step-by-step construction, invests with meaning the mathematical structures defined on differential manifolds. The injunction "to comprehend the world" (i.e., structures on the entire manifold) from "the purely infinitesimal" is just a requirement that sense-constitution of the finite and global mathematical structures of these manifolds ultimately derives from this evidential basis.

Following the injunction, the metric of Weyl's 1918 geometry of space-time, unlike Einstein's Riemannian geometry, allowed for choice of unit of scale ("gauge") at each point; this new degree of freedom, a function $\lambda\ (x_\mu)$ [$\mu = 1,2,3,4$] of the four space-time coordinates of the point, was shown to be mathematically identical to the vector four-potential A_μ of relativistic electromagnetic theory. The end result was that Weyl's geometry yielded mathematical structures incorporating *both* Einstein's gravitation and electromagnetism, the only definitely known interactions in 1918. Hence from a "purely infinitesimal" starting point, global field laws (the "world") are constructed that are required to satisfy not only, as in GR, general covariance (freedom to choose space-time coordinates) but also "gauge invariance" (freedom to choose scale at each point). To be sure, gauge invariance in contemporary physical theory pertains to so-called "internal symmetries" of quantum fields rather than to a factor of scale (of length or duration).[12] On the other hand, the basic idea of gauge invariance, that it involves an arbitrary function of the space-time coordinates, remains. To Weyl, this lends the requirement of gauge invariance, however interpreted, the "character of a 'more general' ('*allgemeiner*') relativity."[13]

V. Symbolic Construction

Around 1925 Weyl began to use the expression "symbolic construction" to underscore the aforementioned distinction between mind-*transcendent* objects, and their symbolic surrogates in physical theory. The term itself originates in Weyl's intervention in a period controversy over foundations of mathematics. His first work in philosophy of mathematics, the predicative analysis of (1918b), drew upon Husserlian phenomenology. By the early 1920s Weyl was an enthusiastic proponent of a version of Brouwerian intuitionism, again largely on phenomenological grounds. Still, he was all-too-aware of the severe limitations intuitionistically acceptable methods placed on classical mathematics, writing in 1925 that "full of pain, the mathematician sees the greatest part of his towering theories dissolve into fog" (Weyl 1926, 534).

Declaring "in the beginning was the sign," Hilbert (1922) entered the lists against Brouwer and Weyl. Hilbert's idea was to begin with the intuitively given but otherwise meaningless signs of "concrete, intuitive" number theory (elementary arithmetic), a finite part of mathematics, including recursion and intuitive induction for finite existing totalities, grounded in "purely intuitive considerations [*rein anschauliche Überlegungen*]," hence acceptable to intuitionism. This finite formal part is to be supplemented by a strict axiomatic formalization of the rest of the mathematical theory (its infinitary part), including its proofs. Questions of the truth or validity of mathematical statements are to be replaced by the metamathematical demand for a consistency proof of the axioms, to be obtained in a "formal proof theory" in which proofs are rule-governed arrays of concrete and displayable formal signs. A formal consistency proof guarantees the reliability of mathematical theory in yielding the conclusion that the permitted arrays cannot yield a contradiction, such as $0 = 1$.[14] A few years later

Hilbert (1926) distinguished between *finitary* and *ideal statements*, justifying appending to mathematics the contentious infinitary part of mathematics (Cantorian set theory) by appeal to a Kantian Idea of Reason, i.e., the regulative demand to complete the concretely given in the interest of totality.[15] To Weyl, Hilbert had demonstrated a necessity to go beyond "the demands of open-eyed [*schauenden*] certainty" for each individual statement of a theory (1931a, 29). Moreover, the Hilbertian shift from truth of individual statements to consistency, a global requirement on theories, prompted an analogy to modern physical theories. While the justification of any mathematical statement pertained only to the theory as a whole, via a proof of the theory's consistency, in theoretical physics evidential justification of particular statements involves the often-complex relation of certain theoretical terms and relations to evidence and observation.[16] Weyl would take as an axiom of "symbolic construction" that evidence can be brought to bear on a theory not statement by statement but only on the theoretical system as a whole. Even so, Weyl dismissed Hilbert's metamathematical "game of formulae" as an adequate philosophical justification of the cognitive worth of mathematics. Instead, he fused mathematics with physics, locating the epistemological value of mathematics via its application in theoretical physics in symbolic construction. Unlike in mathematics, physics posits a mind-transcendent reality, but philosophical reflection (*Besinnung*) on theoretical construction reveals that physics remains content to represent this reality only in symbols ("*das Transzendente darzustellen . . . nur im Symbol*").[17]

The term "symbolic" conveys more than the obvious point that mathematics is the necessary instrument of exact natural science. Far more significantly, the intent of the term is to underscore the conviction that the finite human mind, rooted in "all too human ideas with which we respond to our practical surroundings in the natural attitude of our existence of strife and action" (Weyl 1932b, 6) *can* attain only a symbolic (neither literal nor pictorial) understanding of the infinite (in mathematics[18]) or of the mind-independent real world posited by physical theory.[19] Symbolic representation is essential for the latter, in particular, as the revolutions of twentieth century physics, relativity, and quantum mechanics, demonstrate that "here we are in contact with a sphere which is impervious to intuitive evidence; cognition necessarily becomes symbolic construction."[20] Relativity and quantum mechanics also instructs that "objectivity" is interpreted epistemologically, as "invariance under a group of automorphisms" (Weyl 1948/9). The purely symbolic character of physical theory means that

> A science can only determine its domain of investigation up to an isomorphic mapping. In particular, it remains quite indifferent as to the 'essence' of its object . . . The idea of isomorphism demarcates the self-evident boundary of cognition.
>
> (1927, 22, 1949, 25–26)

The term "construction" also has a polyvalent significance. It echoes Weyl's oft-expressed predilection for constructive vs. axiomatic (i.e., predicative or intuitionist vs. set-theoretic) mathematics (Weyl 1985). It thus reflects an evidential preference for the 'visualizable' or iterative structures of geometry and topology over those of modern abstract algebra, though Weyl insists mathematics (and indeed, differential manifolds) requires both (Weyl 1932a). Above all, it signifies that 'objectivity' in physical theory is constituted as an invariance "for a subject with its continuum of possible positions," and that it arises in step-by-step construction from a basis of what is *aufweisbar*, "something to which we can point to *in concreto*" as demonstrably evident to the constituting consciousness.

> (T)he constructions of physics are only a natural prolongation of operations [the] mind performs in perception, when, e.g., the solid shape of a body constitutes itself as the common source of its various perspective views. These views are conceived as appearances, for a subject with its continuum of possible positions, of an entity on the next higher level of objectivity: the three-dimensional body.
>
> Carry on this 'constitutive' process in which one rises from level to level, and one will land at the symbolic constructions of physics. Moreover, the whole edifice rests on a foundation which makes it binding for all reasonable thinking: of our complete experience it uses only that which is unmistakably *aufweisbar*.
>
> (1954a, 628, 627)

The *fons et origo* of all meaning-constitution, i.e., what is given *in evidence* to 'pure consciousness' remains, but "symbolic construction" is the extra-logical process that substitutes the manipulation of concrete symbols for the data of 'pure consciousness.' It is Weyl's generalization of Husserlian intuition of essences (*Wesensschau*) to step-by-step mathematical construction beginning from the evidentially privileged standpoint of the "purely infinitesimal." "Symbolic construction" is his term of art for philosophical reflection upon the enterprise of theoretical natural science "to comprehend the sense and the justification of the posit of reality [*Wirklichkeitsetzung*]" beginning from the starting point of the "given-to-consciousness [*Bewußtseins-Gegebene*]," recognizing that the starting point in the infinitesimally small requires only elementary and uniform mathematical relations. It is the generic term for *sense-constitution* of the transcendent physical world, a symbolic construct that must not be conflated with that "true real world."[21]

VI. From the "*Raumproblem*" to Lie Groups and Lie Algebras

In a natural development from his 1918 "purely infinitesimal" reformulation of GR, Weyl turned to the new "space problem" posed by the variably curved manifolds permitted in GR. The old "space problem" concerned the group-theoretical characterization of a geometry: A particular geometry is singled out by its continuous group of motions: figures that transform into one another by the operations of the group are considered equivalent. The Helmholtz-Lie solution to the old "space problem" presupposed that any geometric space permits free mobility of rigid bodies, i.e., is a homogeneous space of constant curvature; this is made obsolete by GR, where the metric is no longer homogeneous and the medium itself is space-time, so that the metric is no longer a positive definite quadratic form but one that is indefinite.

Weyl's new solution came about by again drawing from the metaphysics of the "purely infinitesimal." He noted that the old Helmholtz-Lie solution retained its validity in the infinitely small if posed in terms of a point-dependent Lie congruence group G acting transitively in the homogeneous tangent space T_pM centered on each point $P \in M$ (Weyl 1988). The infinitesimal linear action of the group is comparable to rotations about points in Euclidean geometry, yielding "point congruences," generalized metrical relations within each T_pM. Assuming the volume of parallelepipeds formed by basis vectors at P is preserved by these "rotations," these are subgroups of the special linear group $SL(n)$. Metrical relations in the infinitesimally close neighborhood U of P are then given through the "congruent transport" of a vector from P to a neighboring point $P' \in U$ by a single linear connection that intuitively combines parallel transport ("affine connection") and the infinitesimal rotations. This is a "metric connection" A linking infinitesimal congruence relations at P with those at P'. An equivalence class of such connections $\{A\}$ characterizes the same infinitesimal congruence structures at each point differing only by point-dependent infinitesimal rotations. Thus, the "nature" of space at each point P is the same, while the rotation groups at different points may have distinct "orientations." Weyl showed that the local group of isometries must be the (pseudo-)Euclidean group. As required by general relativity, the rotation groups at the various points can have different "orientations" due to variations of matter and energy, but they all share the same infinitesimal Pythagorean (pseudo-Euclidean) metric group structure (1923b, 43–61).

Already in the nineteenth century Sophus Lie had reduced the concept of continuous group to the 'germ' of infinitesimal elements that generated it. In abstract form, these "*groupes infinitésimaux*" had been extensively studied and classified some decades before by É. Cartan, building upon earlier results of W. Killing.[22] But explicit recognition that the simple tools of linear algebra could be brought to bear on this infinitesimal structure was due to Weyl. The concept of "infinitesimal group" was essentially repurposed in Weyl's solution to the "Space Problem" in the light of the variably curved four-dimensional space-times of GR. In an appendix, Weyl showed that the "constitution" of this infinitesimal group structure could be axiomatically expressed in algebraic terms (1923b, 82). In particular, the structure of the tangent space surrounding the point that is the group identity (where the Lie group can be considered homogeneous and linear), is a linear vector space. He could then carry over the linear algebra of the tangent space to a point in a Riemannian manifold to the tangent space of the identity of a Lie group. The infinitesimal structure of a Lie group was first termed a "Lie algebra" in Weyl's lectures ten years later at the Princeton Institute of Advanced Study.[23] In sum, Weyl's solution to the new "Space Problem" crucially rested upon the concept of infinitesimal group, recast in language of linear vector spaces. To Weyl this was compelling evidence that "mathematical simplicity and metaphysical originality [*Ursprünglichkeit*] are narrowly bound together."[24] The purely infinitesimal solution to the new "Space Problem" led in turn to his purely mathematical research on representations of semi-simple Lie groups and Lie algebras.[25] And it is just in the guise of linear vector spaces that the concept of Lie algebra (and its representations) appears in the contemporary gauge theories of the Standard Model.

VII. Weyl's Legacy: Gauge Symmetries and Lie Algebras in Field Theory

A symmetry of a (quantum) field theory is a group of transformations that leave the equation of motion of the field (its Lagrangian) unchanged in form. Symmetries may be discrete or continuous, global (applying in the same way everywhere) or local (applying differently at each point of space). Representations of Lie groups and their respective Lie algebras are used to describe continuous symmetries, both of space-time (the inhomogeneous Lorentz group of spatial rotations, boosts, and translations) and of internal spaces, such as phase changes of the wave function (the commutative (*abelian*) unitary Lie group $U(1)$), or of isospin space where there is an approximate symmetry between nucleons (protons and neutrons), broken by approx. 1% due to the mass difference between the two quark types composing protons (*u,u,d*) and neutrons (*u,d,d*). Isospin symmetry is mathematically identical to rotational symmetry; the physical interpretation is that particle multiplets can be 'rotated' into each other, preserving total isospin ($I = \frac{1}{2}, 1, \frac{3}{2}, \ldots$). Its symmetry group is the non-commutative (*non-abelian*) Lie group of 2 × 2 unitary matrices with unit determinate $SU(2)$.

Our brief remarks here pertain to gauge (local) internal symmetries. A gauge symmetry group is a constraint on the Lagrangian L for any quantum field theory; the first step in constructing a quantum field theory is to ask what gauge symmetry the Lagrangian must obey.[26] The reason for this is two-fold: (1) to reduce the number of possible Lagrangians, and (2) the only known renormalizable quantum field theories contain gauge symmetries. More specifically, the only way to form a relativistic quantum field theory of spin-1 particles (force-carrying *bosons*) is a gauge theory. (Weinberg 1995, 340) The Standard Model (SM) is a spontaneously broken[27] non-abelian gauge theory containing three types of particles: elementary scalars, fermions (spin-1/2) and spin-1 bosons. The spin-1 gauge bosons are called gauge fields. The SM is often designated by its total "gauge group," the direct product group *SU(3)* × *SU(2)* × *U(1)* representing all known fundamental interactions except gravity. Although the SM provides a detailed empirically successful framework in which to analyze and correlate experimental data, the three interactions are not related to each other in any fundamental way except for being derived from local, or gauge, symmetries.

The special unitary Lie groups *SU(2)* and *SU(3)* are of particular significance. *SU(2)* describes the weak interaction in that the Lagrangian describing one form of radioactivity (β-decay) transforms invariantly under *SU(2)* symmetries. As noted previously, the elements of *SU(2)* are 2 × 2 matrices; in the weak interaction matter particles (*up* and *down* quarks; electrons and electron neutrinos) are sorted into doublets such that the two particles in a doublet are interchangeable, indistinguishable in that interaction. The *SU(2)* × *U(1)* group is the gauge symmetry of the "electro-weak" interaction. *U(1)*, by itself (slightly different from *U(1)* above in the product group which is the weak hypercharge) is the phase symmetry of the Dirac Lagrangian of the electromagnetic interaction and quantum electrodynamics (QED, see immediately next).

The quark model of hadrons of Gell-Mann and Zweig was the original justification for the group *SU(3)* which now describes the strong interaction; its elements are represented by 3 × 3 matrices, so the symmetry operations pertain to a triplet of particles. The matter particles of the strong interaction; each quark with its own mass and electric charge comes in one of six *flavors* partitioned into three doublets (*up, down*), (*strange, charm*), (*top, bottom*). *Color*, the strong force analogue to electric charge, itself has three manifestations, *red, green, blue.* Within the same flavor, changing e.g., *red* quarks to *green* quarks leaves the interaction energy of the system unchanged. *SU(3)* plays two roles in the SM: as an approximate global symmetry acting on the (u,d,s) quark flavor indices and as an exact gauge symmetry associated with quark color charge for the strong interactions (the "eightfold way" of Gell-Mann and Ne'eman, an extension of isospin symmetry).

There is a canonical way, the "gauge argument," to "gauge" a field theory. The first and simplest example was found by Weyl himself in 1929 in the course of setting the relativistic equation of the electron (the Dirac equation) into the four-dimensional space-time context of general relativity. The new gauge degree of freedom now appears not as a factor of scale but as a local phase factor in wave function of electron. In more modern terms this happens as follows. One begins with a free electron field $\Psi(x)$ that is determined up to a phase factor θ. The Maxwell-Dirac Lagrangian for the electron field that is the basis of QED transforms invariantly under the *global* phase transformation (applying at each point in the same way),

$$\Psi(x) \Rightarrow \Psi'(x) = e^{i\theta}\Psi(x)$$

where $e^{i\theta}$ is Euler's formula. $\Psi(x)$ is then invariant under a global *U(1)* internal symmetry group; the global invariance of the matter system implies, via Noether's theorem, the existence of a conserved quantity, a matter current j^{μ}.

One then "promotes" the global symmetry to a local phase symmetry; this means that an independent *U(1)* group is associated with *each* space-time point. The requirement of gauge symmetry then demands that the phase parameter θ vary as a function of space-time position x, and the phase invariance becomes *local:*

$$\Psi(x) \Rightarrow \Psi'(x) = e^{i\theta(x)}\Psi(x) \tag{1}$$

Typically, Lagrangians depend not only on the field magnitudes but also on their (at least first) derivatives. However, by imposing a local symmetry, the derivative of the field $\partial_{\mu}\Psi(x)$ picks up an extraneous term $\partial_{\mu}\theta(x)$ in its transformation; as $\theta(x)$ is a function varying with space-time position, it is not a covariant object. In order to cancel this unwanted term, the "gauge covariant derivative" is introduced,

$$\partial_{\mu} \Rightarrow D_{\mu} = \partial_{\mu} - ieA_{\mu}$$

where e is the electric charge of the electron field, expressing a coupling to the newly introduced field $A_{\mu} = A_{\mu}(x)$. The new covariant derivative transforms as

$$D_{\mu} \Rightarrow D'_{\mu} = e^{i\theta(x)}D_{\mu}$$

and $A_{\mu} = A_{\mu}(x)$ is an "invented" vector field required to transform as

$$A_{\mu}(x) \Rightarrow A'_{\mu}(x) = A_{\mu}(x) - \frac{\partial_{\mu}\theta(x)}{e} \tag{2}$$

The resulting Lagrangian is then invariant under joint local transformation of $\Psi(x)$, given by (1), and of $A_{\mu}(x)$, given by (2), the added partial derivative in (2) *exactly compensating the extraneous position-dependent variation of the phase factor*. Moreover, in imposing the requirement of local symmetry in steps (1) and (2), the free electron field $\Psi(x)$ is coupled to the electromagnetic field represented by the Faraday tensor, obtained by taking the derivative of the four-potential $A_{\mu}(x)$,

$$F_{\mu\nu} = \frac{\partial A_{\mu}}{\partial x_{\nu}} - \frac{\partial A_{\nu}}{\partial x_{\mu}},$$

and the conserved current now appears in an interaction of the form $ej^{\mu}A_{\mu}$ in conformity with Maxwell theory. For obvious reasons, the new term $A_{\mu}(x)$ is now called a "gauge field." The "gauge argument" then shows how introducing local symmetries dictates the form of the interaction of matter fields (Yang 1988, 20).

For our purposes, we note that generically one can write the covariant derivative of the field as a linear combination of the ordinary derivative of the field (its infinitesimal displacement in space or space-time) and a field-dependent infinitesimal gauge transformation

$$D_{\mu}\psi \equiv \partial_{\mu}\psi - W_{\mu}\psi$$

where W_{μ} is a Hilbert space matrix whose entries are generated by an infinitesimal gauge transformation. Hence, the gauge field takes values in the Lie algebra corresponding to the gauge group. And this means W_{μ} can be decomposed into its generators t_a,

$$W_{\mu} = W_{\mu}^{a} t_{a}.$$

The generators t_a are Hilbert space matrices satisfying the Cartan structure condition defining Lie algebras.[28] The characteristic of a gauge field is that it carries information regarding the group from one space-time point to another.

The preceding gauge invariance of QED is only a simple example of infinite parameter or Lie group symmetry, and it is not typical as its gauge group *U(1)* is abelian (commutative). The preceding argument was generalized to non-abelian Lie groups by Yang and Mills (1954); Yang-Mills theories arise by generalizing the principle of local gauge symmetry from the *U(1)* local phase rotation invariance of QED to generalized phases associated with the wave functions of multicomponent matter fields. In general, non-abelian Lie groups yield theories of multiple vector particles, whose interactions are strongly constrained by a gauge symmetry.

VIII. Conclusion: Objectivity From the "Purely Infinitesimal"

Weyl's "purely infinitesimal" generalization of GR issued in the demand that fundamental physical theories, in addition to the requirement of coordinate freedom ("general covariance"), should also satisfy the requirement of gauge (more appropriately in its debut, scale) invariance. Both requirements introduce *arbitrary* mathematical degrees of freedom at each point *P* of the four-dimensional differential manifold representing space-time; the arbitrariness can be understood phenomenologically, as each point indifferently can be considered the locus of an experiencing, constructing ego.[29] These degrees of freedom arise from two metaphysical prerequisites: (1) a postulate of transcendental-phenomenological idealism, that "Reality

[*Wirklichkeit*] is not a being-in-itself [*Sein an sich*] but rather is constituted *for* a consciousness," and (2) the *aspiration*, fortified by the successes of differential calculus in physics and indeed of field physics (*Nahewirkungphysik*) itself, that this "reality," constituted as it is by a *situated* consciousness, "can be understood from its behavior in the infinitesimally small," i.e., mathematically comprehended starting from the *evident* simple linear relations within the tangent space. The arbitrary degrees of freedom represent *at each point* particular magnitudes of physical states, *either* as mathematical functions of four independent variables (space-time coordinates) determined by the field laws *or* in terms of an arbitrary function of these space-time coordinates signifying an internal gauge symmetry. Since both coordinate and gauge transformations connect states that cannot be physically distinguished, both symmetries can be understood not as symmetries of nature but of our description of nature. General covariance and gauge invariance are then demands of objectivity, that the constructed physical theory is to be built up from the domain of the "infinitesimal" (point-like interactions) and that the ensuing construction must be independent from any particular starting point from which it is constituted. In a phrase, "physics must not depend on the physicist." In this way we can understand how Weyl's central idea of the gauge principle as a "purely infinitesimal" remnant of sense-constitution is preserved even today in the internal (phase) symmetries of the Standard Model.

Notes

1. Following physicists such as Eugene Wigner, Henry Stapp, Bruce Rosenblum, etc.
2. Redhead 2002, 299: "The gauge principle is generally regarded as the most fundamental cornerstone of modern theoretical physics. In my view its elucidation is the most pressing problem in current philosophy of physics."
3. Husserl 1908, 382:

   > My transcendental method is transcendental-phenomenological. It is the ultimate fulfillment of old intentions, especially those of English empiricist philosophy, to investigate the transcendental-phenomenological 'origins' . . . the origins of objectivity in transcendental subjectivity, the origin of the relative being of objects in the absolute being of consciousness.

   Transcendental sense-constitution of objective nature is founded on one of *empathy*; e.g., Husserl (1931, 92):

   > "a *transcendental theory of experience of the other* [*Fremderfahrung*], the so-called *empathy* [*Einfühlung*]" has within its scope "the founding of *a transcendental theory of the objective world* . . . in particular, of objective Nature to whose existence sense [*Seinsinn*] belongs *there-for-everyone* [*Für-jedermann-da*]."

4. In our opinion, Weyl's misgivings largely concern with what he saw as Husserl's limitation of sense-constitution to the level of 'bodily-given' objects of perception. For example, Weyl 1954b, 640:

> The theoretical-symbolic construction, through which physics tries to grasp the transcendent standing behind the perceptual [*das hinter dem Wahrgenommenen stehende Tranzendente zu erfassen sucht*] is far from stopping at this bodilyness [*Leibhaftigkeit*]. I would say that Husserl merely portrays one of the steps through which the constitution of the external world is completed.

Leibhaftigkeit is Husserl's term for the peculiar mode of conscious awareness pertaining to the perceived spatial thing itself, despite its transcendence (1913, §43). It is argued in what follows that "symbolic construction" is Weyl's mode of continuing sense-constitution to transcendent objects given, if at all, in non-perceptual experience.

5. Weyl 1931c, 49; compare this passage of Husserl 1913, 121: "The existence of a Nature *cannot* be the condition for the existence of consciousness, since Nature itself turns out to be a correlate of consciousness: Nature *is* only as being constituted in regular concatenations of consciousness."
6. About Husserlian *Wesensschau*, Weyl wrote (1954a, 629):

> At the basis of all knowledge lies . . . 1. *Intuition*, the mind's originary act of 'seeing' what is given to it; limited in science to the *Aufweisbare*, but in fact extending far beyond these boundaries. How far one should go in including here the *Wesensschau* of Husserl's phenomenology, I prefer to leave in the dark.

Aufweisbare may be translated as 'evident.'

7. Weyl 1923b, 9: "*Es beruht ja die Leistungsfähigkeit des in der Differentialrechnung, der Nahewirkungsphysik und der* **Riemann***sche Geometrie zum Durchbruch kommenden Prinzips: die Welt nach Form und Inhalt aus ihrem Verhalten im Unendlichkleinen zu verstehen, eben darauf, dass* **alle Probleme durch den Rückgang aufs Unendlichkleine linearisiert werden.**" See also Weyl 1918a, 82, 1927, 61, 1949, 86.
8. "Vorwort des Herausgebers," in Weyl (ed.) 1919.
9. A crucial feature of Einstein's theory of gravitation is that it allows (pseudo-) Riemannian geometry ("pseudo" since time is treated differently than the three space dimensions) to be the appropriate mathematical framework for the concept of "local inertial frame" and so to uphold the "infinitesimal" validity of special relativity in that theory.
10. Weyl 1923b, 34: "*Die Ersetzung der endlichen Gruppe durch die infinitesimal—das ist wieder der 'Rückgang aufs Unendlichkeleine'!—ist einer der Hauptgedanken der Lieschen Theorie.*" Original emphasis. Hawkins (2000, 72) quotes from an 1879 paper of Lie,

> In the course of investigations on first-order partial differential equations, I observed that the formulas that occur in this discipline become amenable to a remarkable conceptual interpretation by means of the concept of an infinitesimal transformation. In particular, the so-called Poisson-Jacobi theorem is closely connected with the composition of infinitesimal transformations. By following up on this observation I arrived at the surprising result that all transformation groups of a simply extended manifold can be reduced to the linear form by a suitable choice of variables, and also that *the determination of all groups of an n-fold extended manifold can be achieved by the integration of ordinary differential equations.* This discovery . . . became the starting point of my many years of research on transformation groups.

11. Husserl, e.g., (1929, 141), is careful to distinguish the usual (and "countersensical") philosophical notion of evidence as the absolute criterion of truth

from evidence as "*that performance on the part of intentionality which consists in the giving of something-itself* [*die intentionale Leistung der Selbstgebung*]. More precisely, it is the universal preeminent form of 'intentionality,' of 'consciousness of something,' in which there is consciousness of the intended-to objective affair in the mode itself-seized-upon, itself-seen—correlatively, in the mode: being with it itself in the manner peculiar to consciousness."

12. Internal symmetries refer to the fact that particles occur in multiplets, members of which can be considered as "the same" under the symmetry of the interaction. Mathematically, the multiplets are realizations of an irreducible representation of some internal symmetry group. See below.
13. Weyl (1929, 246). Cf. Weyl 1931b, 220: "One can in fact take it as a general rule that an invariance property of the kind met in general relativity, involving an arbitrary function, gives rise to a differential conservation theory. In particular, gauge invariance is only to be understood from this standpoint."
14. Gödel's 2nd incompleteness theorem (published in 1931) shows that any such consistency result can only be relative, since a consistency proof can only be carried out in a stronger theory.
15. Hilbert (1926, 190): "The role that remains to the infinite is . . . merely that of an idea—if, in accordance with Kant's words, we understand by an idea a concept of reason that transcends all experience and through which the concrete is completed so as to form a totality." Weyl (1931a, 28):

> Hilbert himself says somewhat obscurely that infinity plays the role of an idea in the Kantian sense, by which the concrete is completed in the sense of totality. I understand this to mean something like the way in which I complete what is given to me as the actual content of my consciousness, into the totality of the objective world, which certainly includes much that is not present to me. The scientific formulation of this objective concept of the world occurs in physics, which avails itself of mathematics as a means of construction. However, the situation we find before us in theoretical physics in no way corresponds to Brouwer's idea of a science. That ideal postulates that every judgment has its own meaning achievable in intuition. The statements and laws of physics, nevertheless, taken one by one, have no content verifiable in experience; only the theoretical system as a whole allows itself to be confronted by experience. What is accomplished here is not the intuitive insight into singular or general contents and a description that truly renders what is given, but instead a theoretical, and ultimately purely symbolic, construction of the world.

16. Weyl 1928a, 147–148:

> [The] individual assumptions and laws [of theoretical physics] have no separate fulfilling sense [that is] immediately realized in intuition [*in der Anschauung unmittelbar zu erfüllender Sinn eigen*]; in principle, it is not the propositions of physics taken in isolation, but only the theoretical system as a whole that can be confronted with experience. What is achieved here is not intuitive insight [*anschauende Einsicht*] into particular or general states of affairs and a faithfully reproduced *description* of the given [*das Gegebene*], but rather theoretical, ultimately a purely symbolic, *construction* of the world.

Weyl goes on to state that if Hilbert's view prevails over Brouwer's, as indeed appears to be the case, then this represents

> *a decisive defeat of the philosophical attitude of pure phenomenology*, as it proves insufficient to understand creative science in the one domain

of knowledge that is most rudimentary and earliest open to evidence, mathematics.

17. Weyl 1926, 540, also 1954b, 645. The latter is a lecture entitled "Erkenntnis und Besinnung (Ein Lebensrückblick)." *Besinnung*, here translated 'reflection,' is a technical term in Husserlian phenomenology, having the meaning of "sense-investigation"; e.g., Husserl, 1929, 8): "Sense-investigation [*Besinnung*] . . . radically understood, is *originary sense-explication* [*ursprüngliche Sinnesauslegung*, orig. emphasis], transforming and above all striving to transform sense in the mode of unclear opinion into sense in the mode of full clarity or essential possibility [*Wesensmöglichkeit*]."
18. Weyl 1932b, 7: "*Mathematics is the science of the infinite,* its goal the symbolic comprehension of the infinite with human, that is finite, means."
19. The influence of Husserl is apparent in Weyl's use of the term "natural attitude."
20. Weyl 1932b, 82. For example, the central underlying theoretical device of quantum mechanics, densities of a complex valued, infinite-dimensional wave function, *can* only be symbolically represented. Dirac, influenced by Weyl (1928b) offers the same philosophical message regarding the necessity of symbolic methods in the first sections of his (1930).
21. Weyl 1954a, 627: "the words 'in reality' must be put between quotation marks; who could seriously pretend that the symbolic construct is the true real world?" The term 'symbolic construct' encompasses not merely the symbolic universe in which physical systems, states, transformations, and evolutions are mathematically defined in terms of manifolds, functional spaces, algebras, etc., but also a symbolic specification of idealized procedures and experiments by which the basic physical quantities or observables of the theory are related to observation and measurement. It reflects an insistence, reinforced by quantum mechanics, that physical quantities (beginning with 'inertial mass') are not simply given, but "constructed" (1931b, 76, 1934, 109ff.).
22. See (Hawkins 2000), and (Eckes 2013). Cartan's "structure theory" for infinitesimal Lie groups (today, Lie algebras) identifies isomorphic groups through their "structure constants." Cartan worked exclusively at the level of abstract groups; Weyl would translate Cartan's structure theory into the language of matrix groups, group representations by matrices, the language of most interest to physics.
23. Weyl, *The Structure and Representation of Continuous Groups*. Based on notes by Richard Brauer taken at Weyl's course at The Institute for Advanced Study, 1934–5; reprinted 1955, 4. In modern terms, a Lie algebra to a Lie group G is usually denoted by the Gothic character $\mathfrak{g}$ and is defined by three properties: (1) the elements X, Y, etc. of $\mathfrak{g}$ form a linear vector space; (2) the elements of $\mathfrak{g}$ close under a commutation relation $[X,Y] =-[Y,X]$, $\forall X,Y \in \mathfrak{g}$; (3) the Jacobi identity $[X,[Y,Z]] + [Y,[Z,X]] + [Z,[X,Y]] = 0$ is satisfied. Using Cartan's "structure theory," the structure of a Lie algebra is completely determined by its "structure constants" c_{ij}^{k} that appear in the commutator of any two basis vectors $\left[X_i, X_j\right] = c_{ij}^{k} X_k$.
24. Weyl 1922, 329: "*auf diesem Felde mathematische Einfachheit und metaphysische Ursprünglichkeit in enger Verbindung miteinander stehen.*"
25. Weyl, 1925–6. A (non-abelian) Lie group whose Lie algebra cannot be factorized into two commuting subalgebras is called *simple*. A direct product of simple Lie groups is called *semi-simple*. In these papers Weyl supplemented Cartan's infinitesimal abstract group viewpoint with global and topological properties of Lie groups, thus in our view, "comprehending the world

beginning from the infinitesimal." In fact, linearizing a Lie group G in the tangent space of the identity to form its Lie algebra $\mathfrak{g}$ destroys G's global properties, i.e., what happens far from the identity. Hence the need for integral and topological methods.

26. The second step is to determine the representations of fermions and scalars under the gauge symmetry; a third step is to postulate the pattern of spontaneous symmetry breaking.
27. A symmetry of a system is said to be "spontaneously broken" if its lowest energy state is not invariant under the operations of that symmetry. This is an extremely important concept in the weak interaction as the bosons introduced by gauge symmetries are massless, like the photon; their masses arise from the "spontaneous breaking" of the SU(2) × U(1) symmetry through couplings to the scalar Higgs field.
28. $[t^a,t^b] = i\, f^{abc}\, t^c$. In general, the covariant derivative associated with the gauge transformation contains one vector field for each independent generator of the local symmetry.
29. Weyl 1918b, 72:

> The coordinate system is the unavoidable residue of the ego's annihilation (*das unvermeidliche Residuum der Ich-Vernichtung*) in that geometrico-physical world that reason sifts from the given under the norm of 'objectivity'—a final faint token in this objective sphere that existence (*Dasein*) is only given, and *can* only be given as the intentional content of the conscious experience of a pure, sense-giving ego.

References

Atiyah, Michael. "Hermann Weyl 1885–1955." *National Academy of Science: Biographical Memoirs*. Washington, DC: National Academies Press, Vol. 82, 2002, 1–14.

Bernard, Julien. *L'idéalisme dans l'infinitésimal: Weyl et l'espace à l'époque de la relativité*. Nanterre: Presses universitaires de Paris Nanterre, 2015

Dirac, Paul Adrien Maurice. *The Principles of Quantum Mechanics*. Oxford: Clarendon Press, 1930.

Earman, John. "Laws, Symmetry, and Symmetry Breaking: Invariance, Conservation Principles, and Objectivity." *Philosophy of Science* 71 (2004), 1227–1241.

Eckes, Christophe. *Les groups de Lie dans l'oeuvre de Hermann Weyl*. Nancy: Presses Universitaires de Nancy/Éditions Universitaires de Lorraine, 2013.

Hawkins, Thomas. *Emergence of the Theory of Lie Groups: An essay in the history of mathematics 1869–1926*. New York: Springer, 2000.

Hilbert, David. "Neubegründung der Mathematik (Erste Mitteilung)." *Abhandlungen aus dem mathematischen Seminar der Hamburgischen Universität* 1 (1922), 157–177.

Hilbert, David. "Über das Unendliche." *Mathematische Annalen* 95 (1926), 161–190.

Husserl, Edmund. *Die Krisis der europäischen Wissenschaften und die transzendentale Phänomenologie: Eine Einleitung in die phänomenologische Philosophie*. Husserliana VI. Ed. by W. Biemel. The Hague: Martinus Nijhoff, 1954.

Husserl, Edmund. "Zur Auseinandersetzung meiner transzendentalen Phänomenologie mit Kants Transzendentalphilosophie (ca. 1908)." In *Erste Philosophie* (1923–24). Husserliana VII, 381–395. Ed. by R. Boehm. Erster Teil. The Hague: Martinus Nijhoff, 1956.

Husserl, Edmund. *Cartesianische Meditationen und Pariser Vorträge* (1931). Husserliana I. Ed. by S. Strasser. The Hague: Martinus Nijhoff, 1963.
Husserl, Edmund. *Formale und Transzendentale Logik* (1929). Husserliana XVII. Ed. by P. Janssen. The Haag: Martinus Nijhoff, 1974.
Husserl, Edmund. *Ideen zu einer reinen Phänomenologie und phänomenologischen Philosophie I* (1913). Husserliana III, 1–2. Ed. by K. Schuhmann. The Hague: Martinus Nijhoff, 1976.
Kant, Immanuel. *Kritik der reinen Vernunft: Erste Auflage 1781*. Berlin: Academy Edition. Vol. 4, 5–252. Ed. by B. Erdmann. Berlin: Georg Reimer, 1911a.
Kant, Immanuel. *Kritik der reinen Vernunft: Zweite Auflage 1787*. Berlin: Academy Edition. Vol. 3, 2–552. Ed. by B. Erdmann. Berlin: Georg Reimer, 1911b.
Price, Huw. "Metaphysics after Carnap: The Ghost Who Walks?" In D. Chalmers, R. Wasserman, and D. Manley, eds., *Metametaphysics: New Essays on the Foundations of Ontology*. Oxford: Oxford University Press, 2009, 320–346.
Quine, Willard van Orman. "On What There Is." *Review of Metaphysics* 2 (1948/1949), 21–38.
Redhead, Michael. "The Interpretation of Gauge Symmetry." In Meinhard Kuhlmann, Holger Lyre, and Andrew Wayne, eds., *Ontological Aspects of Quantum Field Theory*. London-Singapore-Hong Kong: World Scientific, 2002, 281–301,
Ryckman, Thomas. *The Reign of Relativity: Philosophy in Physics 1915–1925*. New York: Oxford University Press, 2005 (Oxford Studies in Philosophy of Science).
Weinberg, Steven. *The Quantum Theory of Fields*. New York: Cambridge University Press, Vol. 1, 1995.
Weyl, Hermann. *Die Idee der Riemannschen Fläche*. Leipzig: B.G. Teubner, 1913.
Weyl, Hermann. *Raum-Zeit-Materie*. Berlin: J. Springer, 1918a.
Weyl, Hermann. *Das Kontinuum: Kritische Untersuchungen über die Grundlagen der Analysis*. Leipzig: Veit, 1918b.
Weyl, Hermann. (Ed.) *B. Riemann: Über die Hypothesen, welche der Geometrie zu Grunde liegen*. Berlin-Heidelberg: Springer Verlag GmbH, 1919.
Weyl, Hermann. "Das Raumproblem." *Jahresbericht der Deutschen Mathematikervereinigung* 31 (1922), 205–221; reprinted in Hermann Weyl, *Gesammelte Abhandlungen* (GA). Bd. II. Ed. by K. Chandrasekharan. Berlin, Heidelberg and New York: Springer Verlag, 1968, 328–344.
Weyl, Hermann. *Raum-Zeit-Materie*. 5 Auflage. Berlin: J. Springer, 1923a.
Weyl, Hermann. *Mathematische Analyse des Raumproblems: Vorlesungen gehalten in Barcelona und Madrid*. Berlin: J. Springer, 1923b.
Weyl, Hermann. "Theorie der Darstellung kontinuierlicher halbeinfacher Gruppen durch lineare Transformationen I, I, II, und *Nachtrag*." *Mathematische Zeitschrift* 23–24 (1925/26); reprinted in Hermann Weyl, *Gesammelte Abhandlungen* (GA). Bd. II. Ed. by K. Chandrasekharan. Berlin, Heidelberg and New York: Springer Verlag, 1968, 543–647.
Weyl, Hermann. "Die Heutige Erkenntnislage in der Mathematik." *Sonderdrucke des Symposion, Heft* 3 (1926), 1–32; reprinted in Hermann Weyl, *Gesammelte Abhandlungen* (GA). Bd. II. Ed. by K. Chandrasekharan. Berlin, Heidelberg and New York: Springer Verlag, 1968, 511–542.
Weyl, Hermann. *Philosophie der Mathematik und Naturwissenschaft*. München und Berlin: R. Oldenbourg, 1927.
Weyl, Hermann. "Diskussionsbemerkungen zu dem zweiten Hilbertschen Vortrag über die Grundlagen der Mathematik." *Abhandlungen aus dem mathematischen*

Seminar der Hamburgischen Universität 6 (1928a), 86–88; reprinted in Hermann Weyl, *Gesammelte Abhandlungen* (GA). Bd. III. Ed. by K. Chandrasekharan. Berlin, Heidelberg and New York: Springer Verlag, 1968, 147–149.

Weyl, Hermann. *Gruppentheorie und Quantenmechanik*. Leipzig: S. Hirzel Verlag, 1928b.

Weyl, Hermann. "Elektron und Gravitation." *Zeitschrift für Physik* **56** (1929), 330–352; reprinted in Hermann Weyl, *Gesammelte Abhandlungen* (GA). Bd. III. Ed. by K. Chandrasekharan. Berlin, Heidelberg and New York: Springer Verlag, 1968, 245–267.

Weyl, Hermann. "Levels of Infinity." In P. Pesic, ed., *Hermann Weyl: Levels of Infinity. Selected Writings on Mathematics and Philosophy*. Mineola, NY: Dover, 2012, 17–31. Translation of *Die Stufen des Unendlichen* (1931a). Jena: Gustav Fischer.

Weyl, Hermann. *The Theory of Groups and Quantum Mechanics*. 2nd ed. Translated by H.P. Robertson. London: Methuen and Co., Ltd., 1931b; reprinted. New York: Dover, 1950.

Weyl, Hermann. "Geometrie und Physik." *Die Naturwissenschaften* **19** (1931c), 49–58; reprinted in Hermann Weyl, *Gesammelte Abhandlungen* (GA). Bd. I–IV. Ed. by K. Chandrasekharan. Berlin, Heidelberg and New York: Springer Verlag, 1968, Bd. III, 336–345.

Weyl, Hermann. "Topologie und abstrakte Algebra als zwei Wege mathematischen Verständnisses." *Unterrichtsblätter für Mathematik und Naturwissenschaften* **38** (1932a), 177–188; reprinted in Hermann Weyl, *Gesammelte Abhandlungen* (GA). Bd. I–IV. Ed. by K. Chandrasekharan. Berlin-Heidelberg-New York: Springer Verlag, 1968, Bd. III, 348–358.

Weyl, Hermann. *The Open World*. New Haven: Yale University Press, 1932b; reprinted in P. Pesic (ed.), *Hermann Weyl: Mind and Nature. Selected Writings on Philosophy, Mathematics, and Physics*. Princeton: Princeton University Press, 2009, 34–82.

Weyl, Hermann. *Mind and Nature*. Philadelphia: University of Pennsylvania Press, 1934; reprinted in P. Pesic (ed.), *Hermann Weyl: Mind and Nature. Selected Writings on Philosophy, Mathematics, and Physics*. Princeton: Princeton University Press, 2009, 83–150.

Weyl, Hermann. *The Structure and Representation of Continuous Groups*. Based on notes by Richard Brauer taken at Weyl's course at The Institute for Advanced Study, 1934–5; reprinted 1955.

Weyl, Hermann. "Similarity and Congruence: A Chapter in the Epistemology of Science." Lecture at ETH, Zurich, 1948/49. *ETH Bibliothek, Hochschularchiv Hs91a* 31, 23.

Weyl, Hermann. *Philosophy of Mathematics and Natural Science*. Princeton: Princeton University Press, 1949. Enlarged translation of Weyl 1927.

Weyl, Hermann. "Address on the Unity of Knowledge Delivered at the Bicentennial Conference of Columbia University." 1954a; reprinted in Hermann Weyl, *Gesammelte Abhandlungen* (GA). Bd. I–IV. Ed. by K. Chandrasekharan. Berlin-Heidelberg-New York: Springer Verlag, 1968, Bd. IV, 623–629.

Weyl, Hermann. "Erkenntnis und Besinnung (Ein Lebensrückblick)." *Studia Philosophica, Jahrbuch der Schweizerischen Philosophischen Gesellschaft*, 1954b, 153–171; reprinted in Hermann Weyl. *Gesammelte Abhandlungen* (GA). Bd. I–IV. Ed. by K. Chandrasekharan. Berlin; Heidelberg; New York: Springer Verlag, 1968, Bd. IV, 631–649.

Weyl, Hermann. "Constructive versus Axiomatic Procedures in Mathematics." Typescript written after 1953. *The Mathematical Intelligencer* 7 (1985), 10–17.
Weyl, Hermann. *Riemanns geometrische Ideen, ihre Auswirkung und ihre Verknüpfung mit der Gruppentheorie*. Berlin-Heidelberg New York: Springer Verlag, 1988. A manuscript written in 1925 for a Russian edition of Lobachevsky's mathematical works; first published in 1988.
Yang, Chen Ning and Mills, Robert L. "Conservation of Isotopic Spin and Isotopic Gauge Invariance." *Physical Review* 90 (1954), 191–195.
Yang, Chen Ning. "Hermann Weyl's Contributions to Physics." A Hermann Weyl Centenary Lecture, ETH Zürich, October 24, 1985. In K. Chandrasekharan, ed., *Hermann Weyl 1885–1985*. Berlin-Heidelberg-New York: Springer-Verlag, 1988, 7–21.

Contributors

Joseph Almog is Professor of Philosophy at the University of Turku, Finland, and Professor Emeritus of Philosophy at UCLA. His recent publications include *Referential Mechanics: Direct Reference and the Foundations of Semantics* (Oxford 2014); *Everything in its Right Place: Spinoza and Life by the Light of Nature* (Oxford, 2014).

Christian Beyer studied philosophy, linguistics, and the history of science in Bielefeld and Hamburg, where he received his PhD in 1999. He was a Visiting Scholar at Stanford University (1994–5), served as Temporary Lecturer in Philosophy at the University of Sheffield (2000), and as Wissenschaftlicher Mitarbeiter at the University of Erfurt (2000–5), where he earned his Habilitation. In 2005 he was awarded a Heisenberg Scholarship of the German Research Foundation, and in 2007 he has been appointed as Professor of Theoretical Philosophy at the Georg-August-Universität Göttingen. He is an elected foreign member of the Norwegian Academy of Science and Letters, and was Associated Fellow at the Lichtenberg-Kolleg (Göttingen) in 2011, as well as Fellow at the Centre for Advanced Study (Oslo) in 2016. Beyer is the author of *Von Bolzano zu Husserl* (Kluwer, 1996), *Intentionalität und Referenz* (Mentis, 2000), *Subjektivität, Intersubjektivität, Personalität* (Walter de Gruyter, 2006), and co-editor of *Philosophical Knowledge* (special issue of *Grazer Philosophische Studien*, 2007), *Edmund Husserl 1859–2009* (Walter de Gruyter, 2011), *Content, Context, and Conversation* (special issue of *Philosophical Studies*, 2014), and *Husserl's Phenomenology of Intersubjectivity* (Routledge, 2019).

Michael Friedman is the Patrick Suppes Professor of Philosophy of Science at Stanford University. Publications include *A Post-Kuhnian Philosophy of Science* (Van Gorcum, 2014); *Kant's Construction of Nature* (Cambridge University Press, 2013); *The Cambridge Companion to Carnap* (ed., Cambridge University Press, 2007); *The Kantian Legacy in Nineteenth-Century Science* (MIT Press, 2006); *Kant: Metaphysical Foundations of Natural Science* (transl., Cambridge University Press, 2004); *Dynamics of Reason* (CSLI, 2001); *A Parting of the Ways:*

Carnap, Cassirer, and Heidegger (Open Court, 2000); *Reconsidering Logical Positivism* (Cambridge University Press, 1999); *Kant and the Exact Sciences* (Harvard University Press, 1992); and *Foundations of Space-Time Theories: Relativistic Physics and Philosophy of Science* (Princeton University Press, 1983).

Leila Haaparanta is Professor Emerita of Philosophy at the University of Tampere and Docent of Theoretical Philosophy at the University of Helsinki, where she earned her PhD in 1985. Previous posts include research and teaching positions at the University of Helsinki and at the Academy of Finland. She is the author of *Frege's Doctrine of Being* (Acta Philosophica Fennica, 1985) and the editor of *Mind, Meaning and Mathematics* (Kluwer, 1994), *The Development of Modern Logic* (Oxford, 2009), and *Rearticulations of Reason* (Acta Philosophica Fennica, 2010). Her co-edited works include *Frege Synthesized* (with J. Hintikka, Reidel, 1986), *Analytic Philosophy in Finland* (with I. Niiniluoto, Rodopi, 2003), and *Categories of Being: Essays on Metaphysics and Logic* (with H.J. Koskinen, Oxford, 2012). She has published articles on the history of logic, early twentieth century analytic philosophy and phenomenology, epistemology, philosophy of mind and language, philosophy of religion, and pragmatism. She has led research projects on the history of logic, on twentieth century philosophy, on metaphysics, and on rationality. She was a member of the Steering Committee of the European Society for Analytic Philosophy 2007–2011. She is a member of the Finnish Academy of Science and Letters, and of Academia Europaea.

Mirja Hartimo is currently University Lecturer at Tampere University and Docent at University of Helsinki, Finland. With a fellowship from the Norwegian University of Life Sciences, she was based at its School of Economics and Business from 2015–17, to participate in the project "Disclosing the Fabric of Reality—The Possibility of Metaphysics in the Age of Science," which was led by Frode Kjosavik and Camilla Serck-Hanssen, at the Centre for Advanced Study at the Norwegian Academy of Science and Letters. She has published several edited volumes and journal special issues, and around 60 articles on Husserl in international journals, such as *Synthese*, *Review of Symbolic Logic*, *Journal of History and Philosophy of Logic*, *Southern Journal of Philosophy*, *Inquiry*, and *Husserl Studies*. She is presently preparing a monograph on Husserl's philosophy of mathematics.

Toni Kannisto is Postdoctoral Fellow at the University of Oslo, Norway. His main focus is on Kantian metaphysics and its application to contemporary metaphysical and metametaphysical questions. His PhD thesis—*From Thinking to Being*—was on Kant's theory of modality. He has published a number of articles in leading international journals, e.g., in *Synthese* and *Kant-Studien*.

Frode Kjosavik is Professor of Philosophy at the School of Economics and Business at the Norwegian University of Life Sciences (NMBU), where he has held a permanent position since 1999. He has been Visiting Scholar at Stanford University (2008–9) and Harvard University (2009; 2017). He was Professor at the Department of Philosophy, Classics, History of Art and Ideas, at the University of Oslo, 2013–14. He was group leader for the project "Disclosing the Fabric of Reality—The Possibility of Metaphysics in the Age of Science" at the Centre for Advanced Study (CAS) at the Norwegian Academy of Science and Letters, 2015–16. He earned his PhD from the University of Oslo in 1999. His research interests include general metaphysics and epistemology; history of philosophy; Kant; Husserl; philosophy of perception; general philosophy of science; philosophy of mathematics; philosophy of biology; and philosophy of the social sciences. He has several publications within these fields. He is editor (together with Christian Beyer and Christel Fricke) of *Husserl's Phenomenology of Intersubjectivity: Historical Interpretations and Contemporary Applications* (Routledge, 2019).

Olli Koistinen is Professor of Philosophy at the University of Turku, Finland. He works on both early modern philosophy and contemporary metaphysics. His publications include *Cambridge Companion to Spinoza's ETHICS* (editor, Cambridge, 2009), *Spinoza: metaphysical themes* (ed., with John Biro, Oxford, 2002) and many papers on Spinoza, as well publications on Kant and Descartes.

Øystein Linnebo is Professor of Philosophy at the University of Oslo. He received his MA in mathematics from the University of Oslo and his PhD in philosophy from Harvard University. Since 2004, he has worked as a philosopher at Oxford University, Bristol University, and at Birkbeck, University of London. Even though his philosophical interests are broad in scope, his research has focused in particular on the philosophy of mathematics, and on metaphysics, philosophical logic, and Frege. Linnebo has headed a research group with five members that was funded by a European Research Council Starting Grant. He is a leader of ConceptLab (together with Herman Cappelen and Camilla Serck-Hanssen), which is funded by the Norwegian Research Council, at the University of Oslo. He has numerous publications in leading journals. His most recent publications include *Thin Objects: An Abstractionist Account* (Oxford, 2018) and *Philosophy of Mathematics* (Princeton, 2017).

Charles Parsons is Edgar Pierce Professor of Philosophy, Emeritus, at Harvard University. He was editor of *Journal of Philosophy* from 1966–90. His most recent publications include *Philosophy of Mathematics in the Twentieth Century* (Harvard, 2014), *From Kant to Husserl* (Harvard,

2012), *Mathematical Thought and Its Objects* (Cambridge, 2008), and *Mathematics in Philosophy* (Cornell, 1983, reissued in 2005). He was also co-editor of Volumes III–V of the *Collected Works* of Kurt Gödel (Oxford, 1995, 2003).

Thomas Ryckman is Professor of Philosophy at Stanford University with a special interest in the philosophy of science and history of twentieth-century physics. He has recently published *Einstein* (Routledge, 2017), and previously *The Reign of Relativity: Philosophy in Physics 1915–1925* (Oxford, 2005) as well as a number of book chapters and journal articles, including, "Analytic and Continental Traditions: Frege, Husserl, Carnap and Heidegger" (with Michael Friedman), Alan D. Schrift, ed., *The History of Continental Philosophy 3* (Chicago, 2011); "Hilbert's 'Foundations of Physics: Gravitation and Electromagnetism within the Axiomtaic Method'" (with Katherine Brading), and *Studies in History and Philosophy of Modern Physics* v. 39, 2008.

Camilla Serck-Hanssen is Professor of Philosophy at the University of Oslo and Scientific Director at the Centre for Advanced Study (CAS), at the Norwegian Academy of Science and Letters. She is a leader of ConceptLab (together with Herman Cappelen and Øystein Linnebo), at the University of Oslo, which is funded by the Research Council of Norway. She has been Adjunct Professor of Philosophy at the School of Economics and Business at the Norwegian University of Life Sciences (NMBU), 2014–18. She was group leader (together with Frode Kjosavik) for the project "Disclosing the Fabric of Reality—The Possibility of Metaphysics in the Age of Science" at CAS, 2015–16. Among her publications are articles on Kant's logic and dialectic, including entries in *Kant Lexikon* (de Gruyter, 2015), and "Fighting Achilles" (In G. Motta and U. Thiel, eds., *Immanuel Kant: Die Einheit des Bewusstseins* (de Gruyter, 2017)), as well as on Kant's metametaphysics, and on Heidegger's reading of the *Critique of Pure Reason* ("Towards fundamental ontology: Heidegger's phenomenological reading of Kant," *Continental Philosophy Review*, 2015).

Houston Smit is Associate Professor of Philosophy at the University of Arizona. He received his PhD from UCLA, and was a postdoctoral Mellon Fellow at Stanford University. His articles on Kant's theoretical philosophy include: "The Role of Reflection in the Critique of Pure Reason," *Pacific Philosophical Quarterly* 80 (1999); "Kant on Marks and the Immediacy of Intuition," *Philosophical Review* 109 (2000); "Kant on Apriority and the Spontaneity of Cognition," Newlands and Jorgensen, eds., *Metaphysics and the Good* (Oxford, 2009); and "Kant's 'I think' and the agential approach to self-knowledge," forthcoming in *Canadian Journal of Philosophy*, and online at https://doi.org/10.1080/00455091.2019.1615682.

David Woodruff Smith is Distinguished Professor of Philosophy at the University of California, Irvine. After studying mathematics and science engineering at Northwestern University, he earned his PhD in philosophy at Stanford University. He has written on systematic and historical aspects of phenomenology, ontology, and philosophy of mind and language. His publications include: *Husserl* (2007, revised 2nd edition 2013); *Mind World: Essays in Phenomenology and Ontology* (2004); *The Circle of Acquaintance: Perception, Consciousness, and Empathy* (1989); and (with Ronald McIntyre) *Husserl and Intentionality: A Study of Mind, Meaning, and Language* (1982). He has co-edited *Phenomenology and Philosophy of Mind* (2005, with Amie L. Thomasson) and *The Cambridge Companion to Husserl* (1995, with Barry Smith).

Index

Note: Page numbers followed by an "n" denote a note on the corresponding page.